Cambridge IGCSE®

Geography

Revision Guide

Second edition

David Davies

CAMBRIDGE
UNIVERSITY PRESS

University Printing House, Cambridge CB2 8BS, United Kingdom

One Liberty Plaza, 20th Floor, New York, NY 10006, USA

477 Williamstown Road, Port Melbourne, VIC 3207, Australia

4843/24, 2nd Floor, Ansari Road, Daryaganj, Delhi - 110002, India

79 Anson Road, #06 -04/06, Singapore 079906

Cambridge University Press is part of the University of Cambridge.

It furthers the University's mission by disseminating knowledge in the pursuit of education, learning and research at the highest international levels of excellence.

www.cambridge.org

Information on this title: www.cambridge.org/9781316635490

First published 2013
Second edition 2017

20 19 18 17 16 15 14 13 12 11 10 9 8 7 6 5 4 3 2 1

Printed in Spain by GraphyCems

A catalogue record for this publication is available from the British Library

ISBN 978-1-316-63549-0 Paperback

Table of contents

How to use this book

Learning summary

By the end of this chapter, you should be able to:

- explain and give reasons for population migration, including internal movements such as rural–urban migration, and international migrations, both voluntary and involuntary
- demonstrate an understanding of the impacts of migration, including the positive and negative impacts on the destination and origin of the migrants, and the migrants themselves
- demonstrate knowledge of an international migration case study.

Learning summary – a summary list of key topics and concepts that you will be looking at in this chapter, to help with navigation through the book and give a reminder of what's important about each topic for your revision.

TERMS

natural resources: any natural resource from water and soil, to wind and minerals (broadest sense)

cumulative causation: certain things will result (be caused) when a group of factors combine (accumulate)

Terms – clear and straightforward explanations are provided for the most important words in each topic. Key terms appear in blue bold type within the main text.

FACT

It is estimated that by 2050, about 66% of the developing world and 86% of the developed world will be urbanised.

Fact – provide additional, quick information to extend knowledge on a topic.

TIP

You do not need to revise dome volcanoes, but they are included here to give a complete understanding of volcanoes.

Tip – quick suggestions to remind you about key facts and highlight important points.

Self-test questions 11.1

1 What is meant by the terms 'weather' and 'climate'?

2 Describe and explain the factors you would take into account in locating a rain gauge and a cup anemometer.

Self-test questions – sets of formative questions to assess knowledge and understanding of content throughout each chapter.
The answers to these questions are at the back of the book.

Case studies

A densely populated area – the Ganges River basin

The Ganges River has the second greatest water discharge in the world, after the Amazon. Its basin, from the Punjab to Bangladesh and Assam, is the most heavily populated river basin in the world, with an average population density of 1000 people per km². The location of the Ganges River basin in India can be seen in Figure 4.2, running across the top of the map, in a north-west to south-east direction.

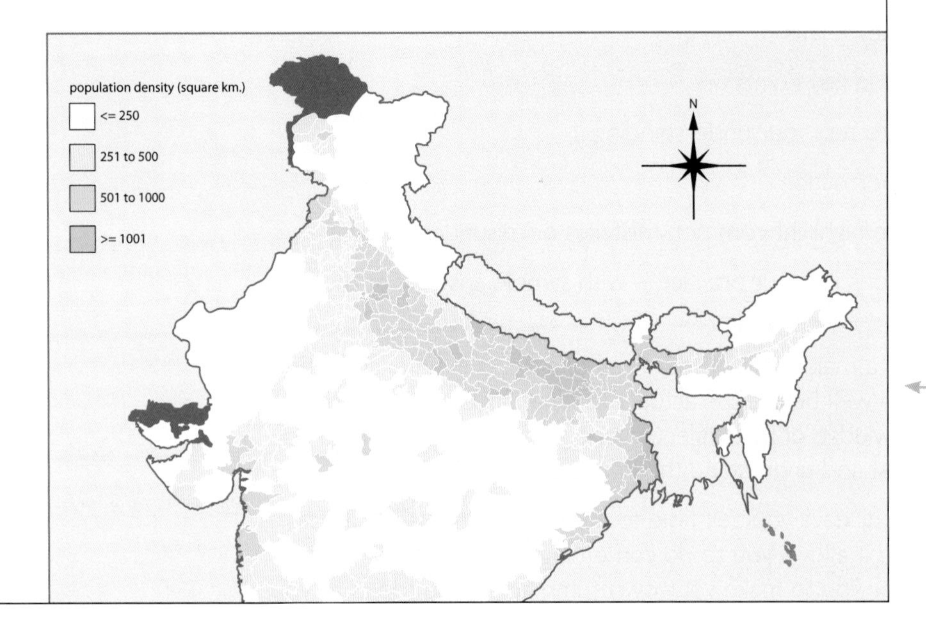

Case study – they illustrate how to work out a particular type of question, emphasising the different stages that students need to work through. This feature also offers students opportunities to practice their case study skills. Suggested answers to these are included at the back of the book.

Sample question and answer – each sample question and answer should contain one exam-style question with a correct answer provided to show students where marks are gained.

Sample question and answer

Explain why a large percentage of the population in many LICs is employed in the primary sector. [3]

Ensure you read the question carefully and refer to LICs only and why people tend to be found in primary sector industries and not in other sectors.

In many LICs, a large percentage of the people work in farming because many people are subsistence farmers **[1]** who have to grow the food that their families eat. They cannot afford machines to do the work on their farms so much of the work is done by people, often the family **[1]**. Also, many of the people do not have the chance to have an education/go to school and so they cannot get jobs that need them to read and write **[1]**. The governments have relatively small budgets which means that there is not much money to invest and establish new secondary and tertiary industries **[1]**.

(A maximum of 3 marks will be awarded.)

Exam-style questions

1 What is meant by the terms nomadic and sedentary farming? [2]

2 a Describe how irrigation and drainage may improve crop production on a farm. [3]

b What problems can be caused by irrigation? [3]

3 Explain the ways in which farmers in LICs might increase the amount of food they produce from their land. [4]

Exam-style questions – Exam-style questions for you to test your knowledge and understanding at the end of each chapter. The answers are provided at the back of the book.

Introduction

This guide has been written to help you revise for your Cambridge IGCSE Geography examination.

The chapters follow the order of the topics and papers in the Cambridge IGCSE syllabus.

Each chapter includes:

- definitions and explanations of key terms
- tips to provide helpful guidance on key points
- self-test questions to allow you to test your understanding
- interesting facts for additional information
- sample questions and answers to highlight common mistakes or misunderstandings
- sample examination-style questions for your practice – with sample answers at the back of the guide.

Revision and exam techniques play a major part in your level of success in an examination, so remember to start your revision well before the actual examination. Use several short sessions to help build up your knowledge of the subject, over a period of time, so that you do not leave too little time to digest and understand the information.

Apart from reading through the text, leave yourself time to attempt the questions. This is also a good method of revision and it allows you to do something practical rather than just reading. It may also be useful for you to make your own summary of the topics as you work through them – this can be done on postcard-sized cards to allow you to quickly reference the information.

Cambridge IGCSE Geography assessment

All students will take Paper 1 and Paper 2 and then either carry out Coursework for Paper 3 or do the written Paper 4 – Alternative to Coursework. You should check the syllabus for the year that you are taking the paper for examination information and details of the individual papers.

Command words

All exam questions ask you precisely what they want you to do and to answer. To make this as precise as possible, they will use **command words** in their questions. It is important to fully understand the meaning of these words in order to answer a question correctly. There are two very simple instructions to follow when doing the exam, i.e., RTQ and ATQ – Read The Question and Answer The Question.

THEME 1: POPULATION AND SETTLEMENT

Chapter 1

Population dynamics

Learning summary

By the end of this chapter, you should be able to:

- describe and give reasons for the rapid increase in the world's population
- show an understanding of over-population and under-population, including the causes and consequences of over-population and under-population
- understand the main causes of a change in population size, including how birth rate, death rate and migration contribute to the population of a country increasing or declining
- give reasons for contrasting rates of natural population change, including the impacts of social, economic and other factors (including government policies, HIV/AIDS) on birth and death rates
- describe and evaluate population policies
- demonstrate knowledge of case studies, including a country which is over-populated, a country which is under-populated, a country with a high rate of natural population growth and a country with a low rate of population growth (or population decline).

A variety of terms are used to differentiate countries with different levels of economic development. Countries with high levels of economic development are termed MEDCs (More Economically Developed Countries) or HICs (High Income Countries). Countries with low levels of economic development are termed LEDCs (Less Economically Developed Countries) or LICs (Low Income Countries). Between these two groups are the MICs (Middle Income Countries). Many of these MICs are also NICs (Newly Industrialised Countries). These terms will be used throughout this section.

1.1 Some important concepts

These terms are a measure of the level of economic development of a country based on its GNI per capita (Gross National Income per person). The value is thought to be a better measure of how 'well off' a country is than the GDP per capita (Gross Domestic Product). However, the levels do change over time. The HIC threshold was originally set in 1989 at $6000. By 2017, a country was classified by the World Bank as an HIC if its GNI per capita was above $12 476, an MIC if its GNI per capita was between $1026 and $12 475, and an LIC if its GNI per capita was $1025 or less. Figure 1.1 shows the global distribution of HICs in 2015.

Figure 1.2 illustrates how the birth rate varies globally and highlights the differences between HICs, MICs and LICs. Globally, the birth rate has been in decline for a long period of time. In 1950, it was 37.2 live births for every 1000 people; in 1970, it was 30.8; in 1990, it was 24.7 and in 2016, it was 18.6 live births for every 1000 people.

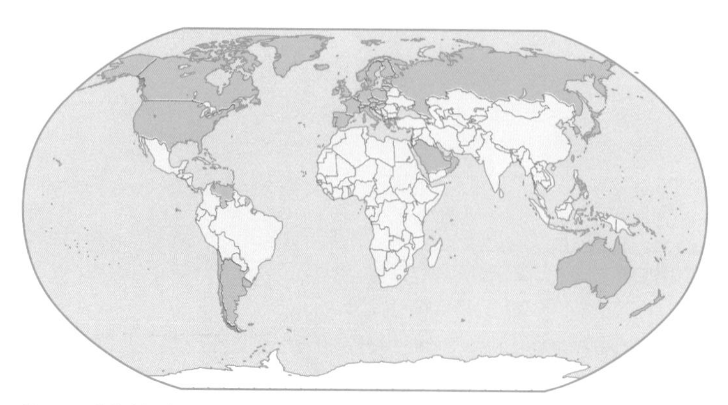

Figure 1.1 High income economies, 2015

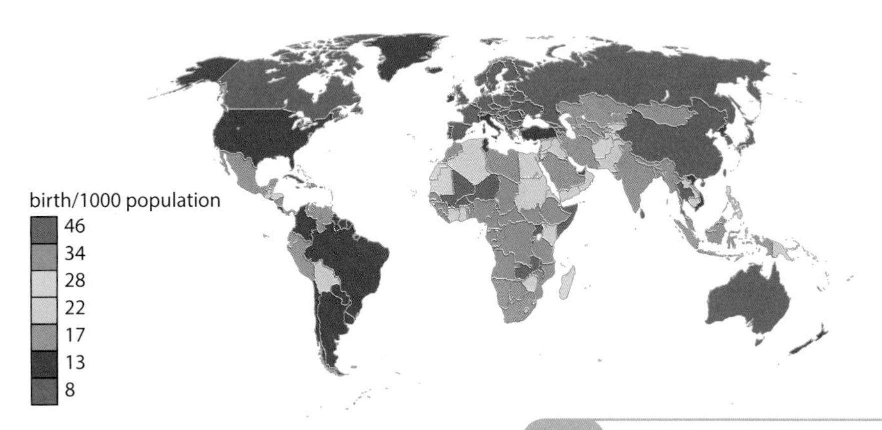

Figure 1.2 Countries by birth rate, 2014

TERMS

GNI per capita: the total value of all the goods and services a country produces (GDP) plus the net income it receives from other countries divided by the population of the country

GDP per capita: the total value of all the goods and services produced in a country in one year by all the people living in that country

birth rate: the average number of live births for every 1000 people in a country per year

TIP

You may be asked to describe the information and trends that you see on graphs. Most of the graphs and diagrams used will be produced specifically for the question and contain information that you should use in your answer. When asked to describe the trends on a graph:

- describe whether the values have gone up/down/stayed the same over the time period
- describe if the change has been rapid or slow
- give the actual values and name the time period
- state whether any increase/decrease has been steady/uniform or whether it has changed in certain periods
- if there are anomalies to the trend, name the years in which it changed and the values (use a ruler to ensure you are precise in finding the exact time/year and recording the exact value on the axes).

1.2 World population increase

Despite the birth rate decreasing, people are living longer, so the world's population in 2017 was estimated to be 7.56 billion and this figure will continue to grow. According to the most recent United Nations (UN) estimates, the world's population is expected to reach 8 billion people in 2024 (Figure 1.3).

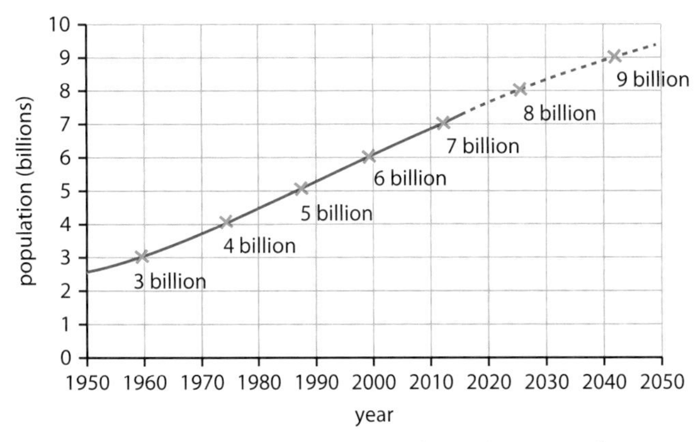

Figure 1.3 World population, 1950–2050 (projected)

However, the rate of growth of the world's population is slowing down – the rate of growth today has almost halved (Figure 1.4) since reaching a peak growth rate of 2.2% per year in 1963.

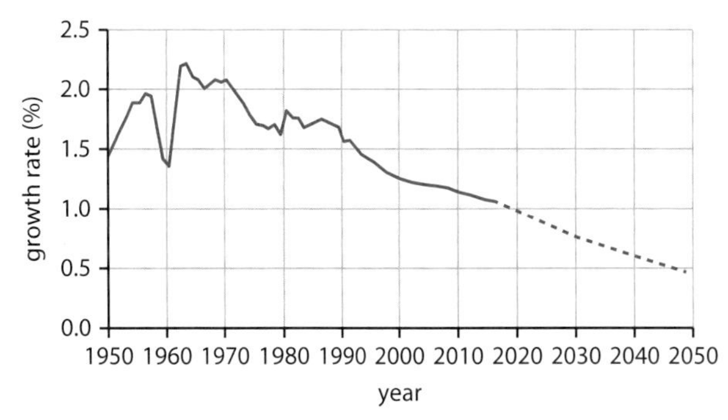

Figure 1.4 World population growth rate, 1950–2050 (projected)

FACT

Figure 1.4 shows that changes in population growth have not always been steady/uniform. A dip in the growth rate from 1959–1960, for instance, was due to the 'Great Leap Forward' in China. During that time, natural disasters and decreased agricultural output in the wake of massive social reorganisation caused China's **death rate** to rise sharply and its **fertility rate** to fall by almost half.

- World births have levelled off at about 134 million per year since the mid-1990s, and are expected to remain constant. However, deaths are only around 56 million per year, and are expected to increase to 90 million by the year 2050.
- Since births outnumber deaths, the world's population is expected to reach nearly 9 billion by the year 2042.
- Population projections/estimates are not always accurate and can vary greatly, a 2014 estimate forecast a population of between 9.3 and 12.6 billion in 2100, and continued growth thereafter. Some researchers have questioned the sustainability of further world population growth, highlighting the growing pressures on the natural environment, global food supplies and energy resources.
- Population growth varies considerably between the continents, Asia has the largest population, with its 4.3 billion people accounting for about 60% of the world population. The world's two most populated countries, China and India, together constitute about 37% of the world's population.
- During the 20th century, the global population saw its greatest increase in known history, rising from about 1.6 billion in 1900 to over 6 billion in 2000.

Reasons for rapid population growth

World population growth, 1750–1900

Up to 1750, both the death rate and birth rate were high. This meant that there was very little growth in the total population. From about 1750, birth rates remained high (around 35 live births per 1000 people), but the death rate dropped rapidly to about 20 deaths per 1000 people. The fall in the death rate was due to:

- improvements in farming techniques, which produced higher crop yields, and improvements in transport; together, they led to food supplies increasing, which reduced starvation and malnutrition

TERMS

death rate (mortality rate): the average number of deaths for every 1000 people in a country per year

fertility rate: the number of live births per 1000 women of the childbearing age group (aged 15–49)

sustainability: the ability of an area or country to continue to thrive indefinitely by maintaining both its economic viability and its natural environment, while meeting the needs of both its present and future generations by limiting the depletion of its resources

life expectancy: the average age to which a person is expected to live, based on the year they were born and their current age

- significant improvements in public health care, including medical breakthroughs (e.g. the development of vaccinations) and improved food handling and general personal hygiene, which came from a growing scientific knowledge of the causes of disease and the improved education
- improved water supply and sewage disposal, so there were fewer deaths from diarrhoea.

This resulted in increased life expectancy. As there was no corresponding fall in birth rates, and more people were living for longer, many countries experienced a large increase in population. This caused the world's population to increase at a much faster rate than previously (a 'population explosion') as the gap between deaths and births grew wider.

TERMS

urbanisation: the increase in the number of people living in towns and cities, causing urban areas to grow

subsistence agriculture: growing enough to feed your family, with little or no extra food to sell for cash

over-population: a country or region that does not have enough resources to keep all of its people at a reasonable standard of living

under-population: when there are not enough people living in a region or country to make full use of the resources at a given level of technology

Factors that contributed to a reduced birth rate	Factors that contributed to a reduced death rate
• access to contraception – family size could be planned • increases in wages – families were better off and no longer required their children to generate income • increased urbanisation – children were not needed as much for work as they were in rural farming families • an improvement in the status and education of women, resulting in women choosing to stay in full-time education for longer and/or deciding to work and to follow careers rather than have children at an early age or have any children at all.	• improvements in health care and nutrition • a reduction in subsistence agriculture • increases in wages.

Table 1.1 Reasons for a reduction in birth rate and death rate

World population growth in the 20th century

In the 20th century, population growth accelerated. This was because as the birth rate fell (to about 20 live births per 1000 people), the death rate also continued to fall (to about 15 deaths per 1000). Life expectancy continued to increase resulting in a rapid increase in population. The reasons behind these changes are shown in Table 1.1.

FACT

Even with all these improvements, the World Health Organization (WHO) estimated that 303 000 women died from complications related to pregnancy or childbirth in 2015. In addition, for every woman who dies in childbirth, dozens more suffer injury, infection or disease.

1.3 Over-population and under-population

Over-population and under-population are relative terms; they express population in relation to the resources in a country or region at a given level of technology. So, a resource-rich HIC with sophisticated technology could be under-populated, but an LIC with few natural resources (having, for example, infertile soils, a challenging climate for crop growing, or no natural resources such as coal or minerals) and traditional technology may be over-populated. Rural areas in LICs may become under-populated where agricultural output has fallen and depopulation has occurred; land has been abandoned due to rural–urban migration (see Chapter 2), or suffered the impact of a natural catastrophe such as drought and floods; because there is a war; or the impact of communicable diseases, such as HIV/AIDs. Most areas considered as under-populated today are large in area and are resource rich, such as Australia, Canada, Kazakhstan and Mongolia.

The two main causes of over-population are an increase in the birth rate accompanied by a decrease in the death rate. The consequences of over-population include:

- **Water** – globally, more than 1 billion people do not have access to clean drinking water and in 2015, 2.7 billion found water scarce for at least one month of the year. Agriculture consumes more water than any other source and aquifers are being depleted faster than they can be replenished. At the current rate of consumption, two-thirds of the global population may face water shortages by 2025.
- **Food** – there is a possibility that the demand for food will overtake food production if the current rate of output continues by 2050. In 2015, around 795 million people did not have enough food to lead a healthy and active life.
- **Environment** – most current research indicates that climate change, due to human emissions is a major consequence of overpopulation. Global warming may result in higher commodity prices, the collapse of fisheries, loss of species' habitats and more extreme climate events.

Self-test questions 1.1

1 Explain why birth and death rates fell in the 20th century.

2 What is the relationship between population growth and resources?

1.4 The main causes of a change in population size

Changes to the population size and structure are caused by social, economic and other factors, such as political policies and decisions.

The **social factors** include healthcare, lifestyle, education and migration. For example, in the UK, the National Health Service set up in 1948 offered free healthcare to the UK population and this has increased life expectancy. Changes in lifestyle has also increased life expectancy through education and advertising the risks associated with alcohol, cigarettes, certain types of food and the advantages of regular exercise.

The **economic factors** include the availability of employment and wage levels which can trigger migration to or from regions and countries. For example, the UK has seen an influx of migrants from east European countries such as Poland, Hungary and Lithuania since 2004 and more recently from Romania and Bulgaria.

The **political factors** include government population policies, as in China, and civil war, such as in South Sudan. A civil war can contribute to lower population densities in a country or region as people become refugees and leave an area.

Changes to the birth and death rates and migration all have an impact on whether the population of a country is growing or declining. Changes in birth and death rates can bring about a natural increase or decrease in population, while migration can impose its own changes on a country's population (see Chapter 2).

In the 21st century, population growth has begun to level off with low birth and death rates in many countries, particularly in the HICs and MICs. In some countries, mainly HICs, birth rates have dropped to well below replacement level, as has happened in countries like Germany, Italy and Japan, leading to a shrinking population. By 2050, the UN projects that three out of every four LICs will be experiencing below-replacement level fertility. Asia's more developed countries are becoming concerned about declining birth rates.

TERMS

natural increase: the birth rate exceeds the death rate and the population grows

replacement level: the average number of children born per woman at which a population exactly replaces itself from one generation to the next

FACT

Replacement level fertility varies globally from about 2.1 children born per woman in HICs to 3.3 in LICs.

It is now common for women in wealthy HICs to delay having their first baby to four years later than they were in 1970. In the UK in 2016, women over 40 years of age were having more babies than women under 20. There were 15.2 births per 1000 women aged over 40, compared with just 14.5 per 1000 women under 20. The reasons for this include:

- advances in fertility treatment
- rising costs of child rearing
- higher levels of women in education leads to more women in the workplace, which can result in fewer births.

Women in the UK aged between 30 and 34 had the highest fertility of any age group – with 111 births per 1000 women.

1.5 Natural population change

Some LIC and MIC countries have both a high birth rate and a high rate of natural population growth, such as in Kenya, Ethiopia, Nigeria and the Philippines. This can have several social and economic impacts.

The social factors and impacts important here include:

- The health service and hospitals will need to be able to accommodate the needs of a young population. High birth rates will require more specialist midwife nurses and clinics to observe, immunise and vaccinate infants.
- More primary schools will be needed.

The economic factors include:

- Food supply will need to increase, either by increasing a country's own food production or by importing food.
- At some point, these children will need jobs and if the economy has not expanded this may lead to higher levels of unemployment.
- However, this population growth will also provide an increased pool of labour which can stimulate economic activity, industrial production and economic growth.

Other factors include the political decision to increase taxation to pay for the expansion of social care, health care and education

In HICs that have gone through the economic transition from manufacturing-based industries into service and information-based industries – a process called deindustrialisation – fertility rates are well below their replacement level, as Table 1.2 illustrates:

	1917	**1967**	**1992**	**2017**	**2040**	**2099**
Australia	3.1	2.8	1.9	1.9	1.9	1.9
China	5.5	5.3	2.2	1.7	1.7	1.9
France	1.3	2.7	1.7	2	2	2
Germany	2.5	2.4	1.3	1.5	1.6	1.8
Japan	5	2	1.5	1.5	1.7	1.8
Sweden	2.9	2.3	2.1	1.9	2	2
UK	2.1	2.6	1.8	1.9	1.9	1.9
USA	3.3	2.6	2	2	2	2

Table 1.2 Total fertility rates (with projections)

This has led to a population decline and can be seen in many countries in Western Europe (UK, Germany, Italy and Spain) and Japan. The population of these countries is falling due to fertility decline, emigration and, particularly in Russia, increased male mortality. The death rate can also increase due to 'diseases of wealth', such as obesity or diabetes, in addition to ageing.

The impact of HIV/AIDS

Although for most of the world, death rates are falling, there are several countries where it is rising. HIV/AIDS is one of the major reasons for this. It is the major cause of deaths in the continent of Africa. In the 12 countries worst affected in Africa, 1 in 10 people between the ages of 15 and 49 are HIV positive. In some parts of southern Africa, such as Botswana, 40% of adults are infected. About 70% of the world's HIV cases live in sub-Saharan Africa.

The AIDS epidemic is concentrated in seven countries in southern Africa, including Botswana (39% of the population) and South Africa (22%). In these seven countries, the population dropped by 26 million (19%) by 2015 and may drop by 77 million (36%) by 2050. For countries affected by HIV/AIDS, there are six main impacts:

- **Labour supply and the economy** – as more people become infected with HIV/AIDS in the economically active 15–49 age group, there will be fewer people available to work and the development of the country and its economy may actually go into reverse. In agriculture, this means less food can be grown and harvested.
- **The family** – the death of parents in many families means that the grandparents and the children are left to look after the family. Where there are no grandparents, it has meant that there are now thousands more orphaned children in southern Africa.
- **Education** – with a lack of government money in poorer LICs to spend on education about HIV/AIDS, many people are not aware of how they can avoid catching HIV.
- **Poverty** – the lack of money in LICs means that most people are unable to afford the cost of the drugs that are now available to treat the disease.
- **Infant and child mortality** – HIV can be passed on to children by their mothers, which means that the numbers of infant and child deaths increases.
- Dependency ratio – the people who have AIDS are normally in the economically active age range, who look after and support the economically inactive group.

TERMS

deindustrialisation: a process of social and economic change caused by the removal or reduction of industrial capacity or activity in a country or region, especially heavy or manufacturing industry

dependency ratio: the ratio between the economically active, 15–65 years old, and economically inactive, under 15 and over 65 years old, population

1.6 Population policies

Several countries have introduced population policies to exert some form of control on their population. Population control normally involves the practice of either limiting population increase, usually by reducing the birth rate, or increasing population by encouraging higher birth rates and/or immigration. France, for example, gives parents money in the form of child benefits and it also gives the parents maternity and paternity leave from work after the birth of a child.

The practice of population control has sometimes been voluntary – as a response to poverty, environmental concerns, or out of religious ideology – but in other times and places, it has been the law of the country. It is generally done to improve the quality of life for a society or as a solution to over-population. Population control may involve one or more of the following practices:

- increasing the access to contraception
- abstinence
- increasing access to abortion
- educating women about family planning
- improving health care so that infant and child mortality rates drop, reducing the need to have more children
- encouraging emigration to other areas in a country
- decreasing the numbers arriving through immigration
- practising infanticide (the intentional killing of infants or children – particularly of females)
- advertising campaigns putting forward the advantages of a smaller/larger family
- offering bonuses to those people who have smaller/larger families.

The methods chosen can be strongly influenced by the religious and cultural beliefs of the community's members. A specific practice, such as abortion, may be allowed or mandated by law in one country while prohibited or severely restricted in another and this generates great controversy between and within different societies.

Population control by governments often involves either anti- or pro-natalist policies.

Anti-natalist policies may be put into place when countries believe that their population is rising too quickly, leading to possible over-population, when their resources may not be able to sustain their population – they may therefore exceed their carrying capacity. Countries with these policies may look at:

- providing free contraception and sterilisation, including financial incentives to encourage people to be sterilised
- legalising abortion – often a controversial policy in many countries
- laws to limit family size
- encourage the education of women.

TERMS

quality of life: the general well-being of individuals and societies, outlining negative and positive features of life, including a person's family, income and access to services

anti-natalist policies: these aim to lower birth rates and encourage lower fertility rates

pro-natalist policies: these aim to encourage higher fertility rates and so increase birth rates

carrying capacity: the amount of resources in a country necessary to support the country's population

Policies favouring natural population growth are often put into operation in countries with a low rate of population growth (or population decline), where there is a stagnant or falling economically active or working population and/or when there is an ageing population. Such policies look at ways that encourage women to have more babies.

Self-test questions 1.2

1. What are the reasons for contrasting rates of natural population change?
2. Describe the ways in which a country may attempt to alter its population growth.

Note that the question does not ask how to **limit** population growth, but how to **alter** population growth. This means that the ways may include either pro- or anti-natal policies or both. In your answer, you should discuss both.

Case studies

Over-population – Kuwait

Causes

In 1938, huge oil reserves (the sixth largest) were discovered in Kuwait. Kuwait experienced a period of enormous economic growth and prosperity driven by the oil revenues from this discovery. The revenue transformed the country and it went from having a very low population density (see Chapter 4) to a country with a high population density (see Table 1.3).

Year	Population
1950	152 000
1960	262 000
1970	750 000
1980	1 384 000
1990	2 059 000
2000	1 929 000
2010	3 059 000
2016	4 348 000

Table 1.3 The growth of Kuwait's population since 1950

Oil accounts for nearly half of Kuwait's GDP and 94% of its export revenues and government income. In 2017, according to the World Bank, Kuwait had the fifth highest per capita income in the world, $73 017 per person. The growth in Kuwait's economy would not have been possible without the use of migrant labour. They accounted for 69% of Kuwait's total population in 2016. In 2016, Kuwait's population density was 236 people per km². However, without its oil revenues, this figure would be nearer 10 per km².

Kuwait has hot summers, very low precipitation (110 mm per year) and high evaporation rates (Figure 1.5).

TERMS

migrant labour: people who move to a country to work

Consequences

There are no permanent rivers and its sources of groundwater would not be sufficient to sustain a population above its 1960 population of 262 000. While the country is one of the most arid on the planet, its water consumption levels per capita are among the highest in the world, with a daily use of around 550 litres per capita in 2015, more than double the average international rate. With extremely limited natural resources, the government has relied on desalination (the removal of salts and minerals) to provide industry and households with freshwater at a cost of over $2 billion annually. In desalinating seawater and air conditioning operating, Kuwait consumes double to quadruple the electricity of other HIC countries (four times that of the UK).

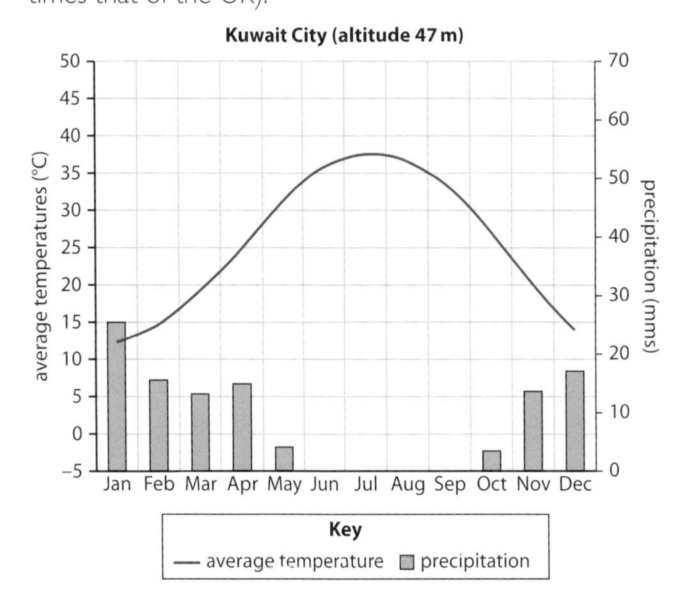

Figure 1.5 Average temperatures and precipitation for Kuwait City, Kuwait

A case study of an under-populated country – Mongolia – can be found in Chapter 4 *Population density and distribution.*

A country with a high rate of population growth – China

One of the most famous examples of government population control is China's **One Child Policy,** introduced in 1979, in which having more than one child was made extremely difficult for a family. China's 'One Child' policy caused a significant slowing of China's population growth, which had been very high before the policy was introduced. The average number of children per woman in China dropped from 6 to 2.5 and, between 1950 and 2005, the birth rate dropped from 44 per 1000 to 14 per 1000 – a figure comparable to many HICs. It has been estimated that at least 300–400 million births were prevented, easing the pressure on China's resources. The policy's impact has now led to China abandoning the policy from January 2016, when the Chinese government decided to allow all couples to have two children as a 'response to an ageing population' and amid concerns over the Chinese economy potentially not having a labour force large enough to allow continued economic growth.

'One Child' measures introduced by the government included:

- couples were allowed only one child
- men could not get married until they were 22 and women 20
- couples had to apply to the authorities to get married and again when they wanted a baby
- couples were rewarded for having only one child by being given a salary/wage bonus (an extra 10%), free education, priority housing, family benefits
- priority in education/health facilities/employment.
- those who did not conform lost these benefits and were given large fines
- women who became pregnant a second time were forced to have an abortion and women who repeatedly became pregnant were sterilised
- a 'workplace snooper' was employed by most factories and businesses, the bosses of these places could grant permission for employees to have a child
- the government advertised the benefits of small families such as having a greater amount of disposable income available.

The Chinese government introduced the policy in 1979 to help solve the social and environmental problems China was facing at that time, including the problem of possibly not being able to feed a fast-growing population, which could have left millions facing starvation and malnutrition. According to Chinese government officials, the policy helped prevent 400 million births. In 1999, the Chinese government started relaxing the policy as the birth rate had fallen from 31 to 19 since 1979. They allowed families in rural areas to have two children.

The policy is controversial both within and outside China because of the issues it raises, the manner in which the policy has been implemented and concerns about the negative economic and social consequences in China. For example, boys are more valued than girls, which has led to female babies being abandoned, and China will now face the problems of an ageing population in the future.

The future of the policy in China

Demographers in and outside China have long warned that its low fertility rate – between 1.2 and 1.5 children per woman – was driving the country towards a demographic crisis. They argued that the human toll has been immense, with forced sterilisations, infanticide and sex-selective abortions that have caused a dramatic gender imbalance, with the result that between 20–30 million men may never find female partners. Also, with China's 1.39 billion-strong population ageing rapidly, and with its labour force shrinking, by 2050 China will have about 440 million people over the age of 60. The working-age population – those between 15 and 59 – fell by 3.71 million in 2015, a trend that is expected to continue.

Since 2013, there has been a gradual relaxation of China's family planning laws that already allowed minority ethnic families and rural couples whose firstborn was a girl to have more than one child, but this was further relaxed in 2016.

A country with a low rate of population growth – Singapore

Since the late 1940s, the government of Singapore has used both anti- and pro-natalist policies. After the Second World War, it attempted to slow and reverse a boom in births. Anti-population policies introduced were accompanied by publicity campaigns urging parents to 'Stop at Two', arguing that large families threatened parents' livelihoods and future security. The Singapore government saw rapid population growth as a potential threat to living standards (widespread unemployment, housing shortages and increasing resource use), education and health care provision, and the political stability of the country.

From the 1980s onwards, however, the government has reversed its policy and has attempted to encourage parents to have more children because birth rates had fallen below replacement levels – seen as one of the country's most serious problems. The old family planning slogan of 'Stop at Two' was replaced by 'Have Three or More, If You Can Afford It'. A new package of incentives for having large families reversed the earlier incentives for small families. It included:

- tax rebates for having a third child
- up to four years' unpaid maternity leave for civil servants
- pregnant women were to be offered increased counselling to discourage 'abortions of convenience' or sterilisation after the birth of one or two children
- a public relations campaign to promote the joys of marriage and parenthood
- a S$20 000 tax rebate for fourth children born after 1 January 1988
- for children attending government-approved childcare centres, parents were given a S$100 subsidy per month regardless of their income
- third-child families were given priority over small families for school registration.

The initial response to the new policy was that the total fertility rate rose from 1.4 in the early 1980s to 1.96 children in 1986, but this was still below replacement level. The government then introduced a Baby Bonus Scheme to remove the financial obstacles associated with having more children. When parents had more than one child, a Children Development Account was set up by the government for the second child of the family, with the government putting S$500 into the account annually and matched, up to another S$1000, for every dollar deposited into the account by the parents. For a third child, the government contributed S$1000 to the account annually and matched, up to another S$2000, in contributions by the parents until the children reached six years of age. This money could be used for the education and development of any child in the family.

Despite these pro-natalist policies, Singapore's population, at 5.77 million, saw its slowest growth in 10 years in 2014. The total fertility rate was 1.24 in 2015, compared to 1.29 in 2013, still well below the replacement rate of 2.1%. The 65 and above age group now forms 12.4% of the population, up from 11.7% in 2013. The total number of marriages and births also fell. Due to the continued low total fertility rate, the government has now changed its immigration policy. As the demand for labour grew with industrialisation, foreigners with key professional qualifications, as well as less-skilled foreign workers, have made up a significant and increasing proportion of Singapore's total population since 2000.

Sample question and answer

Several HICs, such as Japan and Italy, have ageing populations. They now have twice as many people who are over 60-years-old than they have children below 10 years of age. What problems may this cause for these countries in the future? [4]

> **TIP**
> You do not need to know a specific example in this question, but you do need to explain the problems that may occur in HICs in the future and be specific to HICs.

One problem is that HICs such as Japan, will have to find the money and resources to look after the elderly population **[1]** by paying them pensions **[1]**, building retirement homes **[1]** for them to live in, and providing increased geriatric care facilities and specialist treatment in hospitals **[1]**. Also, with so few younger people, they may not have enough people to work in their factories and businesses in the future **[1]**. In the future, the population of Japan will decrease **[1]** because when the older people die there will be fewer people to replace them. Another problem with fewer young people is that some of the schools might be empty and have to close **[1]**. (A maximum of 4 marks will be awarded.)

Exam-style questions

1 Explain why:

a the birth rates are high in many LICs [4]

b the governments of some LICs are attempting to lower their birth rates [4]

c the death rates have started to come down in many countries in recent years. [4]

2 What is meant by the term over-population and what potential problems may be caused by over-population? You should refer to a country or area which you have studied. [6]

Migration

Learning summary

By the end of this chapter, you should be able to:

- explain and give reasons for population migration, including internal movements such as rural–urban migration, and international migrations, both voluntary and involuntary
- demonstrate an understanding of the impacts of migration, including the positive and negative impacts on the destination and origin of the migrants, and the migrants themselves
- demonstrate knowledge of an international migration case study.

2.1 Population migration

Migration is the movement of people from one place to another for economic, social, political or environmental reasons. Migration can be either internal (movement within a country) or international/external (movement between countries). Emigrants are people who *leave* a country. Immigrants are people who *arrive* in a country. Migration can be permanent, where the migrant moves away forever, or temporary, where the migrant returns to their home country at some time in the future.

Reasons for population migration

People migrate due to a combination of both push and pull factors. Push and pull factors are a mixture of natural- and human-caused factors which are summarised in Table 2.1:

TERMS

emigrant: a person who moves out of one country to go to live in another country

immigrant: a person who moves into a country in order to live there

push factors: factors that cause people to move/migrate away from an area

pull factors: factors that attract people to an area

Migration push factors	Migration pull factors
• natural disasters and events (e.g. volcanic eruptions, earthquakes, tsunamis, hurricanes/cyclones/typhoons, floods, drought and rising sea levels) • high unemployment • lack of access to work or progression • lack of health care and other services • poverty due to low incomes • war • racial, political or religious intolerance • lack of safety/high crime • housing shortages • land shortages • crop failure/lack of food/famine.	• higher employment • higher incomes/more wealth • availability of food supplies • better health care and services • better housing and education opportunities • higher standard of living • improved quality of life • greater political stability, racial and/or religious tolerance • more attractive living environment • 'bright lights' syndrome (e.g. when people are attracted by the perceived better lifestyle that cities provide compared to rural communities) • safer/less crime.

Table 2.1 Migration push and pull factors

One of the most common migrations taking place in the world is rural–urban migration – from the countryside to the cities and towns – a process called urbanisation.

Voluntary migrations are often for economic reasons as people look for employment or for improved income.

Involuntary migrations (sometimes referred to as forced migrations) may be as a result of environmental disasters or civil wars that have a political or religious cause. Examples include the recent civil wars in the countries of Syria, Afghanistan, Iraq, Eritrea, Rwanda, Democratic Republic of the Congo and Sudan in Africa. Refugees are people who have been forced to leave their homes because of these reasons.

FACT

In 2016, 65.3 million people around the world were forced from their homes. Among them were nearly 21.3 million refugees, over half of whom were under the age of 18. There were also 10 million stateless people who were denied a nationality and access to basic rights such as education, health care, employment and freedom of movement. In 2016, 34 000 people were forcibly displaced every day as a result of conflict or persecution.

2.2 Impacts of migration

Rural–urban migration

Rural–urban migrations have an impact on both the donor rural area and the receiving urban area.

Rural areas in LICs can become depopulated. It is often the case that the majority of the people who migrate are males. With fewer farmers and people to help with harvesting crops, there can be a drop in food production. Fathers and husbands often leave their families behind for several months or years at a time, leaving the burden of looking after the home and possibly a farm on remaining family members.

Urban areas can find coping with large numbers of migrants difficult with regard to providing housing, health care, education and jobs. Some migrants arrive with very few resources. They may be financially very poor and as a result can end up in squatter settlements.

In some HICs, rural areas are seeing a reduction in their population, such as Snowdonia, an upland area in North Wales, UK. The reasons include:

- few jobs as job opportunities are limited to sheep farming, forestry and water supply, all of which require few workers, and the slate quarries which used to employ many people have closed
- the high land, which is unsuitable for building on and has poor road communications, with narrow, winding roads through mountains.

People are leaving the area as there has been a decline in job opportunities and the area is not attractive for development due to unsatisfactory road and rail links.

TERMS

rural–urban migration: the movement of people from rural to urban areas

standard of living: the factors that affect a person's quality of life and which can be measured; many measures to do with a person's standard of living are to do with material possessions

urbanisation: the increase in the number of people living in towns and cities, causing them to grow

voluntary migration: a migrant chooses to leave their country or a region

involuntary (forced) migration: where a migrant has no choice and has to leave their country or a region

refugee: a person who has been forced to migrate in order to escape war, a natural disaster or persecution but who does not have another country to go to

squatter settlement: an area of makeshift housing that usually develops in unfavourable sites in and around a city; also known as 'shanty' towns, favelas or bustees depending on their location

Urban–rural migration

In some HICs, urban–rural migration (where people migrate from towns and cities to rural areas) is taking place. This is sometimes called counter-urbanisation and normally refers to the movement of people but it may involve the movement of some businesses and economic activities. The push factors for it include an increase in urban areas of the cost of housing, traffic congestion, noise and air pollution, high levels of crime and a poor-quality living environment with a lack of open space. The pull factors of a rural environment include a better living environment, lower cost of housing, less crime and the ability to commute longer distances to work with improved road and rail links to urban areas.

The older, inner city areas can become depopulated, the so-called doughnut effect. It can lead to abandoned housing and services declining as there is a lack of tax revenue coming in from a smaller population. The rural areas can find themselves under pressure to release more land for housing, house and land prices may rise and that can mean local people, especially the young adults, are unable to afford to live in their local area.

TERMS

counter-urbanisation: the process by which an increasing number of people within a country live in the countryside instead of in towns and cities – this could be the result of natural increase in population and/ or migration

doughnut effect: the migration of people from the central parts of cities to the outer suburbs to live and work

economic migrant: a person who emigrates from one country or region to another to seek an improvement in their standard of living (the UN uses the term 'migrant worker')

remittance: transfer of money by a foreign worker to their home country

Voluntary migration for economic reasons

The Gulf States of Qatar, UAE, Bahrain and Kuwait are examples of countries who have received large numbers of economic migrants. The wages earned by migrants in the Gulf States are often sent back to their families, known as remittance. The money received, improves their standard of living and improves their quality of life. Remittances can have a major impact on the local communities in the areas migrants originate from. In Kerala in southern India, for example, 21% of the regions GDP comes from money (remittances) being sent back home by migrant workers. The money is spent in local shops and in the building industry as people either build new homes or improve the quality of their homes. Kerala has over 1 million 'Gulf wives' living apart from their husbands. Of the total population of the UAE (6.4 million), 5.5 million are migrant/ expatriate workers.

Between 2003 and 2015, total remittances worldwide went from $100 billion to $583 billion and involved over 200 million migrant workers. Of this 2015 total, $435 billion went to MICs and LICs. In 2015, India received an estimated $72 billion and China an estimated $64 billion. For some recipient countries, remittances can be as high as a third of their GDP (Table 2.2).

Country	Remittances as a percentage of national economy (2015)
Tajikistan	42
Kyrgyzstan	32
Nepal	29
Moldova	25
Lesotho	24
Samoa	24
Haiti	21
Armenia	21
The Gambia	20
Liberia	19
Lebanon	17
Honduras	17
El Salvador	16
Kosovo	16
Jamaica	15

Table 2.2 Examples of countries in which the remittances from migrant labourers working in other countries play a major role in the national economy

Many economies rely on population growth to provide the workforce for a growing economy. If natural population growth cannot do this, one solution is to fill the gap with migrant labour. This has been done in Germany, for example, where many of the 1 million migrants into Germany in 2015/16 entered the workforce. An increasing proportion of people in many HIC countries are entering retirement age, which creates an economic burden on a shrinking, economically active, working population (they will have to pay more taxes). More migrants entering the workforce of a country can mean more taxes being paid to the government which can be used to support infrastructure projects and/or for social benefit schemes, such as the support of retired people through the provision of, or increase in, pensions and geriatric care.

Between 2000 and 2015, migration into 28 countries (including Canada, Sweden, Denmark, Germany, Italy, Spain, Qatar, Singapore, UAE and UK) either prevented population decline or it doubled the amount of natural increase to produce population growth. However, the population of 51 countries, including Germany, Italy, Japan and most of the countries in the former Soviet Union, is still expected to be lower in 2050 than in 2015.

Effects of migration on migrants themselves

Migrants who move to a new country, whether voluntary or forced can face several problems and difficulties which have consequences. Some of these are shown in Table 2.3.

Problem	Consequences
A lack of qualifications/ skills/education/no experience	Many end up doing low-paid, unskilled jobs or they are unable to obtain employment as there are not enough jobs
Cannot speak the language	They may be exploited by business and factory owners, working long hours and in poor conditions
Higher living costs in the country they move to	Unable to buy homes and end up living in poor, overcrowded conditions
Discrimination; some people may think that their jobs are being lost to the immigrants	Lack of safety, high risk of crime or being hurt or attacked
Migrants may have entered the country illegally	Fear of arrest means they cannot access services for fear of being caught and deported.

Table 2.3 The problems faced by many migrants and the consequences for them

The positive impacts on migrants can include them receiving higher wages in the countries they move to. East European migrants moving to the UK could earn wages five times greater than they could get at home. More qualified migrants may look for medical or education jobs, which are much better paid and increase their skill levels, including language skills, and experience. Families remaining at home can receive remittances from migrants which increases their standard of living and quality of life.

Self-test questions 2.1

1. What are the reasons for people choosing to migrate within a country?
2. What are the positive and negative impacts of migration on the area or country from which people have migrated?

Self-test questions 2.2

1. a With the use of a case study, explain the reasons that have led to an international migration.

   b Explain the impacts of this migration on the destination countries and on the migrants themselves.

Case study

TIP

Ensure you revise the case study you have studied in class. You need to know about an international migration and must be able to give specific place information in your answers to access the highest level.

An international migration

The foreign-born population residing in the European Union (EU) in 2015, including the UK, amounted to 33 million people, or 7% of the total population of the 28 EU countries (about 508 million people). By comparison, the foreign-born population is 1.6% of the total population in Japan, 13% in the United States, 20% in Canada and 27% in Australia.

What became known as the **European migrant crisis** began in 2015, when a rising number of refugees and migrants made the journey to the EU to seek asylum and/or work. Most of the migrants travelled across either the Mediterranean Sea or via Turkey into Greece and South East Europe. The migrants came from areas such as Syria, Iraq, South Asia and Africa (Figure 2.1).

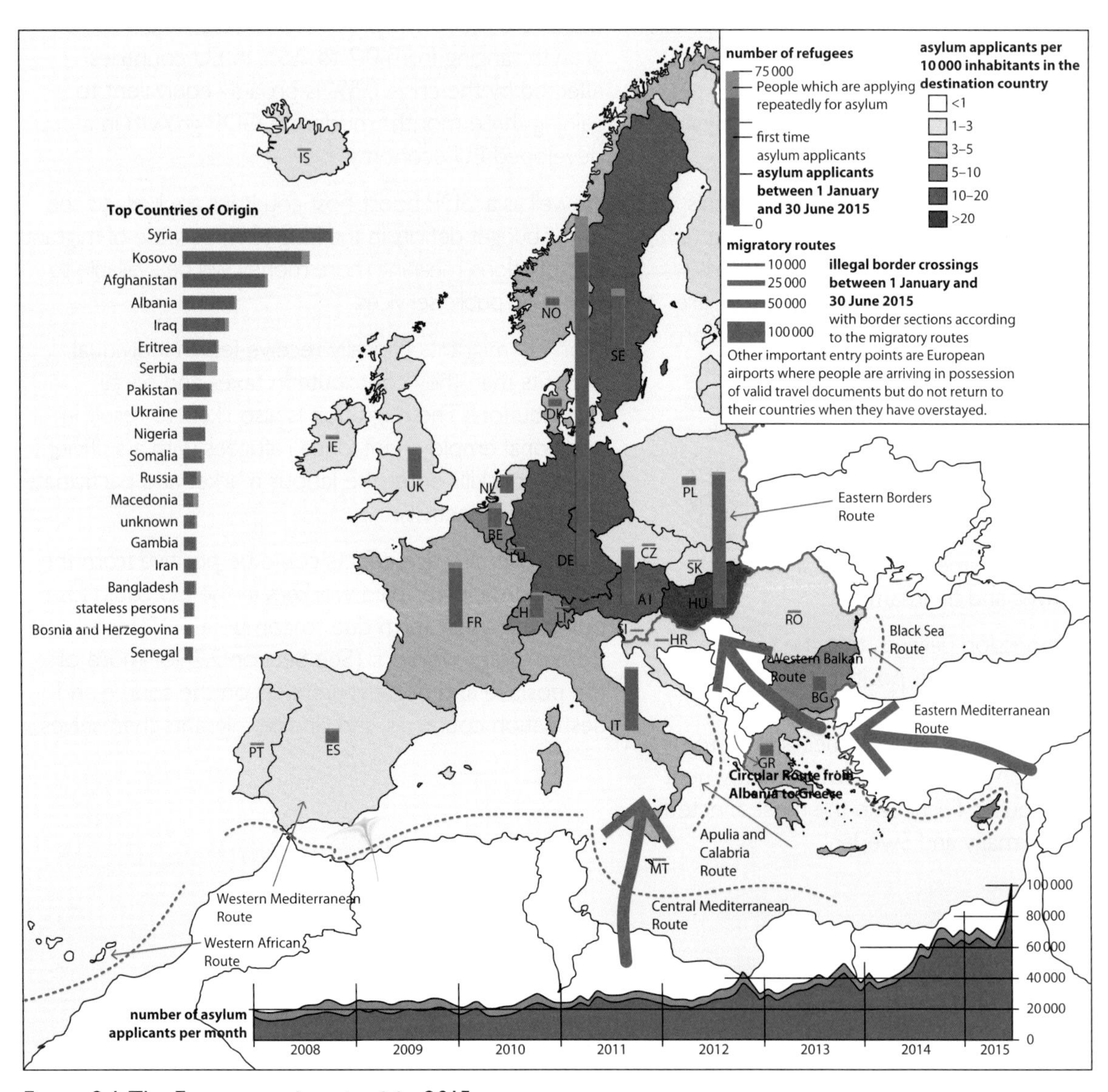

Figure 2.1 The European migrant crisis, 2015

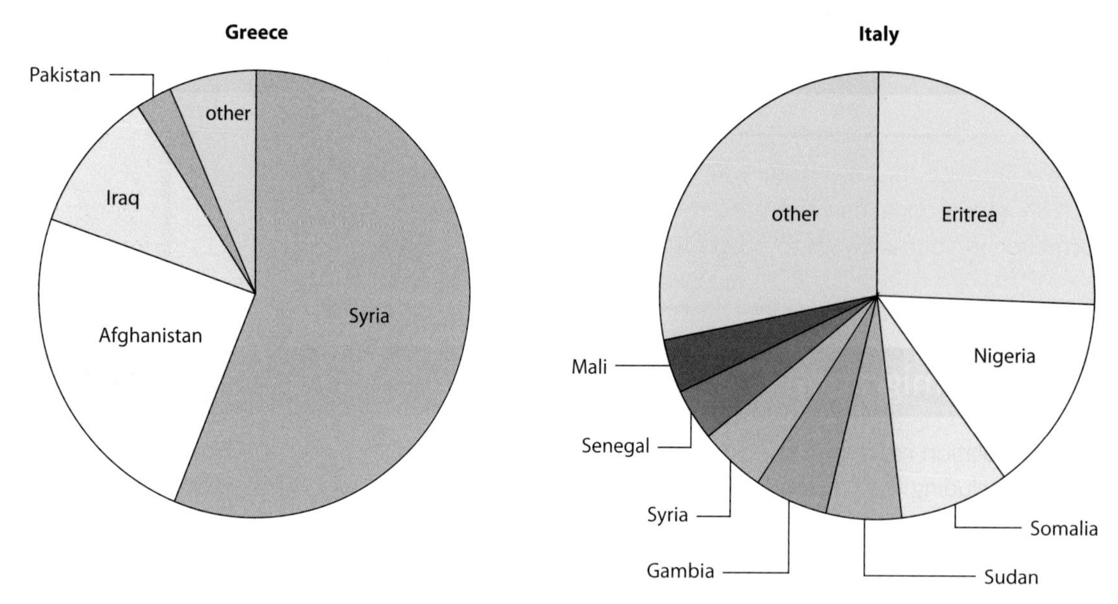

Figure 2.2 Migrant arrivals to Greece and Italy by nationality, 2015

According to the UN High Commissioner for Refugees (UNHCR), the top three nationalities of the over 1 million arrivals in 2015 were Syrian (49%), Afghan (21%) and Iraqi (8%). Figure 2.2 shows the nationalities of the Mediterranean Sea migrants to Greece and Italy in 2015, according to UNHCR data.

There are many push and pull factors that caused this migration. Most of the migrants are refugees, fleeing war and persecution in their home countries of Syria, Afghanistan, Iraq and Eritrea. Refugees from Eritrea, are seeking asylum to avoid indefinite military conscription and forced labour in one of the most repressive states in the world. Migrants from parts of West Africa (such as The Gambia, Nigeria) and South Asia (Bangladesh, Pakistan) are often economic migrants, moving away from poverty and a lack of jobs, many of them hoping to settle in a country with better prospects that will allow them to gain employment and have a better lifestyle for themselves and their families.

The European Commission believes that the refugee crisis is actually having a 'sizeable' positive economic impact on some EU countries. The large influx of people from Syria and other conflict zones is likely to have a positive effect on economic growth, employment rates, and long-term public finances in the most affected countries, such as Germany and Sweden.

It is estimated that the 3 million migrant arrivals by the end of 2016 will produce increases in annual GDP growth ranging from 0.2 to 0.5% in EU countries affected by the crisis. 0.5% is broadly equivalent to gaining three months moderate GDP growth in a developed EU economy.

As well as a GDP boost, host countries are likely to see lower budget deficits in the long term because of migrants' contributions, meaning more money will be available to spend on public services.

Non EU migrants typically receive less in individual benefits than they contribute in taxes and social contributions. The migration is also likely to result in additional employment, once refugees with a sufficient degree of skills enter the labour market and participate in economic activity.

While overall wage effects could be positive from the influx, some lower paid workers in the EU could lose out in the short term due to competition for jobs from migrant workers. (See Section 2.2. for more of the positive and negative impacts on the source and destination countries, and on the migrants themselves.)

Sample question and answer

The number of international migrants in the world is increasing rapidly.
Using named examples that you have studied, explain why people may migrate from one country to another. [6]

TIP

If a question asks for 'named example(s) that you have studied', make sure that you include an example or examples. Note carefully whether just one or more case study examples are needed.

Many people in parts of West Africa (e.g. The Gambia and Nigeria) migrate to a HIC to get a job because they cannot get one in their own country, possibly due to high unemployment **[1]**. Others, such as from Pakistan and Bangladesh, are looking to get a higher standard of living and improved quality of life in a HIC **[1]**. Some people are escaping to get themselves away from war zones **[1]**, such as the war and civil conflicts being experienced in Syria and Afghanistan **[1]**, or from religious or racial conflict, for example the Rohingya people, a Muslim minority group migrating from Myanmar **[1]**. Some people move because they want to escape natural disasters such as earthquakes, volcanoes and droughts **[1]**. (A maximum of 6 marks will be awarded).

Exam-style question

1 Describe the positive and negative effects that the migration of people may have on the area to which they migrate. [6]

Population structure

Learning summary

By the end of this chapter, you should be able to:

- identify and give reasons for and implications of different types of population structure, including age/sex pyramids of countries at different levels of economic development
- demonstrate knowledge of a case study of a country with a high dependent population.

3.1 Different types of population structure

Studying the population structure of a country/region can involve examining several different demographic/population characteristics, including age, sex, fertility (the average number of children born to a woman), mortality (the death rate) and migration (movement of people), race, language, religion, occupation, and so on. The two most studied characteristics are age and sex.

Features of a population pyramid

Age and sex are normally shown together in a population pyramid. These allow population structures to be compared both in a visual and a quantitative way. They are useful in showing changes in population structure over time. They allow national, regional and local government agencies, in areas like health care and education, to project into the future and plan for the future provision of school places and health care services.

Population pyramids are split in half to show males on the left and females on the right. They have a vertical axis which is normally in 5-year intervals, but can be more detailed and have individual years. The pyramids can either use absolute figures in thousands and millions or, more commonly, percentages of the total population. Figure 3.1 shows an example of two population pyramids that contrast the population structures of the world (bordered) and the European Union (EU, coloured in green and orange) in 2013. The EU pyramid reflects the characteristic shape of HIC pyramids, while the world pyramid shows more of the characteristics of LIC pyramids.

Population pyramids can differ considerably in shape from country to country and within any country

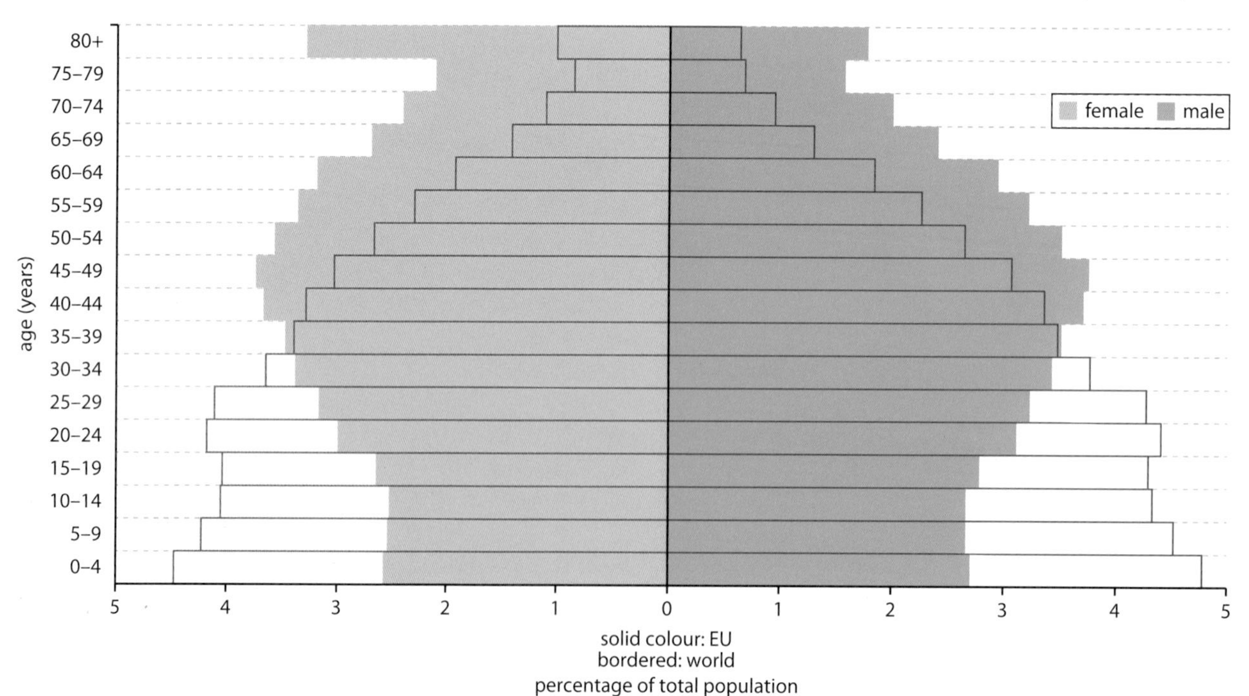

Figure 3.1 The population structures of the world and the European Union, 2013

and over time. They can also be presented in slightly different ways, so in this chapter a variety of ways will be illustrated. They are useful tools in illustrating and analysing changes and trends in population structure.

TIP

If you are asked to compare pyramids, make sure you use comparative statements. For example: 'The pyramid for Country X (an HIC) has a narrower base compared to the broader base of Country Y (an LIC)'.

TERMS

population structure: a term used to describe the structure/composition/make-up of a population in a country or region

population pyramid: a graph that shows the age-sex distribution of a given population

sex ratio: the ratio of males to females in a population

economically active: the 15–65 age group

economically inactive: the age groups below 15 and above 65 who are dependent on the economically active group

dependency ratio: the ratio between the economically active and inactive groups

Ageing population

Globally, the natural sex ratio at birth is estimated to be close to 1.06 males/female – that is, 106 males are born for every 100 females born.

In most countries, however, adult males tend to have higher death rates than adult females of the same age (even after allowing for causes specific to females such as death in childbirth). This is due to natural causes such as heart attacks and strokes, which account for the majority of deaths, and also to violent causes, such as murder and warfare. This results in a higher life expectancy for females.

FACT

In the USA, an adult non-elderly male is 3 to 6 times more likely to become a victim of murder and 2.5 to 3.5 times more likely to die in an accident than a female of the same age.

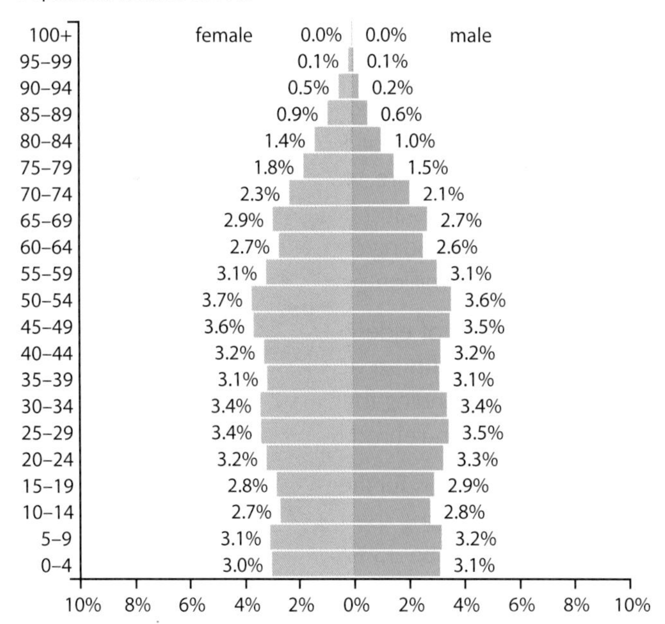

Figure 3.2 UK population pyramid, 2016

The sex ratio tends to reduce as age increases. Among the elderly, there is usually an excess of females, which can be seen for the EU in Figure 3.1. The male-to-female ratio falls from 1.05 for the group aged 15–65 to 0.70 for the group over 65 in Germany, from 1.00 to 0.72 in the USA, from 1.06 to 0.91 in mainland China and from 1.07 to 1.02 in India.

In the UK, the 1:1 ratio stays the same until the 55–59 age group, where both sexes make up 3.1% of the total population, as shown in Figure 3.2. After this, differences in the male and female percentages appear: in the age group 75–79, it is 1.5% and 1.8%; at age 85–89, it is 0.5% to 0.9%.

The issues of an ageing population

In the 20th century, life expectancy rose by 30 years. By 2050, it is estimated that 75% of people in HICs will live to over the age of 75.

In the examples of Germany and the UK, the 15–65 age group are economically active and below 15 and above 65 are economically inactive. The ratio between the economically active and inactive groups is called the dependency ratio.

As people live longer, they place an increasing burden on the economically active group. In many HIC countries, such as Germany and the UK, the number of economically inactive people over the age of 65 compared to economically active people has changed remarkably since 1950. In the UK in 1950, there were 15 retired people to every 100 working people. By

2000, this had increased to 20 in every 100 and by 2040 is estimated to be 40 in every 100.

A large amount of money and resources are spent in HICs on people over working age. In the UK in 2011, for example, the equivalent of $150 billion, or one-seventh of UK public expenditure, was spent. Continuing to provide state benefits and pensions at this level would mean additional spending of $12.5 billion a year for every additional 1 million people over working age.

Growing numbers of elderly people also have an impact on the national health services in countries like the UK, where average spending for retired households is nearly double that for non-retired households.

In Italy, the situation is further advanced and 19% of its GDP is currently being spent supporting the elderly. By 2030, this could be 33%, which is unsustainable in the economic situation that has prevailed in Italy since it became part of the global recession from 2008. Italy may not have the financial resources to support its larger proportion of elderly people.

The effects of an ageing population include the following:

- an increase in proportion of elderly people has taken place at the same time as a rapidly declining birth rate which has resulted in a fall in the total population
- labour shortages – several countries, including Germany, overcome this by importing workers (migrant workers)
- increased spending on medical services for the elderly – geriatric health care and retirement homes to look after the elderly
- the under use and closure of other services such as schools
- more people will require a pension for a longer period of time as life expectancy increases; to counter this, many HICs are putting up the age when people can obtain a state pension, to 67 years and later.

Changes in MICs

Figure 3.3 shows a population pyramid typical of a MIC, India. The wide base of the pyramid indicates that the birth rate is still quite high, as it is still traditional, in rural India in particular, to have larger families and the rapidly sloping sides above the age of 30 show that the death rate was high in the fairly recent past, due to a relative lack of health care for many people.

The base of the pyramid (under 10–14 years) shows that the birth rate is falling rapidly and that the death rate has stabilised at a low level. The fall in the birth rate started with the introduction of India's population policy, which discouraged large families. Much of the reason for the national fertility decline occurred in the southern states, which generally have higher rates of literacy and education, along with greater equality for women.

The percentage of young economically dependent people and young adults is now much bigger than the percentage of elderly economically dependent age groups.

At the base of the pyramid, the male: female ratio is quite large, 1.11 : 1, and indicates the cultural preference for male children in India.

Implications

There are social and economic implications of the fall in birth rates. With so many people under 20, the population of India will continue to rise in the future until at least 2060. Although family size will probably fall, the number of families will continue to increase, so the demand for housing and employment will continue to rise.

After 2070, however, India's population may fall as the impact of a falling birth rate for several decades will start to impact on population numbers. This, in turn, will have implications again for government agencies, with a smaller workforce and increased resources needed to care for the elderly.

Changes in HICs

Population pyramids are capable of revealing anomalies that do not fit the 'normal' patterns seen in most countries. The Gulf States highlight one interesting variation, as seen in the example of Qatar (Figure 3.4).

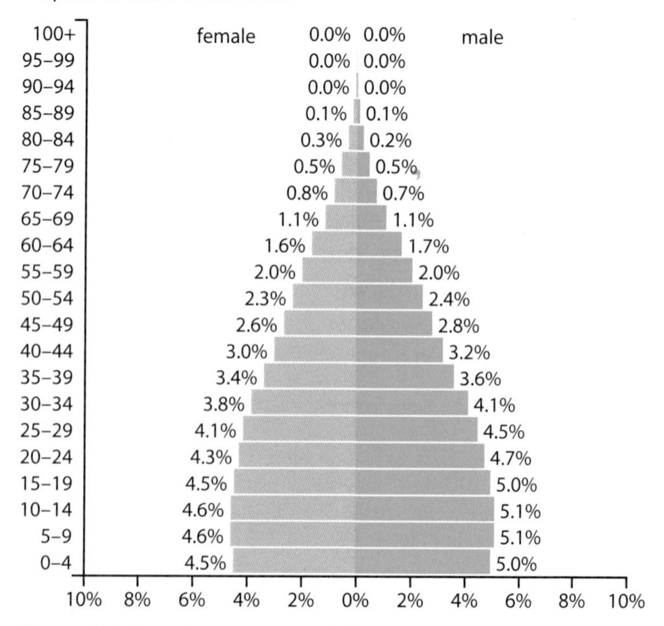

Figure 3.3 Population pyramid for India, 2015

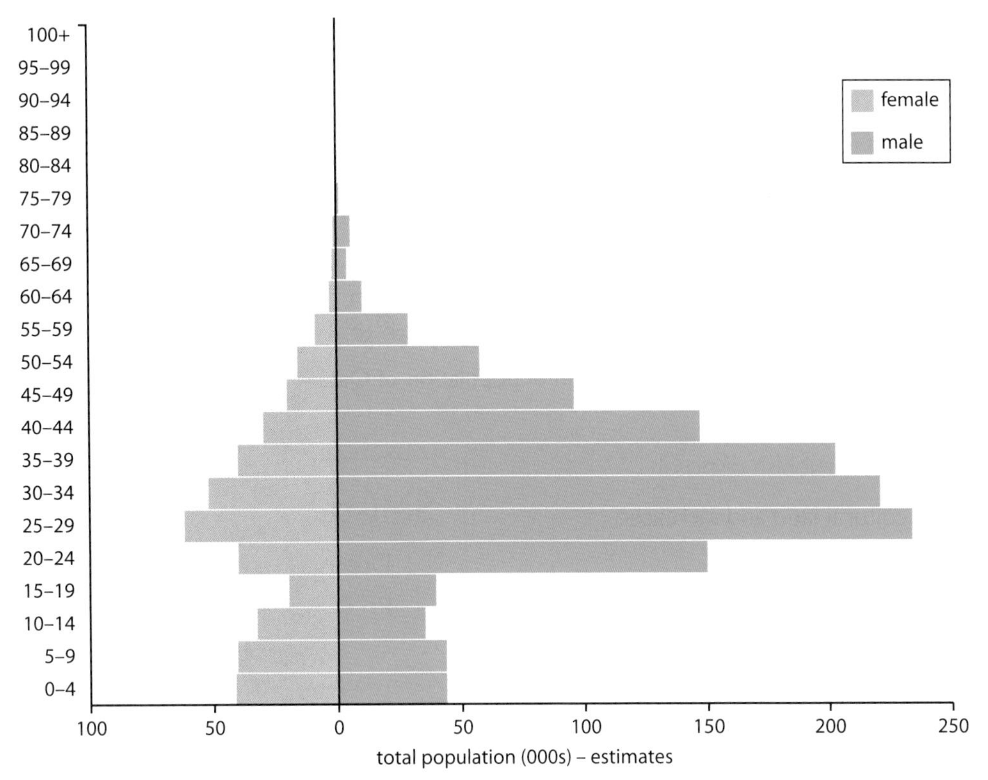

Figure 3.4 Population pyramid for Qatar, 2014

What is very noticeable in Qatar's population pyramid is the massive bulge on the male side of the age-sex structure, between the ages of 20 and 50. This represents the foreign males who have moved to Qatar to work in the construction and oil industries. There is also a bulge in the female side, but to a lesser degree.

With the world's third largest natural gas reserves, Qatar is now the largest exporter of Liquid Natural Gas (LNG) in the world. As a result of huge revenues from LNG sales, Qatar now has the highest GDP per capita in the world, and approximately 14% of Qatari households are dollar millionaires. The country relies heavily on foreign labour to grow its economy, to the extent that migrant workers with temporary residence status compose 86% of the population and 94% of the workforce. Qatar now has the highest ratio of temporary migrant workers to its domestic population in the world.

The exploitation of its gas and oil resources has transformed the country. In 1980, Qatar was a country of just 0.2 million people, making it one of the smallest in the world. Its growth to 2.2 million in 2016 has been through the import of temporary migrant labour rather than natural growth – the fertility rate for Qatari females has fallen to 2.4 compared to the 5.45 children per woman in 1980. Almost all the migrants are in unskilled jobs and most are male, resulting in three out every four residents in Qatar being male.

Qatar and other countries in the area, such as Kuwait, Bahrain and the UAE, have experienced very rapid population growth. This has had a beneficial effect in these countries in allowing them to grow their economies.

Implications

However, in some countries, rapid population growth in a country or region can create challenges, including:

- not enough resources to supply a larger population
- lack of work
- inadequate food supplies
- poor access to education and health care
- overcrowded housing / lack of space for development
- lack of housing leading to the development of squatter settlements (e.g. Nairobi in Kenya or the favelas of São Paulo and Rio de Janeiro in Brazil)
- increased traffic congestion
- increased atmospheric pollution
- inadequate clean water supply and sewage disposal

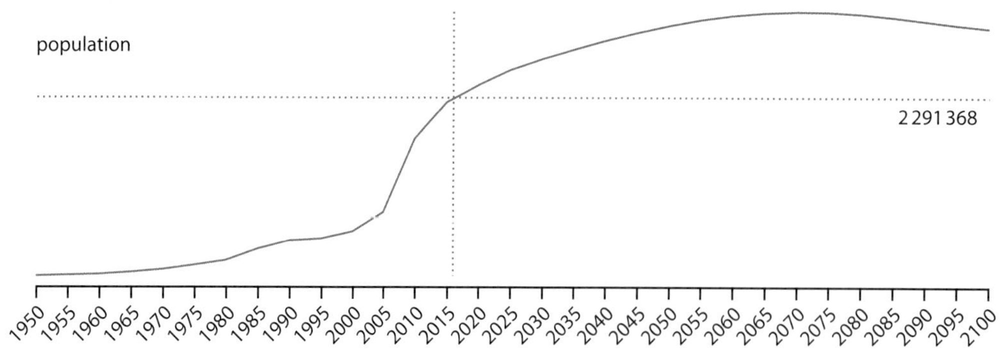

Figure 3.5 Growth and projected growth of population of Qatar, 1950–2100

- lack of facilities for waste/rubbish disposal
- the overuse of agricultural land and increased overgrazing
- the deforestation and loss of natural vegetation and habitat (e.g. Haiti).

Self-test question 3.1

1 Describe the main features of the population pyramid of India (Figure 3.3), an MIC.

Case study

A country with a high dependent population – Somalia

Somalia is located in East Africa, in the so-called Horn of Africa. In 2016, its population was estimated at 11.1 million, up from 10 million in 2013. The country's population is rapidly expanding with almost 3% annual population growth and a high fertility rate of 6.26 children per woman, which is the fourth highest in the world. With its rapid growth and high fertility rate, Somalia's population is expected to reach 13.2 million by 2020.

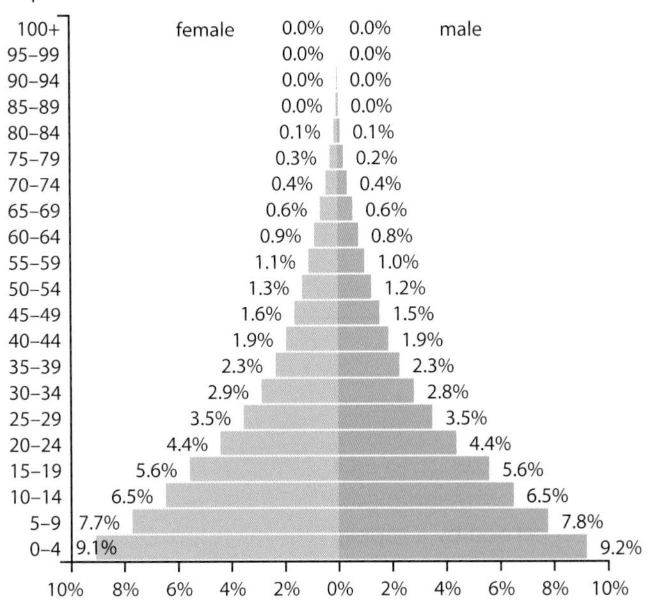

Figure 3.6 Population pyramid for Somalia, 2015

The population pyramid of Somalia (Figure 3.6) is triangular and has the concave shape typical of a LIC. The wide base indicates a high birth rate and the sloping, concave sides show that the death rate is high in all age groups. It also shows that life expectancy is low (54 years) and infant mortality rates are very high (72 deaths per 1000 live births), the latter shown by the marked reduction in population between the 0–9 age groups. Figure 3.4 (Qatar) also shows that the number of young, economically dependent people is very high.

With so many people under 15, the population is almost certainly going to continue to increase for many years. The Somali government has major concerns about whether food production will be able to keep up with this growth in the population (Figure 3.7). There are also concerns for the economy, and whether it will grow fast enough to provide employment for the young people who will enter the 15+ age group shortly and whether it can raise enough revenues from taxation to pay for the improvements in infrastructure needed for both the economy to grow and to provide the education provision and health care needed by such a large, dependent population.

TERM

infant mortality rate: the number of deaths of babies/children under one year of age per thousand live births per year

Educational opportunities have been limited at all levels in Somalia. The schools that do manage to function have low enrolment and only the most basic materials, especially in rural areas. Unclean drinking water, poor sanitation, malnutrition, refugees, along with the high infant mortality rate are all widespread problems. Many preventable diseases are widespread, especially in rural areas where 60% of the population live, or have occasional outbreaks, including intestinal illnesses, malaria, dengue fever, cholera, measles, Rift Valley fever, tuberculosis, leprosy, and bilharzia.

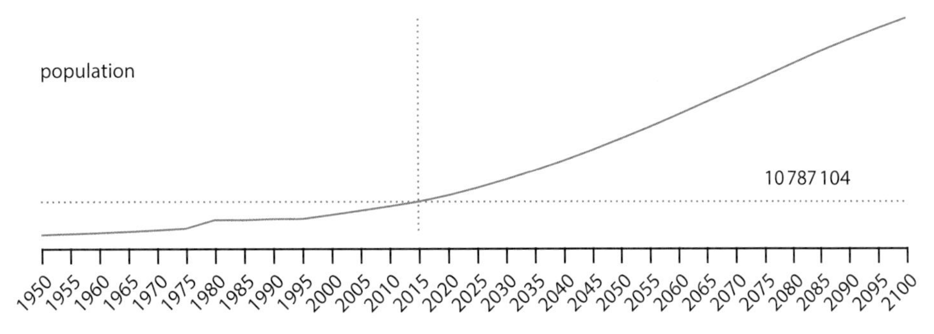

Figure 3.7 The growth and projected growth of population in Somalia, 1950–2100

Although it lacks effective national governance, Somalia maintains an informal economy largely based on livestock, remittances from Somalis working abroad, and telecommunications. Somalia's government, however, lacks the ability to collect domestic taxes and has to borrow heavily from international agencies to keep basic functions operable.

Sample question and answer

Describe the main features of Figure 3.6, the population pyramid for Somalia (an LIC) and suggest reasons for these features. [6]

TIP

You need to read the question carefully and then answer the question that is being asked. There are really two questions here:

i describe the main features of the population pyramid and then

ii suggest reasons for these features.

Many students make the mistake of answering just one of these and so gain only half the available marks.

The pyramid has a broad/wide base **[1]**. There is a large percentage of the population below 15 years; it has a large number of economically inactive young people **[1]**. There is a relatively small population; number of people over 65 years of age; it has a small number of economically inactive old people **[1]**. It has a high dependency ratio **[1]**.

The reasons for these features include a high birth rate **[1]**, a high death rate **[1]** and a low life expectancy **[1]**.

(A maximum of 6 marks will be awarded.)

Exam-style questions

1 Describe the contrasting ways in which the dependent populations may be supported in LICs and HICs. [4]

2 a Suggest reasons why HIC countries, like the UK, have an ageing population. [3]

b Suggest the likely effects of this population trend on the UK economy by 2050. [5]

3 With reference to examples you have studied, explain why some governments may be concerned by a rapidly growing population. [6]

Self-test questions 3.2

1 What evidence from the population pyramids for Somalia (Figure 3.6) and the UK (Figure 3.2) suggests that:

a people have a lower life expectancy in Somalia?

b there is a high birth rate in Somalia?

2 How does the dependent population of Somalia (Figure 3.6) differ from that of the UK (Figure 3.2)? Support your answer with figures from the population pyramids.

Population density and distribution

Learning summary

By the end of this chapter, you should be able to:

- describe the factors influencing the density and distribution of population, including physical, economic, social and political factors
- demonstrate knowledge of a case study of a densely populated country or area and a sparsely populated country or area (at any scale from local to regional).

4.1 Population density

Population density describes the number of people living in a given area, usually per square kilometre. It can be used on a variety of scales, from continents to countries, to regions within a country. It is calculated by dividing the total population by the total area.

$$\frac{\text{total population}}{\text{total area (km)}} = \text{population density}$$

For example, India in 2017, the population density was 408 people per sq km.

$$\frac{1.34 \text{ billion}}{3\,287\,623 \text{ sq km}} = 408 \text{ people per sq km}$$

For the Earth as a whole, with a population of 7.56 billion and a total land area of 149 million km^2, the population density is 50.85 per km^2. Considering that over half of the Earth's land mass consists of areas inhospitable to human habitation, such as deserts and high mountains, and that population tends to cluster around seaports and freshwater sources, this number by itself does not give an accurate measurement or picture of human population density. It therefore needs to be broken down into country, region and area levels to be a useful statistic. Table 4.1 shows a selection of countries to illustrate the great variations in population density.

Rank	Country	Population	Area in km^2	Population per km^2
1	Monaco	37969	1.95	19471
2	Singapore	5760000	699	8240
11	Bangladesh	164305000	143998	1141
33	India	1340000000	3287623	408
38	Philippines	103377000	300076	345
40	Japan	126385000	377873	334
51	United Kingdom	65746000	243000	271
56	Pakistan	192000000	803940	239
84	China	1387000000	9596961	145
90	Indonesia	262703000	1904569	138
118	Egypt	94710000	1001449	95
170	South Africa	55408000	1221037	45
182	USA	325835000	9629091	34
224	Russia	143392000	17098242	8
230	Canada	36533000	9984670	4
236	Australia	24931000	7682300	4
244	Greenland (Denmark)	56227	2175600	0.03

Table 4.1 Countries ranked by population density (number of people/km^2) in 2017

TERM

population density: the number of people per square kilometre

FACT

Macau, a region in China, has the highest population density in the world, with 643 100 people sharing an area of 30.3 km² – a population density of over 21 224 persons per km². Antarctica, in contrast, is 14 400 000 km² in area and with a population of roughly 1000, this results in a population density of 0.00007 people per km²!

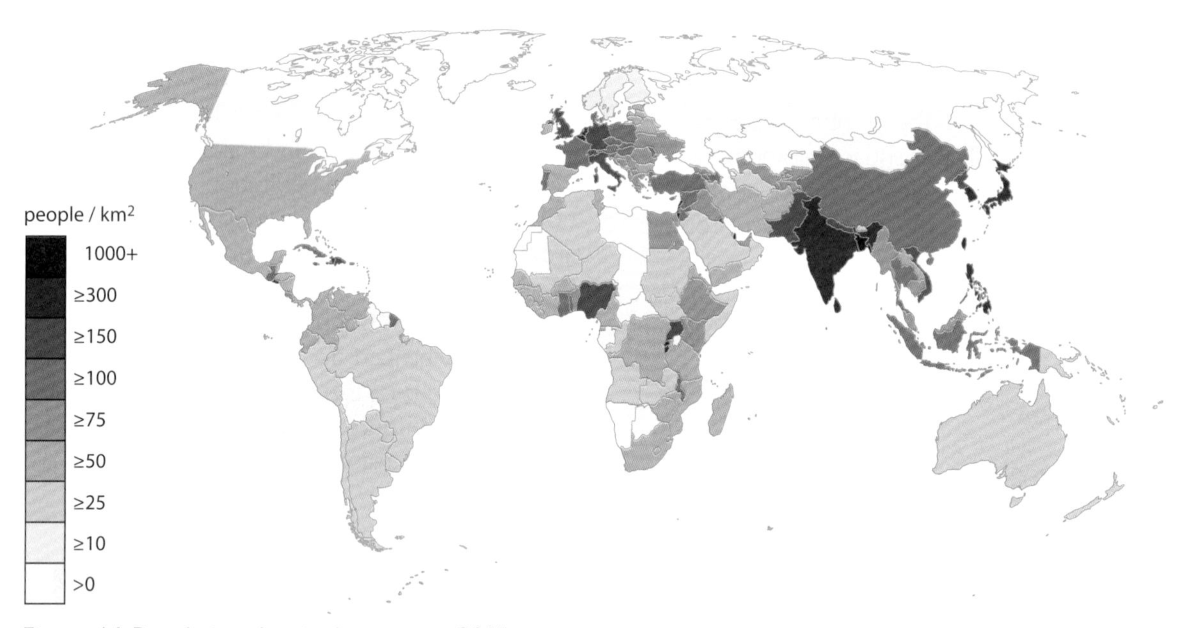

Figure 4.1 Population density by country, 2015

Region	Population	Area (km²)	Density (Pop. per km²)
Taiheiyo Belt (stretching from Tokyo to Nagasaki in Japan)	85 million	60 000	1 417
SE China coast (Guangdong, Hong Kong, Fujian)	140 million	100 000	1 400
Indonesian island of Java	145 million	130 000	1 115
Nile Delta in Egypt	50 million	50 000	1 000
Southern India (Tamil Nadu, Pondicherry, Bengaluru, and Kerala)	120 million	170 000	706
West Indian Coast (Maharashtra and Gujarat Coast)	70 million	100 000	700
BosWash (the area between Boston and Washington in the NE of the USA)	45 million	100 000	450
Northern Europe (Benelux, North Rhine-Westphalia)	44 million	110 000	400

Table 4.2 Areas of the world with high population densities

4.2 Factors affecting population distribution and density

Population distribution is a description of the spread of the human population across the Earth. The distribution is very uneven – most of the world's population live in only one-third of the available land area – and it has changed considerably over time.

TERM

population distribution: the pattern of where people live

The Earth's land surface forms about 30% of the total Earth's surface – the rest being water. However, only about 11% of the land surface is comfortably habitable by people. The reasons for the contrasts in population distribution and density can be divided into two groups – physical (natural) and human.

1 The **physical (natural) factors** involve differences in the natural environment and include climate, water supply, natural resources, relief/topography, natural vegetation and soils.
2 The **human factors** are a result of human activities and these may be economic, political or social (including cultural).

Physical/natural factors

Climate

There are three major climate zones covering the Earth – polar, temperate and tropical.

People tend to avoid living in **polar** areas because of:

- the long periods of freezing temperatures in winter
- the very short growing seasons for plants.

These areas therefore have *very low population densities* and contain people who are traditional hunter/gatherers such as the Inuit of northern Canada, Alaska and Greenland. These people survive by hunting seals, fish and whales at sea and caribou (reindeer) on land. Such cold climates can also be found in mountainous areas of the world, such as the Himalayas, Andes and Rockies.

Other areas of *low population density* are the **tropical** deserts – the Sahara, Arabian, central Australia, Atacama, Kalahari/Namib and south west USA – where the very high temperatures for much of the year and lack of rainfall again combine to produce a very harsh living environment.

Areas of *high population density* tend to be in, firstly, the **temperate** areas where there are not the extremes of temperature and there is adequate rainfall to provide a reliable source of water for both people and farming. For example, Western Europe, North Eastern USA, North East China and Japan. Secondly, the **tropical** areas, but not all – it has to be where year round high temperatures, reliable rainfall and fertile soils produce highly productive areas for farming. For example, the floodplains and deltas of major rivers, such as the Ganges in India and Bangladesh, the Nile in Egypt, the Mekong in Vietnam and the island of Java in Indonesia.

FACT

Around 99% of the Egyptian population of 91 million is found in just 4% of the country's land area – the rest being inhospitable desert.

Water supply

Most people in the world obtain their water for drinking and for farming from two sources – rivers and lakes and from underground storages, called aquifers (see Chapter 18).

Water can flow long distances, both overland and underground, from the original source of the water and this can extend the area where people can live. For example, the Nile flowing across the desert in Egypt and the Colorado in south west USA. Where there is a lack of reliable water, population densities are normally low.

Soils

The fertility of the soil is very important in determining how productive an area will be for crops. The most fertile soils are mineral rich and well drained. These are often found in river floodplains and deltas and in areas of volcanic rocks – which tend to weather down into very fertile soils. Where water is available, either naturally or by irrigation, areas of fertile soil can support high population densities – such as the island of Java in Indonesia.

Relief (topography)

The terms **relief** and **topography** are used to describe the height and shape of the land. Population densities tend to be lowest where land is high and steep, and highest where land is low and gently sloping or flat. The high land found in the world's major mountain ranges tends to have lower temperatures, more frosts and higher rainfall, which often falls as snow.

High land also tends to have steeper slopes, which are more difficult to farm (though this can be overcome by terracing – as in the rice terraces of Indonesia and southern China). As a result, most of the world's population tend to be found in the lower areas of the world – around the coastlines and on river floodplains and deltas.

Natural resources

In population geography, the term natural resources usually refers to minerals such as coal, oil and metallic minerals such as iron ore, bauxite (the natural ore from which aluminium is obtained), gold, silver, tin, copper, and so on. Areas with abundant natural resources provide employment and income.

TERMS

natural resources: any natural resource from water and soil, to wind and minerals (broadest sense)

cumulative causation: certain things will result (be caused) when a group of factors combine (accumulate)

Human factors

Economic factors

Economic factors include transport and money (sometimes called capital). These are extremely important in the location of industry which, in turn, provides jobs which, in turn, affects where people live – that is, population distribution. The term cumulative causation is often used to describe this set of links.

One of the most important economic factors is transport. Where fast, efficient, reliable and cheap transport is available, many industries will have an advantage as it will reduce their production costs and increase the area in which they can sell their products. A large, modern port provides such a location. For example, Europort in Rotterdam in the Netherlands,

Singapore in South East Asia, Shanghai in China and New York in the USA. Motorways/freeways have a similar effect and attract industry.

Route centres where forms of transport, such as road, rail, shipping and air, converge are important locations of industry.

FACT

London and Paris are both good examples of route centres, as they are both at the centre of their country's road, rail and air networks.

Political factors

Political factors include government investment in the infrastructure of an area such as in roads, railways, airports and sea ports, and land reclamation.

National and regional governments, as well as the major trading blocs (see Chapter 13), such as the European Union, have a very important role in deciding where industry, jobs, roads, railways, air and sea ports, housing, hospitals and schools are located and therefore on the distribution of population, for example, in South Wales in the UK.

Social factors

Social factors include housing, health care, education and cultural opportunities.

In the poorest LICs, the birth rate is often high to compensate for the high rates of infant mortality. In some parts of sub-Saharan Africa, where there is no government social care provision in old age, a woman may have 8 or 9 children to increase the possibility of having a surviving child to look after them in old age.

In contrast, the HICs/MEDCs have low birth rates as there is no need to compensate for high rates of infant mortality and there are higher levels of government social care provision in old age. Also, social improvements in healthcare and sanitation and diet have taken place.

Cultural factors include the traditional expectation in some countries of large families and to marry at an early age and start families right away. Some groups of people also prefer to live together for security and friendship.

Areas of the world where many physical and human factors combine together can either be densely populated or sparsely populated. For example, the Sahara desert, which stretches across several countries of North Africa, is very sparsely populated because its climate is too hot and dry for people and animals to survive comfortably. Its soils are too dry, sandy or rocky. It has a poor water supply. Lastly, the countries in the Sahara are all poor LICs whose governments do not have the money to invest in improving their infrastructure in transport, housing, education and health care or industrial development.

At the other extreme, one of the world's most densely populated countries, Bangladesh, has rich, fertile soil. It has a hot, wet climate with easy access to water supplies. All of this means that it is ideal for growing crops and can support a very large population on a relatively small area of land.

Sometimes large urban areas may grow up in places which are otherwise sparsely populated. For example:

- around an oasis in a desert
- near rivers where they flow through arid desert areas
- mining settlements on coal or iron ore fields, or in Kuwait (where there is a high production of oil)
- growing tourist destinations and resorts such as Dubai in the UAE and Sharm El Sheikh in Egypt
- market towns where the products of a large rural area can be brought to sell
- route centres and the junctions of major highways often in openings or gaps in mountain and hill ranges – gap towns
- towns of strategic importance controlling access to a region
- new towns created by government policies
- dry areas in otherwise waterlogged marshy land
- a sheltered, fertile valley in a highland area.

Self-test question 4.1

1 What factors affect the distribution of population?

Self-test questions 4.2

1 Explain the factors that can lead to an area or country having a low population density.

2 How may physical factors encourage a high population density?

Case studies

A densely populated area – the Ganges River basin

The Ganges River has the second greatest water discharge in the world, after the Amazon. Its basin, from the Punjab to Bangladesh and Assam, is the most heavily populated river basin in the world, with an average population density of 1000 people per km^2. The location of the Ganges River basin in India can be seen in Figure 4.2, running across the top of the map, in a north-west to south-east direction.

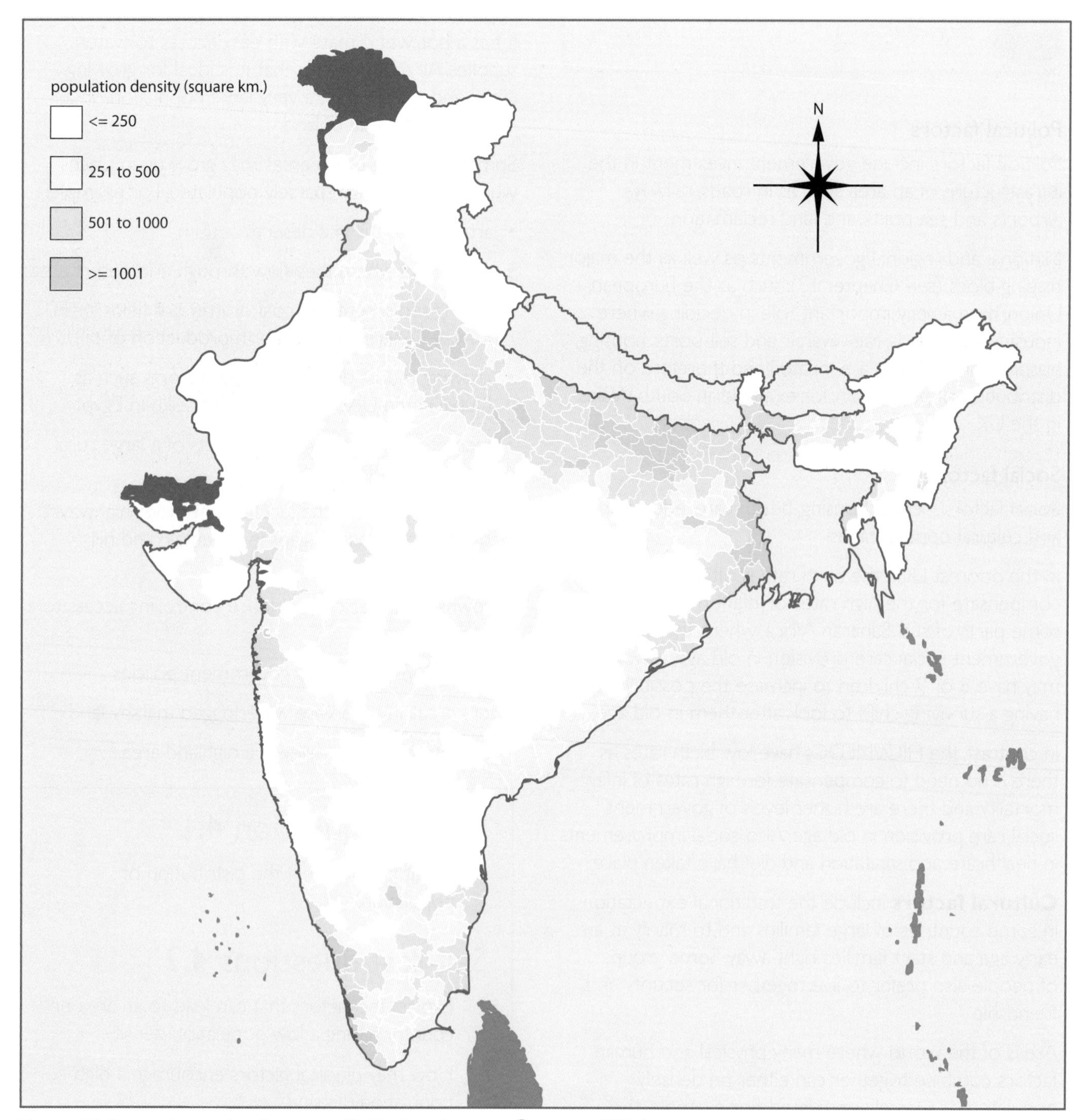

Figure 4.2 A population density map of India, based on the Census of 2001

TERMS

census: a procedure of systematically acquiring and recording information about the members of a given population; it is a regularly occurring, official count of a particular population

hydro-electric power: electricity generated by using moving water to turn turbines

The Ganges River basin is extremely important to the people of India and the other countries – Nepal, China and Bangladesh – which it runs through, as most of the people living beside it use it for daily needs such as bathing and fishing.

Physical and human factors

The Ganges River basin is also important to the agricultural economies of India and Bangladesh. The huge floodplain of the river is rich in alluvium and produces very fertile soils, while the climate and access to the river for irrigation make the Ganges River basin one of the most intensely farmed and densely populated areas of the world, capable of supporting a very large population. About 580 000 km^2 of the Ganges River basin is arable land and it contains 29.5% of the cultivable land area of India. Precipitation in the Ganges River basin arrives with the south-west monsoon winds from July to October, but it also comes with tropical cyclones that originate in the Bay of Bengal between June and October.

The Ganges and its tributaries provide water for irrigation – 90% of the annual water withdrawn from the river is used for agriculture and 8% is for domestic use. A wide range of crops are cultivated in the basin including rice, sugar cane, lentils, oil seeds, potatoes and wheat. These crops support millions of people in both the rural areas and urban centres of the basin.

The hydro-electric power potential of the Ganges and its tributaries is enormous, of which India currently uses only 12%.

Pollution of the Ganges River

The deterioration in both surface water and groundwater quality, as a result of the high population density, is now a matter of serious concern. Pollution of the Ganges is caused by both human and industrial waste due to the very large population and its rapid growth, as well as religious events. Much of the waste, including raw sewage, is dumped into the river. In addition, many people bathe and use the river to clean their laundry. Fecal coliform bacteria levels near Varanasi are at least 3000 times higher than what is considered safe (by WHO).

These impacts may have a major influence on agricultural production and food security, ecology, biodiversity, river flows, floods, droughts, water security and human and animal health.

A sparsely populated country – Mongolia

Mongolia, situated in the centre of the continent of Asia, is one of the world's most sparsely (Figure 4.3) *and* under-populated countries (see Chapter 1). With a land area of 1 564 000 km^2 and a population of just over 3 039 000 people in 2017, along with a combination of an extreme continental climate with long, cold winters and short summers and its remoteness, it has a population density of just 1.94 km^2, even though the country has large mineral resources, which could be developed to support a larger population. The country is bounded to the north by Russia and on the east, south, and west by China. It is also one of the largest countries in the world with no access to the sea.

Mongolia's relief mainly consists of a high plateau, ranging from 900 to 1500 metres in height, with vast areas of semi-desert and desert plains, grassy steppe, mountains in the west and south-west and the Gobi Desert in the south-central area.

Mongolia is classed as a MIC, although about 20% of the country's population lives on less than $1.25 per day.

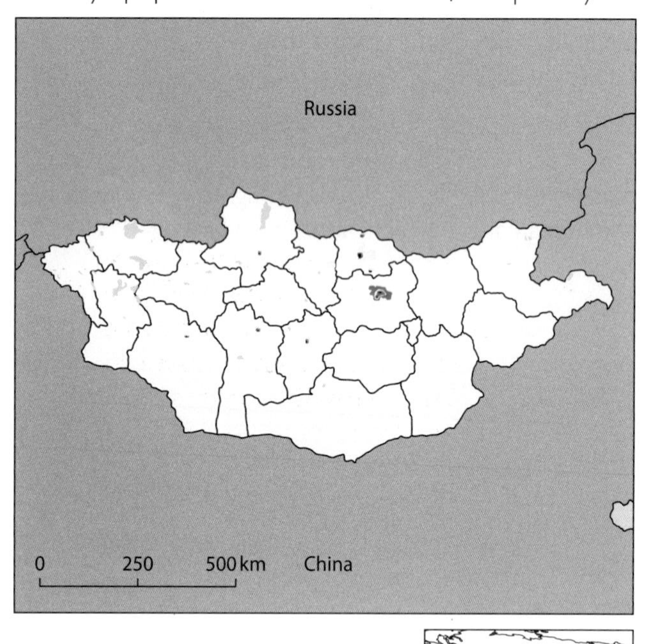

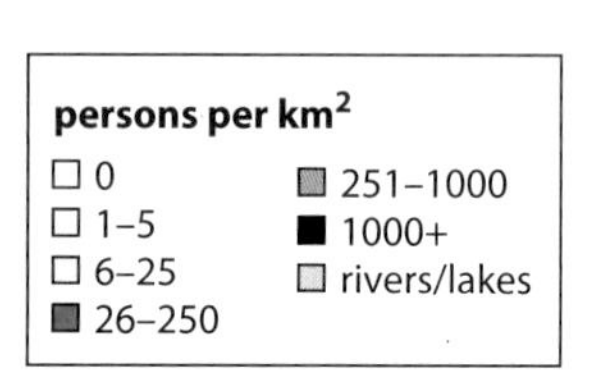

Figure 4.3 Population density of Mongolia, 2000

The influence of climate

Mongolia has an extreme continental climate with long, cold winters and short summers, during which most of its annual precipitation falls. Most of the country is hot in the summer and extremely cold in the winter. Temperatures fluctuate from as low as −50 °C in the steppe in winter to 40 °C in the Gobi desert in the summer, with January average temperatures dropping as low as −30 °C. The growing season is very short, 90 days, and the population has to survive the long, freezing winters, when any outside human activity is very limited. The annual average temperature in the capital city of Ulaanbaatar is −1.3 °C, making it the world's coldest capital city (Figure 4.4).

Remember to study and interpret the resources such as maps, graphs, diagrams and extracts carefully, using appropriate facts and statistics derived from them to back up your answer.

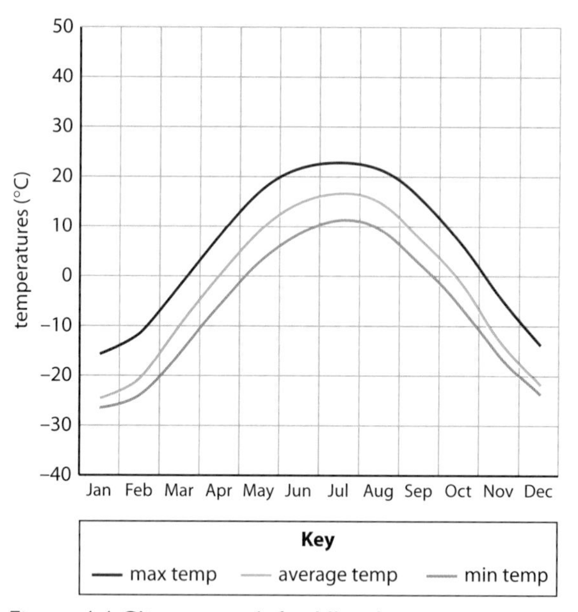

Figure 4.4 Climate graph for Ulaanbaatar

Ulaanbaatar is home to about 45% of Mongolia's population. Climate has had a huge impact on the population density of the rest of the country. The period of grazing, the timing and methods of performing many animal and crop production tasks, all depend directly on climate and weather conditions. The short summers, short growing season and sharp falls in temperature and frosts during the warmer spring and summer period make agriculture very difficult.

In 2002, about 30% of the families in Mongolia made a living from breeding livestock. One of the country's key economic activities, employing almost half of the population, rearing livestock accounts for 85% of Mongolia's agricultural output. Most herding families are self-sufficient in meat and milk products and also earn a cash income from selling live animals, milk, meat, skins and hides, sheep's wool and cashmere wool.

Most livestock herders and their families in Mongolia follow a pattern of nomadic or semi-nomadic pastoralism, where livestock are herded and moved in order to find fresh pastures on which to graze (see Chapter 14).

The severity of the climate in Mongolia was highlighted by a severe winter in 2009–10, when Mongolia lost 9.7 million animals, or 22% of its total livestock. The effect of this was to force many Mongolian herding families, who had lost all their livestock and means of living, to leave their traditional areas and migrate to Ulaanbaatar.

Human factors

Dzud, a Mongolian term, refers to a range of severe weather conditions, including severe summer droughts and exceptionally cold winters, which can prevent access to or destroy pasture, causing significant loss of animal life and, consequently, devastating the livelihoods of herding families. Consecutive *dzuds*, from 1999 to 2002, killed about one-quarter of the livestock, forcing many people to migrate to urban areas. Herd numbers did recover, but there was an increase in the number of goats reared for cashmere, as their wool gained a higher price. This resulted in widespread overgrazing as too many animals were kept grazing too long in too small an area – a contributing human factor.

Other human factors contribute to the low population density in Mongolia. For example, inadequate rural infrastructure and a lack of markets make life difficult for rural families.

The government has attempted to initiate a programme to protect animals from climatic hazards such as droughts and *dzuds*, which includes both pre-emptive measures and weather warnings as well as a system of compensation. There has been greater investment in rural areas as a result, in basic infrastructure, such as building wells and watering points.

Economic activity in Mongolia has traditionally been based on herding animals, although the development of extensive mineral deposits such as copper, coal, molybdenum, tin, tungsten and gold has increased industrial production. Workers and their families are migrating from rural areas into the capital city and mining areas to work in newly established mines and industries. Minerals now represent more than 80% of Mongolia's exports. However, their jobs are dependent on the world prices of the minerals and new investment in industry, and any changes in these can have destabilising effects on the nation's economy and its population.

The influence of remoteness

Moving around the country is not easy. Many of the roads in Mongolia are only gravel roads or simple, cross-country tracks. The only sealed/paved roads are from Ulaanbaatar to the Russian and Chinese borders. A number of road construction projects are currently underway. Mongolia only has 4800 km of paved roads, with 1800 km of that total built in 2013 alone.

Sample question and answer

Using a named country or area that you have studied, describe the problems that may be caused by it having a high population density. [6]

TIP

You should know how to use the mark allocations in brackets and the space provided as a guide to the amount of detail or number of responses required. Normally there will be two lines available for each mark.

State your named example, such as Bangladesh, at the start of your answer. Then you need to develop several themes. The example shows more answers than are needed, but remember you can only gain a maximum of only 6 marks, so any additional answers will not gain a mark.

Answers could include:

- There may be a lack of work/employment **[1]**, those jobs that are available may be part time or provide a small income **[1]**, many people may have little or no income, leading to poverty for many people **[1]**.
- There may be inadequate food supplies **[1]**, which may lead to starvation/malnutrition **[1]**.
- People may not have access to education, so they remain unqualified **[1]**; poor access to health care **[1]**, leading to high death rates **[1]**.
- Housing may be overcrowded **[1]**, often with inadequate basic amenities such as water supply and sanitation **[1]**.
- Traffic congestion **[1]** as there are too many vehicles on the roads, especially in cities such as the capital city of Dhaka in Bangladesh **[1]**.
- High levels of air and water pollution **[1]**.
- There may be widespread deforestation **[1]** for people obtaining firewood for fuel, leading to soil erosion **[1]** and an increased number and size of floods **[1]**.

(A maximum of 6 marks will be awarded.)

Exam-style questions

1 Using a named area that you have studied, explain why it has a low population density. [6]

2 Explain why an urban centre may become established in an area of low population density. [4]

Settlements and service provision

Chapter 5

Learning summary

By the end of this chapter, you should be able to:

- explain the patterns of settlement, including dispersed, linear, and nucleated settlement patterns
- describe and explain the factors which may influence the sites, growth and functions of settlements, including the influence of physical factors (such as relief, soil, water supply) and other factors (such as accessibility, resources)
- give reasons for the hierarchy of settlements and services, including high-, middle- and low-order settlements and services, sphere of influence and threshold population
- demonstrate knowledge of a case study of settlement and service provision in an area.

5.1 Patterns of settlement

Settlements take on certain shapes when they form (Figure 5.1). The following are the most common.

- **Dispersed:** where individual buildings are spread out across a landscape.
- **Linear:** where a settlement occurs along either side of a road and looks like a long line.
- **Nucleated:** circular in shape with the buildings mostly concentrated around a route centre.

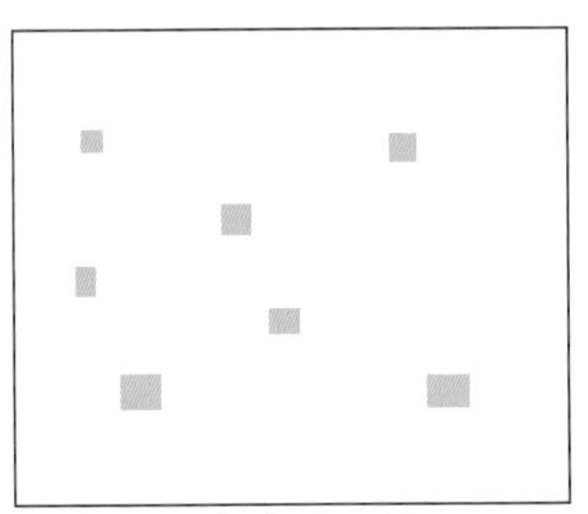

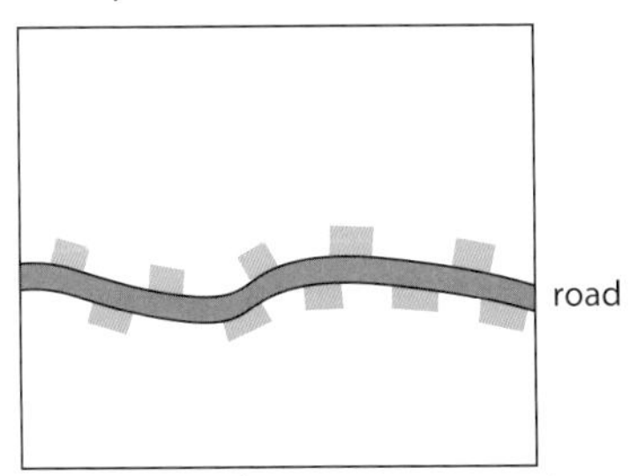

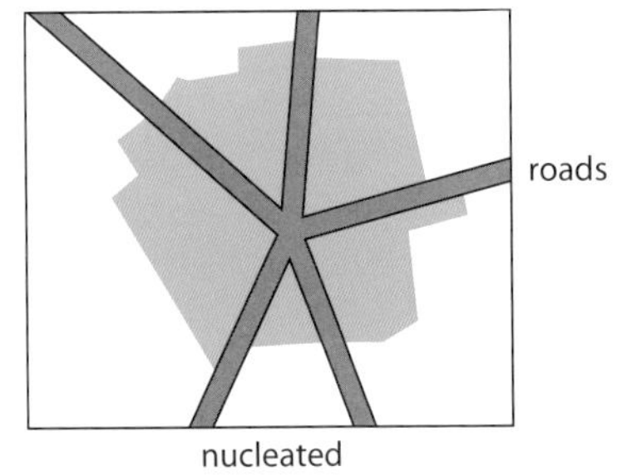

Figure 5.1 Patterns of settlement

5.2 The sites, growth and functions of settlements

The site, growth and function of a settlement often depends on a combination of factors. These factors often change with time, with the result that many settlements can become more or less important with time. A topographic map (or similar) can give a great deal of basic information about a settlement.

TERMS

site: the area of land actually covered by the buildings in a settlement

topographic map: a detailed and quantitative representation of relief – the height and shape of the land, usually using contour lines

situation: a description of a settlement in relation to the other settlements and physical features that surround it

settlement function: the term given to the functions that take place in a settlement. For example a tourist resort or market town.

After initially being established, the further growth of a settlement may then depend on the situation of the settlement in relation to the surrounding area, the size and functions of surrounding settlements, the presence of large physical features such as valleys and hills, and access to natural resources such as fertile soils, fuel, minerals and route ways.

In the past, the main factors influencing the sites, growth and functions of settlements included the following physical and other factors:

- flat or gently sloping land that was easy to build on
- a good defence site – this was often a high point, such as a hill top, surrounded by steep slopes or inside a meander
- a wet point site to be near a reliable source of water – beside a river, stream or spring or oasis
- a dry point site to avoid flooding
- having building materials nearby
- a supply of fuel – firstly wood and then coal
- fertile land for growing food
- a sheltered site – often in the lower parts of valleys – which often meant being on warmer, south-facing slopes compared to colder, north-facing slopes in the northern hemisphere and the opposite in the southern hemisphere
- good transport links – often where several tributary valleys joined, which became a route centre (sometimes called a nodal point), a bridging point of a river (particularly the lowest bridging point of a river before it enters the sea) or a port.

Some of these factors are no longer important for many settlements, especially in HICs. For example, there is no need for a defence point or a reliable source of water as most people have piped water in HICs, food is bought from shops and supermarkets, and flood prevention schemes stop or reduce the threat of flooding.

Some settlements may have one or several functions. These include defence, administration (containing several government functions, some of which may be national, state or local), commercial (shops and offices), industrial, mining, route centre, market town, education (may contain a university), tourism and residential.

A successful port, such as Singapore, Shanghai in China and Europort in Rotterdam in the Netherlands, for example, can be found where there is:

- a deep water estuary such as the mouth of a large river
- a deep water anchorage so that it can accommodate large ships
- a sheltered anchorage so that ships can anchor, load and unload safely
- a large, flat area beside the port for the building and the expansion of warehouse and industrial activities
- access to important sea routes
- a large hinterland of nearby cities and industrial towns
- a well-developed road and rail network of communications, from the port to its hinterland.

After their initial siting and building, many settlements have continued to grow and expand in size because they also provide excellent sites for industry and factories to locate. For example:

- A large river nearby will provide a water supply for a factory – the water can be used for cooling, or as a raw material in processing materials. The river would also provide any easy means of disposing waste.
- Relatively cheap, flat land is likely to be found on the floodplains beside rivers. This can be drained and factories could be built with room to expand.
- Near excellent transport routes/rivers/railways/air transport for the easy, fast, efficient and cheap access for either obtaining or sending out raw materials, component parts and finished goods.
- Nearby work/labour force.

Self-test questions 5.1

1. List the factors that describe the site of a settlement.
2. Explain how the situation of a settlement might determine its future growth.

5.3 Hierarchy of settlements and services

Settlement hierarchies

A settlement hierarchy is found by putting the settlements in a region or country into a rank order, using either their population size or the type and range of services that they provide, such as retailing (shops), health care, banking, legal, education, entertainment and leisure, government functions and religious. The position of a settlement in the hierarchy will depend on the number and type of services and functions it provides. The larger a settlement is, the more services and functions it will have and the higher it will normally be on the settlement hierarchy. Figure 5.2 shows a typical settlement hierarchy.

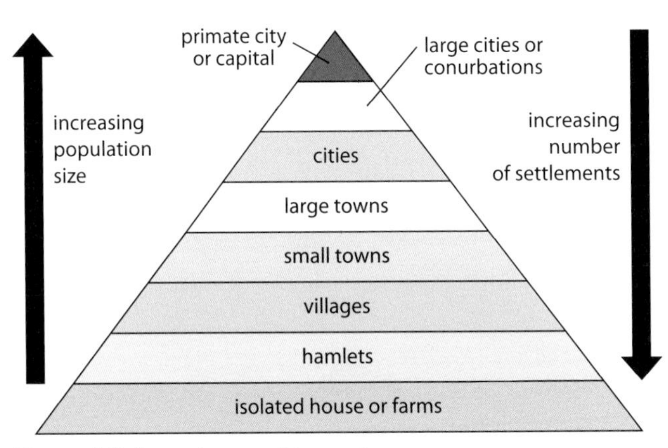

Figure 5.2 A settlement hierarchy

TERMS

settlement hierarchy: a way of arranging settlements into a rank order based upon their population or some other criteria

service hierarchy: settlements can be ranked according to the type of services they provide, such as in health care or education. A high order settlement may have large hospitals; while a low order settlement might have a small health centre.

convenience goods: goods that people need to buy perhaps two to three times a week such as water, vegetables, fruit, milk and newspapers

sphere of influence: the area surrounding a settlement that is affected by the settlement's activities

A city, or high-order settlement, will have many **high-order services,** including:

- retailing, such as specialist shops/department stores, including international chain stores such as Coach, and Tiffany from the US and Gucci and Louis Vuitton from Europe
- leisure, such as cinemas/theatres
- educational institutions, such as schools/university
- medical facilities, such as a hospital
- financial and insurance services, such as banks
- estate agents, selling houses in the local area
- large supermarkets for weekly shopping.

Middle order settlements, such as large towns, will have a more restricted range of **high-order services,** such as some specialist shops and services, with some national chain stores and supermarkets, a cinema, secondary schools and banks.

A village will only have **low-order services**, such as a small shop which provides convenience goods. Below a village in the hierarchy, a hamlet may have no shops or services and people who live there may have to travel to a nearby village or town to obtain the goods and services that they need.

Spheres of influence

A sphere of influence is the area surrounding a settlement that is served by the settlement in terms of its services such as shops, jobs, health care, education and government services. The size of a settlement's sphere of influence will depend on a number of factors, such as:

- the number and type of services it provides
- the transport facilities available to the settlement
- the level of competition from surrounding settlements.

Higher-order settlements will have several advantages over lower-order settlements because of the high-order services they provide. This will result in several outcomes:

- People will often be prepared to travel further for higher-order services. For example, to buy specialist goods, such as furniture, computers, cars and 'white goods' such as washing machines, and cookers.

- People may also travel further to buy comparison goods (goods where you would like to compare quality and price as they are expensive – such as the specialist goods mentioned above) rather than convenience goods (inexpensive items that people will buy two or three times a week).
- High-order settlements offer more shops / wider range / opportunity to shop around / get cheaper prices.
- People may also travel further for some services to find a better/higher quality service, or to buy/obtain specialist services, such as the financial and legal services provided by banks, insurance and law firms.
- People who live in settlements with low-order services will have further to travel than people who live in high-order settlements for many services.

In this process, the range of a good or service is important. This will have a different size hierarchy depending on the good or service that is needed. Each service or good will have a threshold population. In the UK, for example, it has been estimated that a village shop needs 350 customers, 2500 patients for a doctor, 10000 people for a secondary school or a Boots Chemist, 50000 for a Marks and Spencer store and 60000 people for a large national supermarket chain such as Sainsbury's or Asda.

TERMS

range: the maximum distance people are prepared to travel to purchase a service or product

threshold population: the minimum number of people necessary before a particular good or service can be provided in an area

Self-test questions 5.2

1. What factors will determine the size of the sphere of influence of a settlement?
2. Why is it useful to determine the nature and characteristics of the settlements in an area such as Cornwall?

Case study

Settlement and service provision in an area – Cornwall, UK

Cornwall is a county in the south-west of England. It has a population of 566 750 and covers an area of 3563 km^2 with a population density of 159 people per km^2, making it a relatively rural county. Cornwall is also one of the poorest parts of the UK in terms of per capita GDP and average household incomes. At the same time, parts of the county, especially on the coast, have high house prices, driven up by demand from relatively wealthy retired people and second-home owners from the rest of the UK. Cornwall has a relatively high retired population, with 23% of pensionable age, compared with 20% for the UK as a whole. The GDP per head for Cornwall is 64% of the EU average.

The Cornish economy depends heavily on its tourist industry, which makes up around a quarter of the economy and contributes up to 24% of Cornwall's GDP, bringing almost $2.8 billion into the Cornish economy annually. Cornwall has a unique culture and a spectacular landscape. It has a large number of Areas of Outstanding Natural Beauty (AONB) and all these factors, along with its mild climate, make it a popular tourist destination, despite being somewhat distant from the UK's main centres of population.

Why determine and analyse a settlement hierarchy in an area?

There are many hundreds of settlements in Cornwall, ranging from those that are strategically significant such as Truro, through small towns and local centres to the smallest of villages and hamlets. Analysing the nature and characteristics of the settlements in an area helps the planning and organisational processes for the future provision of services in an area.

The settlement hierarchy in any area is usually scored according to the services and facilities available in any settlement, but it is normal to 'weight' those services and facilities that are considered necessary to meet residents' day-to-day needs higher than those that are 'nice to have but not essential'. Three different levels of 'scoring' were identified in order that those services and facilities that are essential to everyday life are scored higher than those that are used less frequently.

- **High** – a settlement that has the services and facilities that people need to use on a regular basis that are essential to everyday life and which is a sustainable settlement, such as Truro (population 21 000), should contain: supermarket/food store, primary school, doctors' surgery, magistrates courts, good broadband connection, travel to work bus/rail service, and a high work self-containment (i.e. the proportion of the resident population employed locally).
- **Medium** – for example, Helston (population 11 000), with services and facilities that people would only expect to be available in larger settlements or those that are not needed on a day-to-day basis: secondary school, bank/building society, Post Office, pre-school provision, chemist/pharmacy, pub/social club, petrol filling station, community hall, leisure centre/swimming pool, formal sports area/playing pitch, children's equipped play area.
- **Low** – for example, Breage (population 3000), with services and facilities that are either so strategic that people would expect to travel a considerable distance to access them or that they are 'nice to have' but not essential – some of these services, however, may be very important in settlements with few other facilities: hospital, permanent library, mobile library, place of worship, other bus/rail service (not for travelling to work).

The scores for different facilities and services are shown in Table 5.1.

Facility or service	Score
Primary school	8
Doctors' surgery	8
Travel to work bus/rail service	8
Pre-school provision	7
Chemist/pharmacy	7
Food shop	6
Post Office	6
Bank	6
Secondary school	6
Pub/social club	6
Petrol station	6
Community hall	6
Leisure centre	6
Formal sports area	6
Children's equipped play area	6
Hospital	4
Place of worship	4
Other bus/rail service	4
Library (permanent)	3
Library (mobile)	1

Table 5.1 Scores for facilities and services

This scoring system can then be used to determine the functionality of a settlement (rather than its size in terms of population). It was used, in the case of Cornwall, to group similar settlements into the same category. Six categories were identified (Table 5.2). It is worth noting that, outside of this, were 96 smaller settlements which did not even meet the criteria for Category F.

Category	Level of function	Number of settlements
A	strategically significant towns and the main employment and service centres	7
B	market and coastal towns that are locally significant and contain a good range of housing, employment, and community facilities and services	8
C	small towns and villages that meet local needs for some services and facilities and employment	41
D	smaller settlements that perform an important role in their local areas, i.e. 'rural service centres'	66
E	settlements that do not meet the criteria for category D but are considered important in their immediate local area and contain a primary school or general store	66
F	settlements that do not meet the criteria for category E but contain a travel-to-work bus or rail service and either a general store or a meeting place	46

Table 5.2 Functionality of settlements

Sample question and answer

Explain why people travel further for some shops and services than for others. [5]

TIP

Underline command words in the question and also the words or terms that identify the content and/or context of the question.

People will travel further for higher-order services **[1]**, which are likely to be used less frequently than low-order services **[1]**. High-order settlements offer people more opportunities to buy specialist goods or comparison goods rather than convenience goods **[1]**. People travel further for some higher-order goods and services to seek better quality or to find precisely what they want **[1]**. Higher-order services may have a larger sphere of influence **[1]**. Those people who live in settlements with fewer services will have further to travel for higher order goods and services than the people who live in towns or cities **[1]**.
(A maximum of 5 marks will be awarded.)

Exam-style questions

1 What is meant by the term 'hierarchy of settlements and services'? [4]

2 a What is the difference between a nucleated and a dispersed settlement? [3]

b List three low-order and three high-order services. [3]

Chapter 6

Urban settlements

Learning Summary

By the end of this chapter, you should be able to:

- describe and give reasons for the characteristics of, and changes in, land use in urban areas, including land use zones in countries at different levels of economic development and the effect of change in land use and rapid urban growth in an urban area
- explain the problems of urban areas, their causes and possible solutions, including different types of pollution, inequality, housing issues, traffic congestion and conflicts over land use change
- demonstrate knowledge of a case study of an urban area, including changing land use and urban sprawl.

6.1 Characteristics of and changes in land use in urban areas

The characteristics of land use zones in urban areas

Over time, towns and cities have grown, but not in a haphazard way. Urban geographers began to recognise patterns in the land use appearing in urban areas. Several urban geographers produced theoretical models of urban land use that could be applied to HIC and LIC cities. LIC cities have been characterised by rapid urban growth, as a result of rural–urban migration due to perceived opportunities in the growing LIC cities and changing agricultural economies. This rapid growth in LIC cities has led to a number of problems, including the uncontrollable growth of shanty towns, unemployment, pollution and social unrest. Although each urban area (town or city) is unique and has its own distinctive shape and land use patterns, these models attempted to show that they all shared certain common characteristics. Two of the earliest and simplest models were put forward by American researchers Burgess and Hoyt.

FACT

The Burgess and Hoyt models of urban land use were based on US cities.

The **Burgess Concentric** model (figure 6.1), devised in 1925, saw a city displaying a pattern of concentric rings around one centre, the Central Business District (CBD). A typical CBD will have quite distinctive features:

- high cost of land – which can only be afforded by shops and offices
- high rise buildings – to make full use of the expensive land
- few houses/residences – due to the high cost and rents charged for the land
- a lack of open space – the land is so expensive and valuable
- a transport focus – most roads and railways will focus on the centre of the city or town and make it the most accessible part of the urban area; it will have transport links with all parts of the town/city.

Outside the CBD was a ring/zone of light industry, followed by three rings/zones of low-, middle- and upper-class residential housing. The better off/richer you were, the further away from the crowded city centre (CBD) and industry you could afford to live. As the urban areas grew, so did the rings.

TERM

Central Business District (CBD): the main commercial and shopping area of a town or city

inequality: the extreme differences that exist within many urban areas in poverty and wealth, access to employment opportunities or access to services such as healthcare and housing provision.

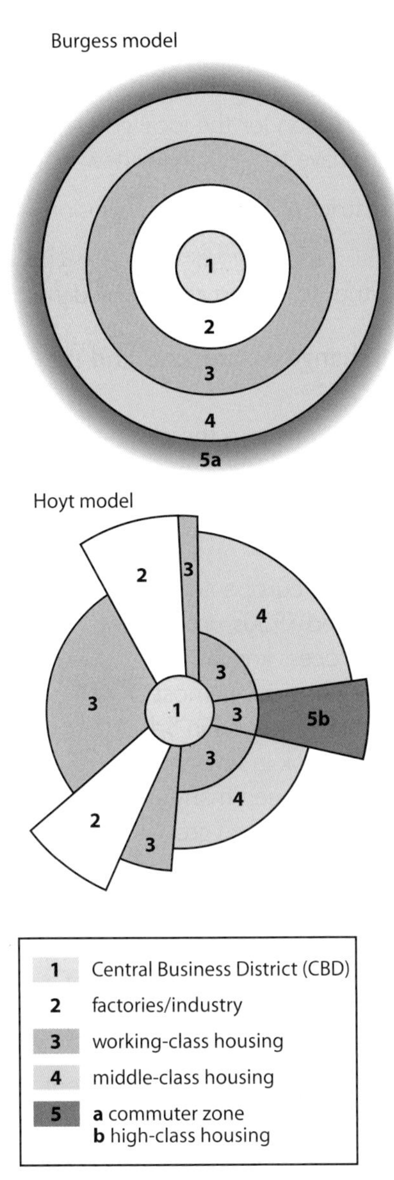

Figure 6.1 The Burgess and Hoyt models

The **Hoyt Sector** model (figure 6.1) came later, in 1939, when public transport had developed in the cities of the USA. In Hoyt's model, land use developed along the main transport routes going out (radiating) from the CBD. This formed a series of wedges or sectors. He identified industrial sectors in the urban areas and adjacent to these areas, low-class housing developed as the workers could not afford to be far from their place of work. A high-class housing sector developed furthest away from the sectors of industry with their potential air, noise and water pollution.

The value of land also changed with the changes in land use. The most expensive sites were located in the CBD, as it was the most accessible part of the city for most people and so businesses competed with one another for these sites. With competition came higher land prices. The major land users were banks, shops and offices, as they needed access to as many customers as possible and they could afford the land prices – unlike industry and housing. The low-class housing had the cheapest land because there was very little competition for these sites.

By the 1970s, urban land use and both cities and towns had further evolved and changed. Another model, the **Composite** model, appeared and this combined the concentric and sector models. Mann developed this model after studying medium-sized UK cities like Nottingham.

In LICs and many MICs, it was found that urban areas were developing in different ways. Consequently, the theoretical models differ in several ways from those in the HICs:

- Many develop a commercial spine which extends out from the CBD and an elite upper class residential sector on either side of this.
- Surrounding the central area is a mixture of older traditional housing. This used to house the elite high income group who now live in the elite sector, mentioned above. These buildings are in relative decline and have often been subdivided. Some have been replaced by self-built housing.
- Most of the better housing is located in or near the city centre compared to being on the edge of the city.
- The quality of housing decreases as you go towards the edge of the city instead of improving.
- Surrounding the edge/periphery of the city is a zone of squatter settlements which is home to new migrants.
- Industry tends to locate along the main roads leading into the city centre rather than being in a distinct zone or sector on the edge of the city. Industries tend to be located here because this area has both power and water supplies.

TERM

rural–urban fringe: where the urban area meets the rural countryside at the edge of a town or city

Changes in land use in urban areas

Over time, several factors have changed and new patterns and trends of development and inequality have occurred in cities across the world. For example, CBDs have become increasingly crowded, congested and less accessible as cities grow and people have found themselves living much further away from the centre of cities. As a consequence, there have been several important developments on the rural–urban fringe. The mixture of land uses on the fringe often causes conflict as different groups have different needs and interests. For example, the decision to build a new runway at Heathrow airport in 2016 on the outskirts of London is a source of controversy.

Out-of-town shopping centres began to develop with shopping malls which had easy access, free parking and 'one-stop shopping' – you could get all your shopping done in one large, covered mall as well as have a meal and entertainment in multiplex cinemas. Shops started to leave the congested city centre sites and set up in the new, out-of-town locations. This has produced the **doughnut effect** – a run-down city or town centre abandoned by shops and businesses – an urban settlement with a hole in the middle!

Many cities and towns have now put in place regeneration schemes to attract businesses back in to the centre – a process called reurbanisation. These schemes have involved some new initiatives:

1 Creating **pedestrianised zones** or precincts, where selected roads/streets in the centres have been closed off to traffic or only allow public transport vehicles, like buses, and sometimes delivery trucks/vans to enter. This has improved the environment, with less traffic pollution of the air and less noise.

2 Old office blocks and shops have been pulled down and redeveloped as shopping malls on brownfield sites.

3 Gentrification – where older houses and buildings in or near the city centre, which are often large and well-constructed, are renovated and brought up to modern standards. Gentrification has several advantages:

- The buildings can be used for a range of uses, including shops, offices and apartments.
- It is often a cheaper option than pulling down the buildings and constructing new buildings.
- Older houses add character/retain culture/image.
- They provide a convenient residential/housing location close to people's workplaces/CBD.

There are several **social advantages** of improved housing rather than building new flats:

- People have lived there for many years, so there is an established community spirit as many of the people will know each other.
- Nowadays, people cannot afford to move to a costly new house.
- The area will already have convenient small corner shops/cafes/pubs etc.
- It is a cheaper option for the local government authority who have to pay for any new developments.
- It helps to restrict the outward expansion of the town or city.
- It minimises the disruption caused by demolition.

The effect of change in land use and rapid urban growth

In addition to the changes in the central parts of cities and towns, there has been a great deal of change over time on the edge of the urban areas where they meet the rural countryside.

In the UK, the rest of Europe and in North America during the 1920s and 1930s, urban areas spread out dramatically – a process known as urban sprawl. This process is still very obvious in most countries, such as in Jakarta in Indonesia, Colombo in Sri Lanka, Kuala Lumpur in Malaysia and Bangkok in Thailand. In the UK and elsewhere, there have been many attempts to limit and control this development and growth. One of the first attempts was the setting up of green belts from 1947.

TERMS

regeneration scheme: the use of public money to reverse the decline of a city or town by improving both the physical structure and the economy of those areas by encouraging private investment into an area

reurbanisation: the movement of people back into an area that has been previously abandoned

brownfield site: land that was previously used for either industrial purposes or some commercial uses

gentrification: renovation and revival of deteriorated urban areas to attract more affluent residents

urban sprawl: the expansion of an urban area away from central urban areas into low-density and often car-dependent communities on the edge of existing urban areas

green belts: areas of land surrounding an urban city area where any new housing or industrial development was to be stopped or severely restricted

The pressure on this rural–urban fringe has not disappeared, however, and they have proved very attractive sites to many land users, including:

- new housing
- science and business parks – often containing footloose businesses (further explained in Chapter 15)
- retail parks and hypermarkets
- new hotels, conference centres
- new road developments
- new sports centres and stadiums
- landfill (rubbish) sites and sewage works.

The reasons for these developments at rural–urban fringe locations include the following:

- The land is much cheaper than in the centre of towns and cities.
- There is much less traffic congestion.
- There is room to expand the development in the future.
- There is a larger area for car parking.
- There is less pollution and a much more attractive place for people to live, being surrounded by countryside.

These new sites in the countryside/rural areas are called greenfield sites.

TERMS

footloose businesses: businesses that are not tied to a particular location

greenfield site: an area of undeveloped land, often being used for agricultural needs, amenity or forest use, or some other undeveloped site that has been identified for commercial development or industrial projects

counter urbanisation: when large numbers of people move from urban areas into the surrounding countryside or rural areas

suburbanised villages: villages which have adopted some of the characteristics of urban areas

commuter hinterland: the rural area around large urban areas/cities that are economically tied to the urban area

Historically, people have always been migrating into cities. However, in HICs there has been a move in the opposite direction – a process called counter urbanisation. As people have wanted to live in better environments, they have moved into the countryside, often to new housing developments attached to existing villages – producing expanded villages called suburbanised villages. These villages and the land on the rural–urban fringe have many attractions for people who want to move out of town and city centres as they provide:

- houses with a modern design, services and amenities/electricity/water
- space to park cars
- relatively easy access to local services
- relatively low cost compared to houses and property in a town or city centre
- gardens/space/close to open space/play areas
- no air pollution as there are no nearby factories
- close proximity to their workplaces which are now also built on the urban fringe
- good public transport / main road access to the CBD if they still work or want to go into the centre for higher-order shops and services.

Many of the people who live in these settlements, however, do not work in them as they travel/commute to work in the cities and towns. This has given these villages the name of **commuter** or **dormitory settlements**. Many of the commuters use large supermarkets on the edge of towns and so there is a lower demand for a village shop and services, which has forced many to close. There may also be social impacts, as once tight-knit village communities begin to lose their community spirit as more and more people move in. The people in these villages form part of an area called a commuter hinterland. The size of a hinterland usually depends on the ease, speed, and cost of transportation between the urban area and the hinterland.

Self-test questions 6.1

1. Explain the possible changes that may occur in rural communities due to counter urbanisation.
2. What are the pressures on land in the rural–urban fringe?

6.2 Problems of urban areas, their causes and possible solutions

Many cities and towns had large areas of old, traditional industry just outside and adjacent to the CBD. As industry has changed and developed, these inner-city sites became unsuitable for modern industry which needed large, flat, easily accessible sites. Many areas became derelict and abandoned, but have since been regenerated and their functions completely changed. Many were older river and coastal industrial sites which have been cleaned up to provide excellent residential and business sites.

Urban Development Corporations (UDCs) were setup by the UK government to develop and improve areas of the inner cities. The first two UDCs were started in 1981 in London's Docklands (LDDC) and Merseyside (MDC) in Liverpool. These UDCs were given the means, by the government, to buy and acquire land, reclaim it and then attract private investment into the areas to set up new businesses, industry, housing and community developments.

Ten more UDCs were set up in England and one in Wales by 1993. After they had served their purpose, they were all wound up in 2000.

Challenges of urban areas in HICs and LICs

In addition to the problems of urban decay and dereliction, many city centres in both HICs and LICs suffer from a wide variety of challenges, summarised in Table 6.1. Many cities in LICs also suffer from squatter and shanty settlements. Most of the growth in cities in the last 30 years has been in LICs.

Challenges	Causes	Possible solutions
traffic congestion	inadequate road infrastructure to meet the number of vehicles using the road network	increase provision of urban transport options, using buses, trains, trams and light railways
housing shortages and overcrowding	lack of both public and private housing as a result of inadequate planning	planned urban housing expansion
unemployment	urban economy failing to expand to provide employment for the expanding urban population, especially where rural–urban migration is taking place	increase employment opportunities by attracting new businesses and expand jobs in the service sector
deprivation	lack of employment and education opportunities	increase educational opportunities by expanding school and training facilities
pollution – air, water and noise	lack of environmental pollution laws and regulations	provide better water and sewage facilities and enforce stricter environmental pollution controls

Table 6.1 Challenges faced by urban areas

For more problems associated with urban areas and strategies to reduce them, see Chapter 7 *Urbanisation*.

Self-test questions 6.2

1 What are the causes of the problems of urban areas in LICs and HICs and what are the possible solutions?

2 Using a case study you have studied, explain how the transport problems of a large urban area can be addressed.

Case study

6.3 An urban area – Dubai, United Arab Emirates

Dubai, with a population of 2.7 million, is the most populous city in the UAE. It is a rapidly developing urban area. About 50% of the emirate of Dubai's 4114 km^2 of land has been brought under urban development. Very large new residential areas have been built on the edge of the city, separate to the city's major employment areas.

Transport issues

This expansion has created urban sprawl in Dubai, which has not been matched by an expansion of public transport. This has meant that many people have to travel to work by private car, which has led to traffic congestion, increased air pollution and traffic-related injuries and fatalities. For a city of its size, relatively few people in Dubai use public transport – only 11% compared to Hong Kong's 63%, New York's 52%, London's 42% and Madrid's 30%. In an attempt to reduce transport problems in Dubai, a number of solutions have been used. These solutions aim to:

- reduce congestion
- save time for those people travelling to the centre, which, in turn, will save on transport costs
- reduce accidents
- reduce personal stress levels
- create a more pleasant urban environment by reducing atmospheric pollution from exhaust fumes and noise pollution
- save space – fewer car parks needed.

The solutions include the following:

- Providing more frequent, modern public transport systems so that people do not have to come to work in their car. This may include increased use of buses and **mass transit systems,** such as the combined over and underground railway system – the **Dubai Metro** system.
- **Transport hubs** (transport interchanges) served by adequate car parks to allow Park and Ride schemes to operate where people can leave their cars in car parks and be taken into their place of work or to shop by frequent buses and the Metro system.
- **Congestion charging** – as used in London, Milan, Stockholm, San Diego and Singapore – where people who wish to drive in to the centre of cities are charged to do so. Certain central areas in Dubai have been highlighted as areas where this could be introduced.
- Providing **discounted travel tickets** for commuters who can use them for all forms of public transport in city or town centres by bus, metro or water – the Nol Silver cards in Dubai.
- **Restrict parking** in the city centre and fine or tow away offenders.
- **Increase parking charges** in city centre.
- **Provide less polluting forms of public transport** – electric and gas-fuelled vehicles and trains.
- Major companies/employers settled in Dubai should provide to workers a **free (or subsidised) public transport seasonal-ticket** as a standard employee benefit.
- Public transport services should operate at least 18 hours per day.

It is hoped that by 2020, 30% of households in Dubai will be within 400 metres of public transport services with a peak service frequency of at least one service every 10 minutes. By increasing the efficiency of public transport and road networks, the average journey time to work should be within a range of 25 to 30 minutes for every worker.

Public open space

The urban sprawl of Dubai has meant that there is little public access to its waterfront and a lack of public open spaces within the city. Public parks account for only 2.3% of the urban area, compared to 42.5% in Seoul, Korea, and 8.5% in Colombo, Sri Lanka. It is hoped that access can be improved and more open space created in future urban expansion projects.

Loss of natural habitat

Dubai has also lost wildlife sanctuaries and conservation areas on the rural–urban fringe during its urban expansion. It is planned that future urban expansion should compensate for this loss with the creation of an inland protected area.

Affordable housing

Much of the new urban residential area in Dubai is not affordable for lower-income earners. The average monthly rent for a three-room apartment is $2500 while the average wage is $3300. People who live in these areas also lack easy access to public transport. To alleviate this problem, 450 000 new residential units for low-income households are due to be built, with easy connection to public transport.

Waste treatment

The urban expansion of Dubai has not taken into account any recycling of waste so that over 90% of waste is currently sent to official landfill sites in the city, which already cover about 10.6 km^2 on the edge of the expanding urban area. At present rates of urban expansion, an additional area of 57 km^2 will be needed by 2020. At the moment, less than 1% of waste is recycled. It is essential that Dubai develops an adequate recycling policy and looks at **waste-to-energy systems** that could use waste to generate energy.

Sample question and answer

Explain why many new shopping and entertainment centres are being built in suburban areas rather than in the centre of cities. [5]

Write as clearly and precisely as possible, including examples and developing your ideas wherever possible, especially where extended writing is required.

It is much easier to build new shopping and entertainment centres in a suburban area rather than in the centre of a city because they will be much easier for people to drive to **[1]** as they will be nearer the motorway, on the edge of the city, and people will not have to get through the crowded streets of the city centre **[1]**. Supply lorries/trucks for the shops will be able to get to them more easily **[1]**. The land in the suburbs will be cheaper than the land in the city centre **[1]** and so large shopping malls, which require large areas of land for retail outlets and parking/delivery areas, can be built **[1]**. Parking will be easier because there will be more land for car parks and it might be free **[1]**. People will be able to do one-stop shopping **[1]** and won't have to walk long distances between shops **[1]**. The shopping malls will be covered and their inside temperatures regulated – warm in winter and cool in the summer **[1]**. It will also be away from the noise and air pollution of the city centre **[1]**.
(A maximum of 5 marks will be awarded.)

Exam-style questions

1 The area that surrounds many towns and cities is called the rural–urban fringe. Why may urban authorities and councils want to carefully monitor and control any new development that takes place in these edge-of-town areas? [5]

2 Why may a national supermarket chain want to locate a new superstore/hypermarket on the edge of a large urban area? [4]

Urbanisation

Learning summary

By the end of this chapter, you should be able to:

- identify and suggest reasons for rapid urban growth with reference to the physical, economic and social factors which result in rural depopulation and the movement of people to major cities
- describe the impacts of urban growth on both rural and urban areas, including the effects of urbanisation on the people and natural environment and the characteristics of squatter settlements
- describe strategies to reduce the negative impacts of urbanisation
- demonstrate knowledge of a case study of a rapidly growing urban area in a developing country and migration to it.

7.1 Rapid urban growth

Urbanisation refers to a population movement from rural to urban areas, resulting in the gradual increase in the proportion of people living in urban areas. It is the process by which towns and cities are formed and become larger as more people begin living and working in urban areas.

According to the UN, half of the world's population was living in urban areas by the end of 2008 (Figure 7.1). By 2017, the urban population accounted for over 54% of the total global population, up from 34% in 1960.

FACT

It is estimated that by 2050, about 66% of the developing world and 86% of the developed world will be urbanised.

TERM

urbanisation: the process by which an increasing proportion of people live in towns and cities instead of the countryside – this could be the result of natural increase and/or migration

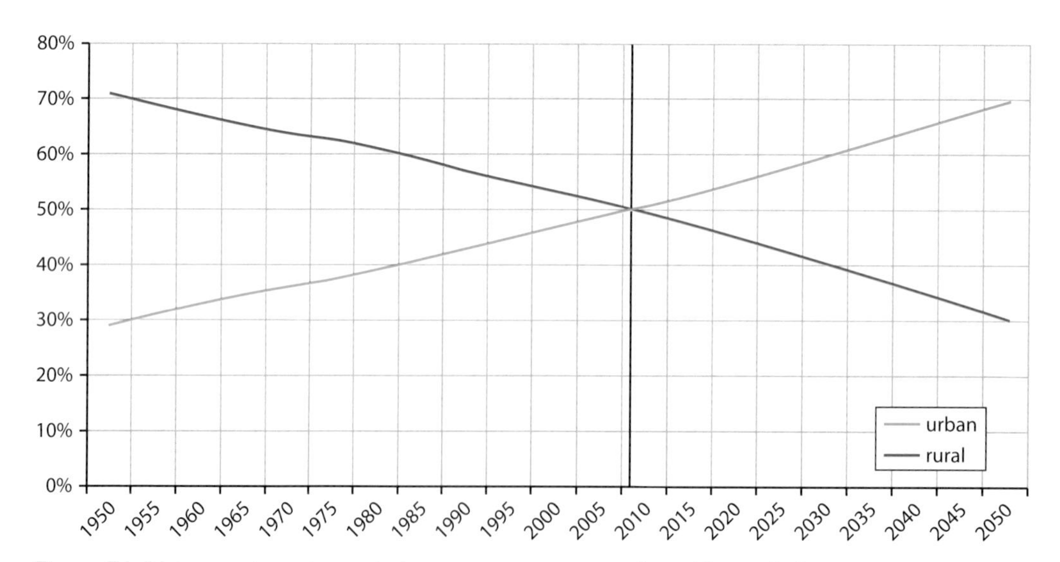

Figure 7.1 Urban and rural populations as a percentage of world population

The global and regional growth in the urban population

In real terms, the urban population of the world has grown rapidly from 746 million in 1950 to over 4 billion in 2017. Asia has 53% of the world's urban population, in contrast, Europe has 14% and Latin/South America and the Caribbean 13%. The UN has projected that nearly all the global population growth from 2016 to 2030 will be absorbed by urban areas, accounting for a growth of about 1.1 billion people in this period, and most of this will be in Asian cities.

The need for urban planning

Due to the vast numbers of people involved in this growth, careful **urban planning** will need to be given to urban settlements of all sizes. If cities are sustainably managed in the future, they will provide important opportunities for economic development and for expanding access to basic services, including health care and education, for very large numbers of people. Providing public transportation, as well as housing, electricity, water and sanitation for a densely settled urban population is typically cheaper and less environmentally damaging than providing a similar level of services to a dispersed rural population.

Cities are important catalysts in the development process of most countries and in the reduction of poverty in both urban and rural areas, as they concentrate much of a country's economic activity, government, commerce and transportation. Urban living is often associated with higher levels of literacy and education, better health, greater access to social services, and enhanced opportunities for cultural and political participation.

The implications of rapid and unplanned urban growth

Rapid and unplanned urban growth may threaten the sustainable development of many urban areas when the necessary infrastructure is not developed or when policies are not implemented to ensure that the benefits of urban life are equally shared.

Today, despite the advantages of urban areas, they often have greater inequalities than rural areas and hundreds of millions of the world's urban poor live in sub-standard conditions. In some cities which are unplanned or inadequately managed, urban expansion has led to rapid urban sprawl, pollution and environmental degradation, together with unsustainable production and levels of consumption.

TIP

For more on the problems of urban areas, see Chapter 6 *Urban settlements*, Section 6.2 *Problems of urban areas, their causes and possible solutions*.

In 1990, there were ten mega-cities containing 153 million people or slightly less than 7% of the global urban population at that time. In 2014, there were 28 mega-cities worldwide, containing 453 million people or about 12% of the world's urban population. Tokyo is the world's largest city with 38 million inhabitants, followed by Delhi with 25 million, Shanghai with 23 million, and Mexico City, Mumbai and São Paulo, each with around 21 million inhabitants. Osaka has just over 20 million, followed by Beijing with slightly less than 20 million. The New York-Newark area and Cairo complete the top ten most populous urban areas with around 18.5 million inhabitants each. Overall, nearly half of the world's 3.9 billion urban dwellers reside in relatively small settlements with fewer than 500 000 inhabitants, while only around one in eight people live in the 28 mega-cities.

TERMS

mega-cities: cities with 10 million inhabitants or more

rural depopulation: the fall in population of rural areas by migration or by a fall in birth rate as young people move away

The rural population of the world has grown slowly since 1950 and is expected to reach its peak around 2020, as a result of the increasing urbanisation of the world's population. Urban areas are growing faster in the developing world than anywhere else in the world, but this growth brings problems and challenges, all of which require careful management and solutions.

Rural depopulation and the movement of people to major cities

Push factors cause people to move away from rural areas, leading to rural depopulation. **Pull factors** cause people from the rural areas to move to urban areas. Table 7.1 gives examples of push and pull factors.

Push factors	Pull factors
• The mechanisation of farming activities and a shortage of alternative jobs resulting in a lack of employment opportunities in rural areas and high levels of rural unemployment • The trend for large landowners to take back the land of their tenant farmers to grow cash crops for export. • High infant mortality in many rural areas due to a lack of clean water, electricity, sewerage and health care. • Natural disasters, such as periodic droughts in NE Brazil. • Lack of infrastructure, such as adequate roads, education facilities, retail opportunities/shops.	• More schools, doctors and health care services are available in the urban areas. • Successful migrants encourage their extended families and other members of their rural communities to join them in the large urban areas. • In Brazil, a series of advertising campaigns were carried out in rural areas in the 1950s and 1960s to attract workers from rural areas to move to the city.

Table 7.1 Push and pull factors

Why is the urban population increasing so fast at the expense of rural areas?

The rapid growth of urban areas is the result of two factors:

- a **natural increase in population** (an excess of births over deaths)
- **migration to urban areas**.

In general, cities are perceived as places where people can have a better life, because of better opportunities, higher salaries, better services and better lifestyles. These perceived 'better' conditions attract poor people from rural areas.

To further illustrate the causes of rural migration, it is useful to consider the policies that have led to migration in many developing countries. At a national level, in trying to pay foreign debts and to be more competitive in international markets, national governments have encouraged the export of national resources and agricultural products. In order to produce agricultural products quickly, efficiently, and to maximise profits, national governments often look to decrease the number of small producers, and turn agricultural production and resource extraction over to larger enterprises (often large multinational/transnational companies (TNCs) or nationally important companies), with larger production facilities, and a lower per-unit cost of production.

This trend has the effect of turning the land into a commodity that can be bought and sold, and it is viewed only in terms of its productive capabilities. Many countries now encourage free market economics which looks to increase economic efficiency to deliver goods at the lowest possible price, and the supporters of this form of economics maintain that any government intervention will reduce this efficiency. Other government policies can reinforce this. For example, in order to increase the production of cheaper food, some governments have maintained artificially low food prices in urban areas to reduce the cost of urban living. This policy has sometimes resulted in no form of adequate compensation being given to rural farmers for the costs they have incurred in producing their food products and this has increased levels of rural poverty.

At the same time, these policies have also made urban life more attractive and pulled people from rural areas. In Mexico, as a result of these policies, an average of 270 000 rural migrants have been arriving in Mexico City annually over the last ten years, transforming it into one of the largest cities in the world.

TERMS

free market economy: an economy where the government imposes few, if any, restrictions on buyers and sellers

favelas: housing areas of 60 or more families in houses that often lack the basic services of running water, sewerage and electricity and the residents have no legal right to the land on which they live

7.2 Impacts of urban growth

The impacts of rapid urban growth can include the following:

- A large proportion of housing that is inadequate housing and services. In cities, like São Paulo in Brazil, as many as 40% of the population live in the shanty towns or favelas.

- The infrastructure in the shanty towns is often almost non-existent or incapable of maintaining a basic standard of living. They lack a clean water supply, rubbish collection and sanitation, increasing the risk of disease.
- What infrastructure that does exist is poor, with governments either not providing or not having sufficient money to either build or even maintain the existing infrastructure.
- A shortage and lack of affordable formal housing.
- The shanty town/favelas are often constructed on inappropriate land – more prone to flooding, steeply sloping, and prone to landslides. It may badly polluted by industrial activities.
- Houses are often made of wood, plastic and their high population densities increase the risk of fire and higher casualties.
- Poor transport infrastructure.
- A lack of employment means that many people work in the informal sector. The employment available is usually in the form of low paid, menial jobs in construction industries, hotels and restaurants.
- Many cities in the LICs and MICs have very high levels of air, land and water pollution. The increasing amounts of industry brings with it inevitable pollution problems, especially as laws to protect the environment are either non-existent or rarely enforced.

The effects of urbanisation on the people and natural environment

The impacts are seen across HICs, MICs and LICs. In HICs, such as the USA the negative impacts include:

- increased traffic and traffic congestion
- increased air, water and noise pollution
- destruction of agricultural land, parks and open space
- huge infrastructure costs for new water and sewer systems, transport projects – urban railway and underground systems, new schools, and increased police and fire services
- crowded schools in the suburbs and empty, crumbling schools in city centres.

The impact on rural areas of rural depopulation

In HIC and MIC especially, small labour-intensive family farms have often been merged together and replaced by heavily mechanised and specialised industrial farms. This has meant that employment opportunities are severely reduced and many workers and their families have migrated away to find work in urban areas. Unfortunately, many of their skills are not relevant to the jobs available in the urban areas and so they find themselves in unskilled or low skilled jobs. The large 'mega' farms have also consolidated their supply and processing industries in feed, seed, processed grain and livestock industries, which has meant that there are fewer small businesses in rural areas. Many have therefore closed and their owners and workers have been forced to migrate away.

FACT

Rural areas, which used to be able to provide employment for all young adults, now provide much fewer employment opportunities.

The decline in rural population has led to a decrease in services such as schools, business, health care and cultural opportunities and the increasing age of the remaining population puts further strain on the social service system of rural areas. Many small rural communities, therefore, witness decreased populations, decreased incomes, increased income inequality, fewer retail outlets and less retail trade.

Characteristics of squatter settlements

Over 100 million people living in the cities of the developing world have no shelter of any kind and over one-third of the population of these cities live in squatter settlements.

TERMS

squatter settlement: an area of makeshift housing that usually develops in unfavourable sites in and around a MIC or LIC city; they are also known as 'shanty' towns or 'bustees'

self-help schemes: small-scale schemes which allow local residents to help improve their local area

Within the squatter settlements, the residents can face many hardships and challenges, including:

- overcrowding – these settlements have high population densities

- devastating fires – fires can spread quickly as there is little or no room between the tightly packed buildings and there are very few points of access for fire engines and their equipment
- often being overpopulated – the settlements do not have enough resources to support their large and growing populations
- intense competition for any form of employment
- poor sanitation and limited health care services and facilities can lead to the spread of disease
- poor access to education services and schools
- lack of space in the settlements – the newest and often poorest arrivals may be forced to live on the worst quality land, often the steeper, less stable slopes, wetland/marshland, or land most prone to flooding
- lack of infrastructure – access to services can be poor; public transport is limited and connections to the water, sanitation and electricity supply can be limited and sometimes dangerous.

Most city authorities would like to remove these squatter settlements, but very few authorities have the resources to replace the housing that would be lost. As a result, many of these settlements become permanent homes, though authorities try to offer support and help to set up self-help schemes and community housing projects in which the residents can improve their houses and the infrastructure of the areas they live in. The government provides the basic building materials which enables residents to improve their own homes. This often has the added advantage of developing a community spirit and a sense of common ownership for their area and community, as many families may come together to make the improvements. In the favela of Rocinha, in Rio de Janeiro, such self-help schemes have improved the area from slums to low-quality housing where the majority of homes have basic services like electricity. There are now also many services available to the residents including cafés and shops. Some people have also been granted legal ownership of the land on which their houses are built.

Self-test question 7.1

1. Describe the factors causing people to move away from rural areas.

7.3 Strategies to reduce the negative impacts of urbanisation

Possible solutions to reduce the negative impacts of urbanisation include:

- creating urban growth boundaries, parks and open space protection
- improved planning and increased expenditure to promote public transportation
- reversing and changing government policies that help create sprawl
- revitalising already developed areas through measures such as attracting new businesses, retailing, reducing crime and improving schools.

Other initiatives to improve the quality of life of the residents include the following:

- **Site and service schemes** which give people the chance to rent or buy a piece of land. The city has provided new areas of development land that is connected to the city by transport links and has access to essential services – water/sanitation and electricity).
- **Self-help schemes**, a cheaper option than the site and service schemes, where people are provided with the materials, the tools and the training to improve their homes. Low-interest loans are made available to help people fund these improvements. Such schemes are relatively cheap and give the migrants a sense of control over their future and encourage a community spirit.
- **Rural investment** where attempts and schemes have been set up to improve the quality of life, and in so doing, creating greater opportunities in rural areas which persuade and prevent people in rural areas from migrating to urban areas.
- **Slum and squatter settlement clearance** where authorities have tried to clear the settlements, but have found that the inhabitants have just moved elsewhere to other slum and squatter areas. In some cases, new residential tower blocks are built on the cleared land and the former slum residents re-housed in them.

Self-test questions 7.2

1. In the cities of the developing world, a large proportion of the population may live in slums (squatter settlements). What are the possible reasons for this?
2. What are the problems faced by the people who live in favelas?

Case studies

Managing rapid urban growth in a sustainable way in Curitiba, Brazil

Curitiba is the capital and largest city of the Brazilian state of Paraná, with a population of 1 879 000. The city expanded greatly from the late 1960s, with innovative urban planning that increased the population from a few hundred thousand to more than a million. Its economic growth occurred in parallel with the inward migration of people from surrounding rural areas and other parts of the country; almost half of the city's population was not born in Curitiba.

At its peak, the state of Paraná produced almost one-third of the world's coffee. However, after a series of frosts killed a large proportion of the coffee trees, many of the workers lost their jobs and migrated to Curitiba in search of employment. However, Curitiba's infrastructure was unable to cope with the large number of migrants. Only a third of the families living in Curitiba had access to clean water and sewers. Traffic congestion was beginning to become an increasing problem and the growth in population and unplanned development threatened to damage the character of the city and the city was forced to make swift decisions to avoid possible chaos in its urban development and to meet future demands

In response, the **Curitiba Master Plan** was initiated which included a number of social, economic and environmental programmes:

- A **planned transportation system** consisting entirely of buses. On major streets, lanes devoted to a rapid-transit bus system were built. Its huge popularity saw the population of the city shift from car travel to bus travel.

TERM

corridor of growth: a planned strategic corridor, usually along a major road, which allows the continued growth of an urban area – the planning of the corridor involves the provision of infrastructure such as the upgrading of the road and managed open space, as well as possible business, industrial, retail and residential development adjacent to the route

- A **new road design** to minimise traffic. Urban growth is restricted to corridors of growth – along key transport routes. Tall buildings are allowed only along bus routes.
- Under the '**Garbage? That's not garbage!**' programme, 70% of the city's waste and rubbish is recycled by its residents.
- A **green exchange programme** means that poor families in squatter settlements that are unreachable by trucks bring their rubbish bags to neighbourhood centres, where they can exchange them for bus tickets or for eggs, milk, oranges and potatoes, all bought from outlying farms.
- **Entrepreneurial 'sheds'** – business centres designed to help small companies get established and prosper.
- An **Open University**, created by the city, allowing residents to take courses in many subjects such as mechanics, hair styling and environmental protection for a small fee. Retired city buses were used as mobile schools or offices.
- **Pedestrian streets in central areas**, including a 24-hour mall with shops, restaurants and cafés, and a street of flowers with gardens tended by street kids.
- A **Municipal Housing Fund** to house 50 000 poor families.
- To help solve the city's flood problems, floodwater was diverted into lakes in newly created parks and green space beside the rivers. Young people were hired to keep the parks clean.
- Orphaned or abandoned street children are found all over Brazil. Industry, businesses and shops were encouraged to 'adopt' a few children, providing them with a daily meal and a small wage in exchange for simple maintenance gardening or office chores.
- The city has 200 km of **bicycle routes**, used by around 30 000 people every day.
- Creation of **green park areas**. Nearly one-fifth of the city is now parkland, and volunteers have planted 1.5 million trees along the streets. Builders get tax breaks if their projects include green space.

A rapidly growing urban area – Rio de Janeiro, Brazil

Rio de Janeiro in Brazil is typical of many MIC and LIC cities in that there is an enormous inequality between the rich and the poor. With a population of over 6.4 million people in 2016, Rio accounts for 6% of Brazil's GDP, however, it has half a million homeless people who live on the streets with no shelter. Its public services, such as education and health care, struggle to cope with such growth. Physically, it is trapped between high steep mountains and the coast and has little room for growth of buildings or to improve transport.

Rio has over 750 **favelas** (Figure 7.2). Most modern favelas appeared in the 1970s due to rural–urban migration, when many people left rural areas of Brazil and moved to its cities. Unable to find places to live in the cities, many of the migrants ended up in favelas. Due to overcrowding, unsanitary conditions, poor nutrition and pollution, communicable diseases can be prevalent and spread rapidly in the poorer favelas, and infant mortality rates are high. Poverty, inadequate water supplies and weak public health systems are major factors in the spread of other mosquito-borne diseases, such as dengue, Chikungunya and, in 2016, the potential for a Zika epidemic.

The largest favelas, such as Rocinha, have a population of over 100 000 people. The favelas tend to be located either on the rural–urban fringe, or on land that building developers would not normally use, such as steep hillsides or marshland beside rivers. Rio has a problem with crime and the favelas can be centres of organised crime, drug trafficking and violence. There is not enough available and affordable housing to meet demand from the incoming migrants, so that many are forced into the favelas.

There are four main types of favela in Rio de Janeiro:

1. Dense areas with self-built housing lacking any infrastructure such as roads, electricity, water or sewerage.
2. Areas where land or housing has been illegally subdivided into very small plots or houses.
3. Invasions of risky areas such as beside railway lines, electricity lines or streets.
4. Corticos – areas of old and decaying housing that is illegally rented out to the poor.

Strategies to manage the housing challenge in LIC and MIC cities

There are several possible solutions to the lack of available and affordable housing in LIC and MIC cities, like Rio de Janeiro, including the following:

- The government buying land and building houses that are cheap to rent – very few people will be in a position to buy.
- Providing interest-free or low-interest loans for people to either build their own homes or to improve their current homes or neighbourhoods.

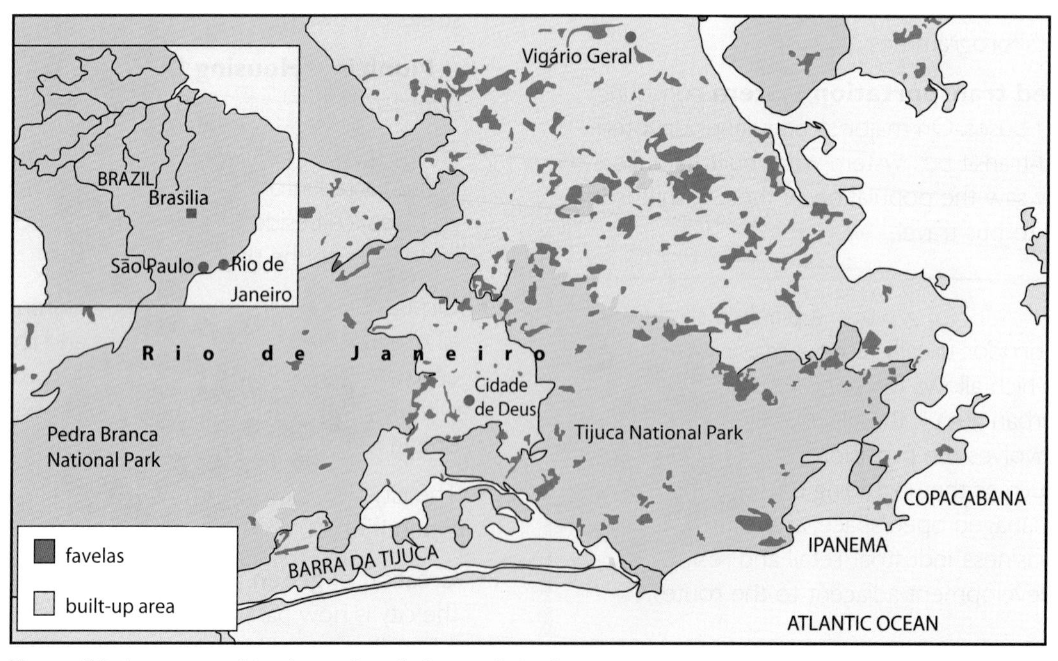

Figure 7.2 Location of favelas in Rio de Janeiro's built-up areas

- Provide the infrastructure and services like clean drinking water, sewage systems, refuse/rubbish disposal, health care and schools to improve the quality of life of the people in the favelas.
- Supporting small businesses in the favelas to provide employment and incomes, often through microfinance schemes.
- Teaching basic building and construction skills to the people and providing low-cost building materials so that they can improve their houses.

TERMS

microfinance: a form of banking/financial service that is provided to unemployed or low-income individuals or groups who otherwise have no other access to financial services – microfinance offers them an opportunity to become self-sufficient by providing a way to save money, borrow money and get insurance and set up in business.

edge towns or cities: concentrations of businesses, shopping and entertainment venues outside a traditional city centre in what had previously been a residential or rural area

The local communities, charities, national and city government departments are working together to improve conditions in the favelas. An example of a successful scheme is the **Favela Bairro/ Neighbourhood Project**, which was set up in Rio de Janiero in 1994. This scheme greatly improved the quality of life of the people living in the squatter settlements and areas of poor housing by:

- increasing the availability of clean water from communal taps
- reducing the incidence of disease – helped by the building of new toilets to prevent the contamination of water by sewage
- building surface drains to reduce the amount of standing water, which is likely to smell and be polluted
- constructing paved roads, which will mean people can travel more easily in cars or on bikes; this could lead to better quality and more regular and more widespread public transport
- putting in security lighting in housing areas which will improve personal safety, so that there is less danger of muggings
- providing free education for street children
- providing low-cost building materials.

FACT

Communities form in favelas over time and often develop social organisations and community associations to obtain such services as running water and electricity. Sometimes the residents manage to gain title to the land and then are able to improve their homes.

People who can afford to have often moved away from the congestion of the favelas and moved to cleaner, safer environments further along the coast from Rio, to new edge towns or cities such as Barra de Tijuca. This new city is 20 km along the coast, linked to Rio by a four-lane highway, which goes through tunnels in the mountains and is raised up along the coast. It now has a population of over 300 000, stretching 5 km along the coast, with a beach that stretches for 20 km. Three-quarters of the housing is in 10–30 storey apartment blocks. Although representing only 4.7% of Rio's population, Barra is responsible for 30% of all tax collected in the city.

Sample question and answer

How can the people in squatter settlements improve the quality of their homes and the areas they live in? [5]

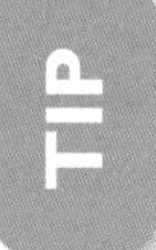

Use the mark allocation as a guide to the amount of detail or number of responses required, and the amount of time you should spend on a question.

The question refers to squatter settlements and how individual homes and the area can be improved, so make sure both are covered. Points to make include: provide low-cost building materials **[1]**; the installation of mains electricity/water **[1]**; infrastructure development including installation of sanitation systems, such as the laying of sewage pipes / the building of sewage works **[1]**; increase in the provision of local authority housing **[1]**; improvement of communications to/from and within the settlements **[1]**; the regular disposal of rubbish/refuse **[1]**; set up self-help schemes/site and services schemes **[1]**; provide skill/education in building skills **[1]**; build schools/health care facilities/clinics/hospitals **[1]**; support charitable initiatives, such as food banks, working with street children **[1]**.
(A maximum of 5 marks will be awarded.)

Exam-style questions

The Curitiba and Rio de Janeiro case studies can be used to answer the following questions.

1. Most large urban areas have made attempts to solve the problem of traffic congestion faced by the people who live there. For a named urban area, describe how attempts have been made to solve the problem. [6]
2. Using examples that you have studied, explain the strategies that can be used to reduce the negative impacts of urbanisation. [6]
3. In the cities of the developing world, a large proportion of the population may live in slums (squatter settlements). What are the possible reasons for this? [4]

THEME 2: THE NATURAL ENVIRONMENT

Chapter 8

Earthquakes and volcanoes

Learning summary

By the end of this chapter, you should be able to:

- describe the main types and features of volcanoes and earthquakes, including the types of volcanoes (such as strato-volcanoes and shield volcanoes)
- describe and explain the distribution of earthquakes and volcanoes, including the global pattern of plates, their structure, and an awareness of plate movements and their effects
- describe the causes of earthquakes and volcanic eruptions and their effects on people and the environment
- demonstrate an understanding that volcanoes present hazards and offer opportunities for people
- explain what can be done to reduce the impacts of earthquakes and volcanoes
- demonstrate knowledge of case studies of an earthquake and a volcano.

8.1 The main types and features of volcanoes

A volcano is a landform, often the size of a mountain or hill, typically conical in shape, having a crater or vent through which lava, rock fragments, hot gases and steam are, or have been, erupted through the Earth's crust. Volcanoes normally form in three possible types of location:

1 Where two tectonic plates move away from each other – i.e. the plates diverge, at constructive/divergent boundaries, and a gap/line of weakness is created in the Earth's crust. Molten magma emerges through the gap. When magma appears on the Earth's surface, it is called lava. The lava solidifies to form a volcano.

2 Where two plates move towards each other – i.e. the plates converge at destructive/convergent boundaries, where a heavier, denser oceanic plate moves down under a less heavy continental plate (the process of subduction). Friction and intense heating will take place, which results in the destruction of the oceanic plate, causing the rocks in the plate to melt and turn into molten magma. The build-up of magma creates enormous pressures on the crust above it and the magma rises through lines of weakness to emerge on the surface of the Earth. The lava then solidifies to form a volcano.

3 Some volcanoes form over thinner, weaker areas towards the centre of a tectonic plate, called hot spots. At these weak spots, magma can force its way to the surface to form a series of volcanoes, such as the Hawaiian Islands.

TERMS

volcano: cone-shaped mountain formed by eruptions of lava at the surface of the Earth

crater: a depression on the surface of a volcano, formed by volcanic activity, often circular in shape with steep sides

vent: the natural pipe or fissure that links the magma chamber to the crater or opening on the Earth's surface through which lava, ash and gases flow

lava: magma that has escaped from beneath the Earth's crust and has flowed onto the surface

tectonic plates: huge pieces of the Earth's crust that float and move on top of the much denser mantle below them

constructive/divergent plate boundary: where two plates move apart, allowing magma to come to the surface as lava.

magma: molten rock found beneath the Earth's crust

TERMS

magma chamber: a large natural underground chamber of magma found within the surface of the Earth beneath a volcano.

destructive/convergent plate boundary: where two plates move towards each other

subduction: when one plate sinks below another

hot spot: a central part of the Earth's crust where plumes of magma rise to the surface

Apart from lava, the magma may come out on to the Earth's crust in a two other ways:

1 It may explode out and fall as **volcanic bombs**. Fragments of molten rock between 60 mm and 5 m in diameter are ejected into the air; they cool into solid fragments before they reach the ground up to 5–600 m from the vent.

2 It may appear as very fine **ash** – fragments of pulverised solid lava, measuring less than 2 mm in diameter. Once in the air, ash can be transported by wind up to thousands of kilometres away.

Some volcanic eruptions are also accompanied by:

- **pyroclastic flows:** clouds of very hot, poisonous gases mixed with ash that flow down the sides of volcanoes at speeds up to 200 km per hour
- **lahars:** melted snow and ice from the top of the volcano combined with rainwater, which mixes with ash and runs off the volcano, flooding valleys and flatter areas surrounding the volcano with mud
- **earthquakes:** as the magma moves upwards, the enormous pressure on the crust reaches breaking point, causing violent shaking (see Section 8.2).

FACT

Over time, varying from a few weeks to several thousand years, all this ejected material may build up to form a volcano.

The main features of a volcano are shown in Figure 8.1.

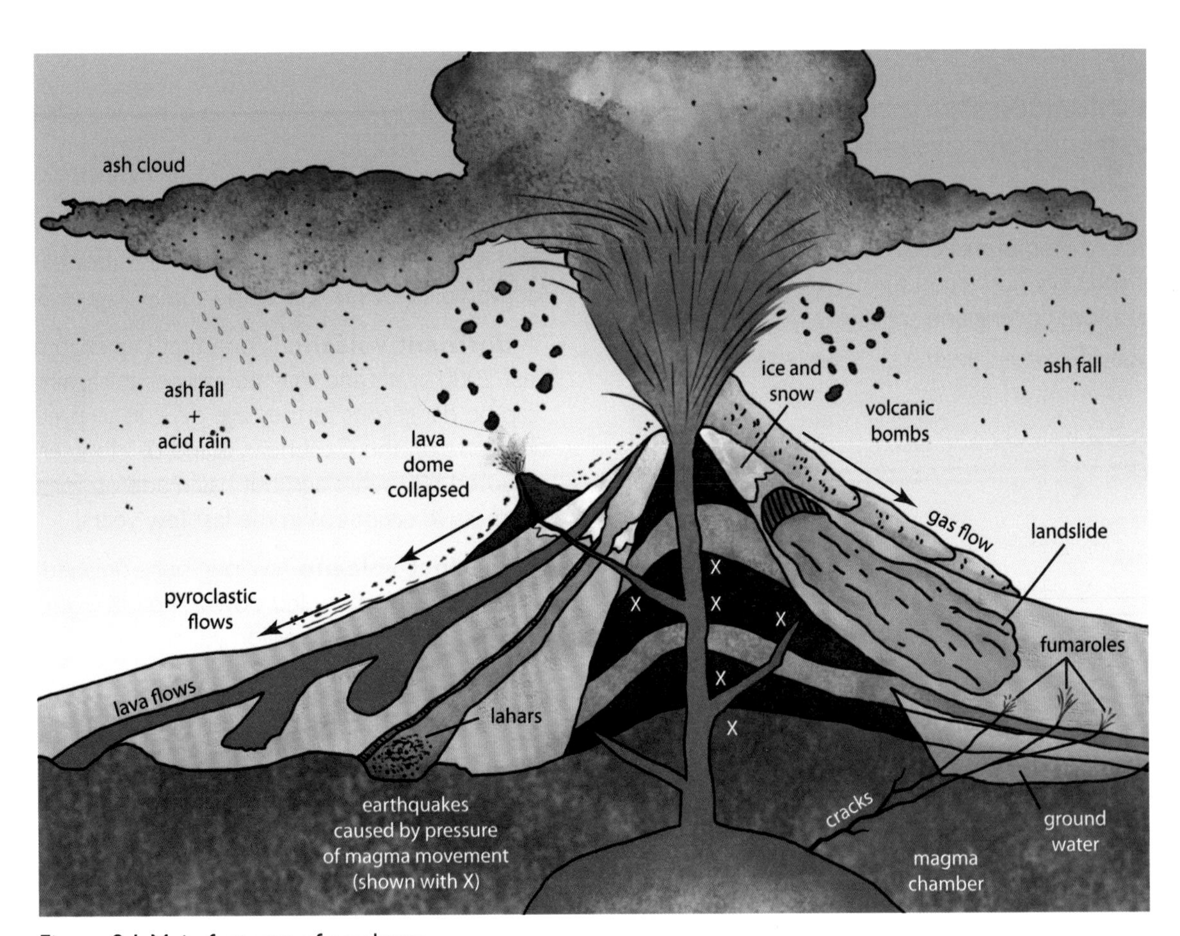

Figure 8.1 Main features of a volcano

There are three main types of volcano based on their shape and what they are made of.

1 **Composite or strato-volcanoes:** these are made from alternating layers of lava and ash as both come out of the vent during an eruption (Figure 8.2). They form on destructive plate boundaries where oceanic crust has melted as it is subducted. The lava forces its way up through the crust and emerges as a violent explosion. An example is Mount Etna in Sicily, Italy.

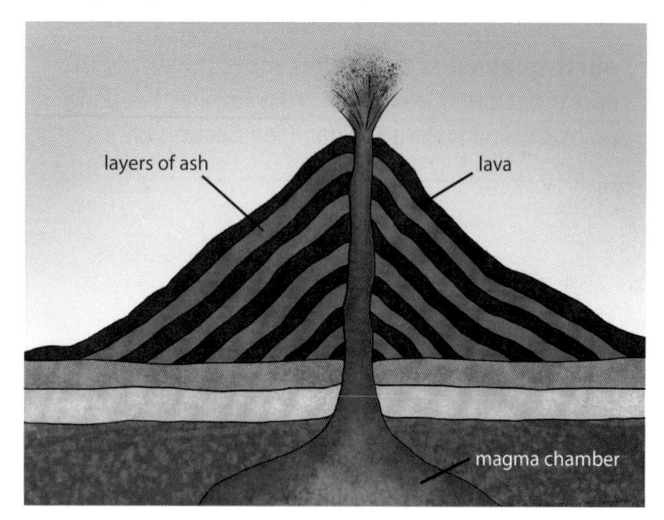

Figure 8.2 A composite volcano

2 **Shield volcanoes:** these are made from lava and form on constructive plate boundaries or at hot spots, such as Mauna Loa in Hawaii. They form large volcanoes, with gently sloping sides, sometimes hundreds of kilometres across because the lava is alkaline and very runny and travels a long way on the surface before cooling and solidifying (Figure 8.3).

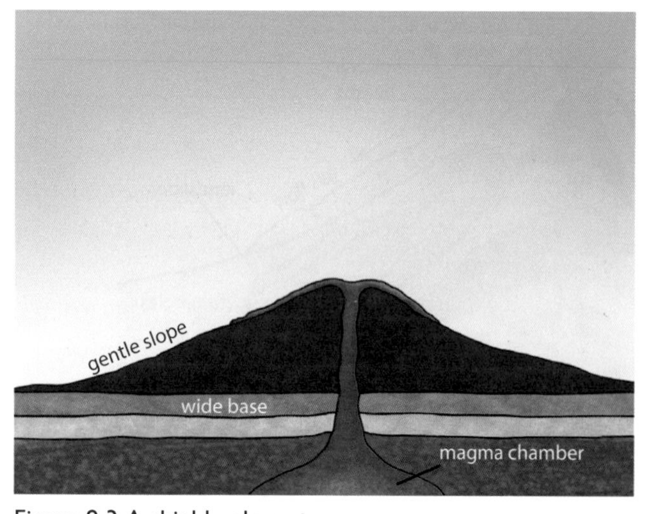

Figure 8.3 A shield volcano

3 **Dome volcanoes:** these are also made from lava but the lava is acid and thicker and cools quickly. It does not flow very far before becoming solid, and so these volcanoes are steep-sided and high (Figure 8.4). An example is Mount St. Helens in the USA.

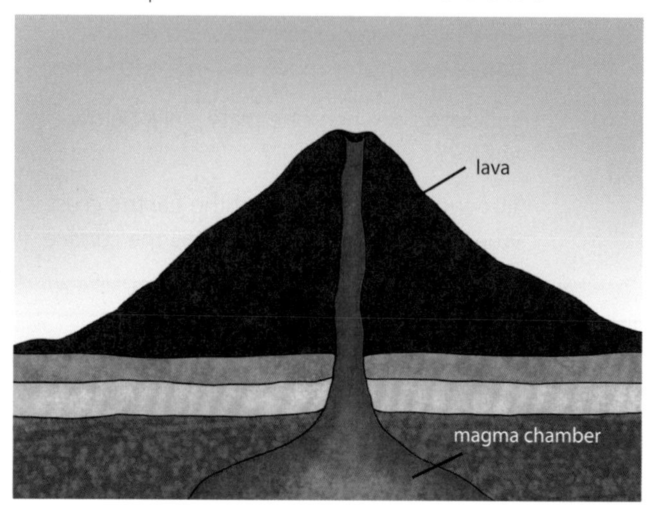

Figure 8.4 A dome volcano

You do not need to revise dome volcanoes, but they are included here to give a complete understanding of volcanoes.

Volcanoes can be active, dormant or extinct:

- An **active volcano** is one that has recently erupted and is likely to erupt again. There are about 1700 active volcanoes in the world today (Figure 8.5).
- A **dormant volcano** is one that has erupted in the last 2000 years and may possibly erupt again. These can be dangerous as they are difficult to predict. The one on the Caribbean island of Montserrat last erupted 500 years ago but has made up for this with its massive eruptions in the last few years!
- An **extinct volcano** has long since finished erupting and has cooled down. The UK's volcanoes last erupted over 50 million years ago.

8.2 The main features of earthquakes

An earthquake is the result of a sudden release of energy that causes the Earth's crust to shake, sometimes violently. Each day, there are at least 8000 earthquakes on the Earth. In a typical year, about 49 000 earthquakes are actually strong enough to be felt and noticed by

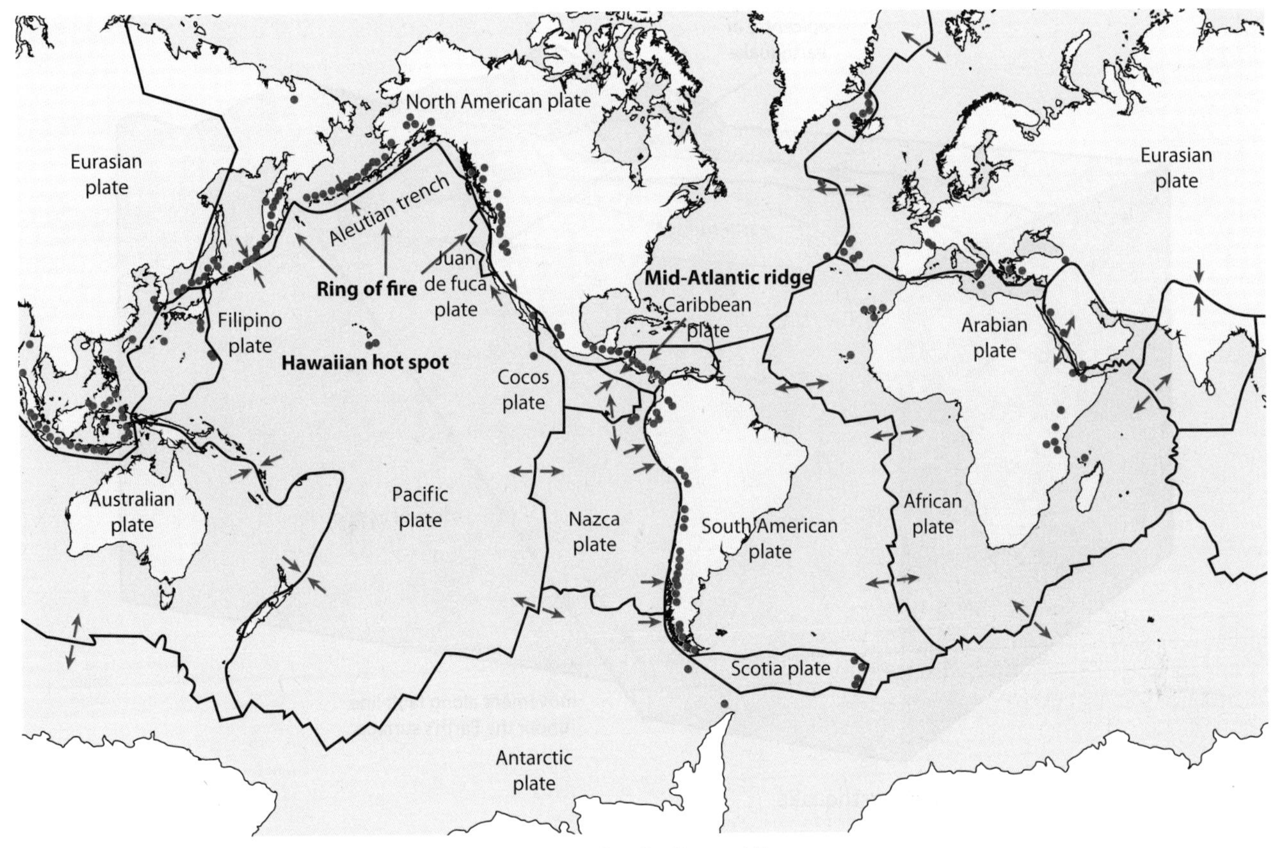

Figure 8.5 Active volcanoes, plate boundaries and the Pacific Ring of Fire

people and an average of 18 of these can cause serious damage to buildings and possibly injure and kill people.

Most of the world's earthquakes (90% of them and 81% of the largest) take place along the **Pacific Ring of Fire** – a 40 000 km long, horseshoe-shaped zone found along the edge of the Pacific Ocean (Figure 8.5). As plates move, the rocks on their edges may become locked together until, at the weakest point along a plate boundary – a fault line – they tear apart, or rupture, and this releases the strain.

Earthquakes can also be produced by the movement of magma inside volcanoes. These earthquakes can serve as an early warning of volcanic eruptions. Some human activities can also trigger earthquakes, such as underground nuclear explosions and the creation of large reservoirs behind dams, the weight of which compresses and depresses the crust below them.

An earthquake's point of initial tearing, or rupture, inside the Earth's crust is called its focus (Figure 8.6). Earthquakes with a deep focus are found in the subduction zones beneath destructive plate boundaries. Those with a shallow focus are found along conservative and constructive plate boundaries. Shallow earthquakes can sometimes be very violent, as in Haiti in 2010, as their energy is not absorbed by the relatively thin crust above them and appears on the surface very quickly.

TERMS

earthquake: a sudden and often violent shift in the rocks forming the Earth's crust, which is felt at the surface

fault line: a fracture or break in the Earth's surface along which rocks have moved alongside each other

focus: the location of the actual source of an earthquake **below** the ground surface; also called the origin

conservative plate boundary: where two plates are sliding alongside each other

epicentre: the location **on** the surface of the Earth above the focus or origin of the earthquake

The epicentre (Figure 8.6) is the point on the Earth's surface directly above the focus, where the energy bursts onto the surface and spreads out in a series of shock waves. Earthquakes cause two major effects on the Earth's surface – shaking and slipping of the crust. Surface movement by slipping in the largest earthquakes can be more than 10 metres. A slip that occurs underwater can create a huge wave called a tsunami (see Section 8.4).

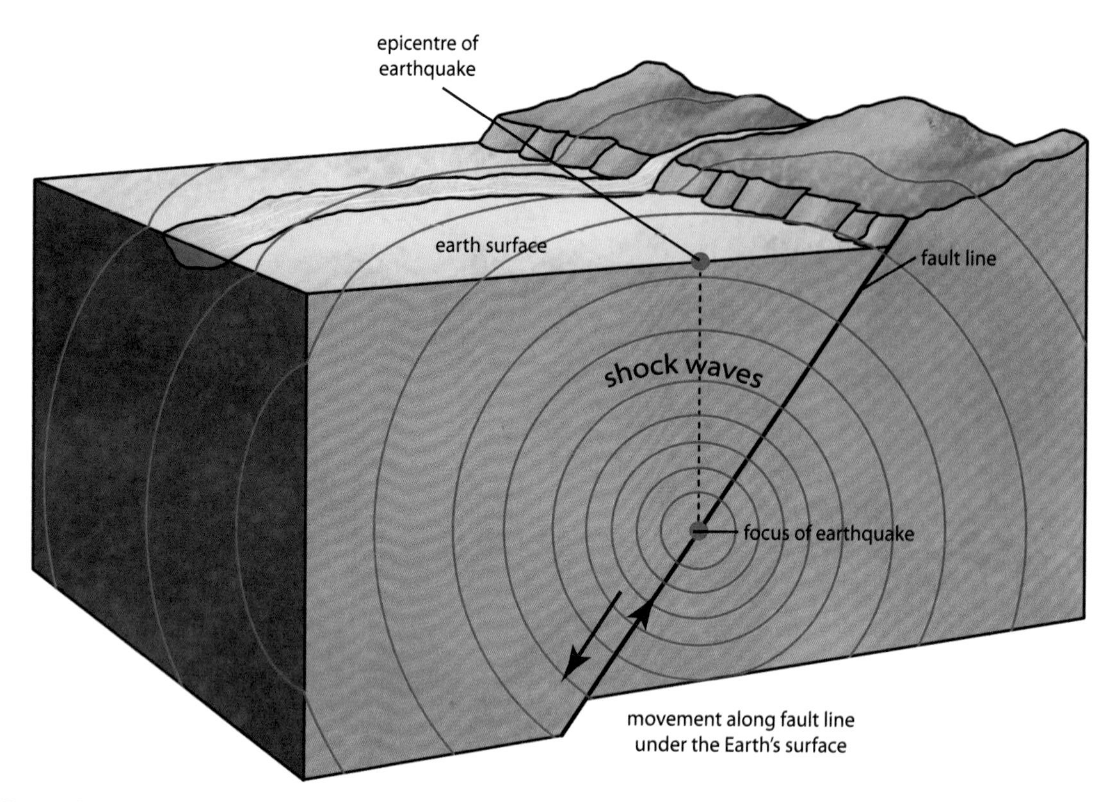

Figure 8.6 The features of an earthquake

Earthquakes are recorded with a seismometer or seismograph, with the results displayed on a seismogram. Records of the seismic waves allow seismologists to map the interior of the Earth, and locate and measure the size of earthquakes.

There are two common scales used to measure the size of an earthquake:

1 The size of an earthquake is reported by the moment magnitude scale and quantifies the magnitude (the amount of energy) released by an earthquake. This scale was developed to replace the Richter scale. Both are logarithmic scales, so that an earthquake of 7.0 releases about 32 times as much energy as one of 6.0 and 1000 times that of 5.0.

2 The moment magnitude and Richter scales only give a rough idea of the actual impact of an earthquake. An earthquake's destructive power varies depending on the composition of the ground in an area and the design and placement of human-made structures. The Mercalli scale is rated on the extent of the physical damage caused by an earthquake. Mercalli ratings are given as Roman numerals and they are based on largely subjective interpretations of the physical damage caused by an earthquake.

TERMS

tsunami: powerful, devastating waves at the coast caused by an undersea earthquake or volcanic eruption that displaces the water lying above it

seismometer: an instrument that measures movement of the ground, including the seismic waves generated by earthquakes and volcanic eruptions

seismic waves: waves of energy that travel through the Earth's layers as a result of earthquakes, volcanic eruptions, magma movements and large landslides

moment magnitude scale and Richter scale: numerical scales showing the size or magnitude of an earthquake based on readings from a seismometer

Mercalli scale: a scale showing the effect of an earthquake on the Earth's surface

aftershock: a smaller earthquake following after the main earthquake

After a major earthquake, it is common for there to be aftershocks. An aftershock is an earthquake that occurs after a previous earthquake. Aftershocks are formed as the Earth's crust around a fault adjusts to the effects of the main shock.

8.3 Distribution of earthquakes and volcanoes

The layers of the Earth

Integral to understanding the global pattern of plates and their movements is a knowledge of the structure of the Earth. The Earth, when seen in cross section, has four main layers (Figure 8.7)

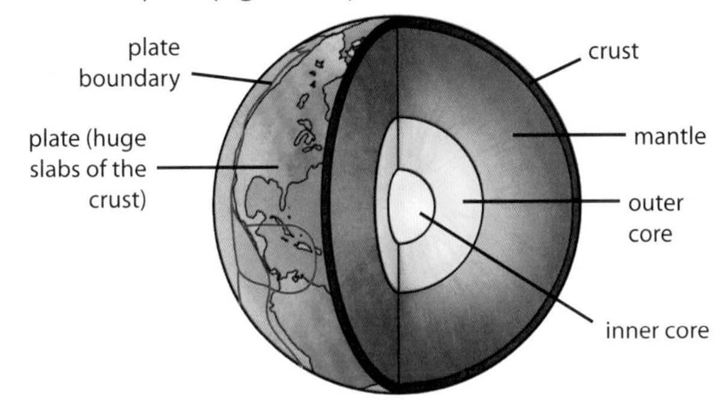

Figure 8.7 The structure of the Earth

1. At the centre is the **inner core**, which is about 1400 km across. Here, the rocks are solid and extremely dense – they are about five times denser than the rocks at the surface. They are made out of iron (about 80%) and nickel and are extremely hot, about 5500 °C.
2. On the edge of the inner core is the **outer core**, which is made up of dense, semi-molten metal with a temperature of 5000–5500 °C. It is about 2100 km thick.
3. Outside the core is the **mantle**. This is also semi-molten, but is less dense and makes up a layer which is about 2900 km thick.
4. At the surface is the **crust**. It varies in thickness and there are two types – oceanic crust and continental crust (Table 8.1).

Oceanic crust	Continental crust
Relatively thin – 5 to 10 km thick	Relatively thick – 25 to 90 km thick
Younger	Older
Heavier/denser	Lighter/less dense
Continually being formed and destroyed	Cannot be destroyed

Table 8.1 The differences between oceanic and continental crust

If oceanic and continental crusts collide, the oceanic crust, being denser, will go under the continental crust, a process called subduction. Relative to the other layers, the crust is actually quite thin.

Plate tectonics is the study and explanation of the global distribution of tectonic plates, fold mountains, earthquakes and volcanoes.

The Earth's crust is split into many huge pieces called tectonic plates (see Figure 8.5). These pieces of crust float and move on top of the much denser mantle below them. The movement of these crustal plates is due to huge convection currents in the mantle below the crust. These convection currents move heat from the interior core of the Earth towards the surface. This movement of heat causes the mantle and, in turn, the crustal plates on top of the mantle to slowly move. The movement of the plates produces several distinct landforms, as well as causing earthquakes.

TERMS

fold mountains: a long, high mountain range formed by uplifting and folding of sediments

convection currents: differences in temperature of material beneath the plates of the Earth's crust leads to the creation of currents to transfer the heat. These currents move the plates above them

Types of plate boundary

Where two plates meet is called a **plate boundary** or **plate margin**. There are four types of plate boundary.

Constructive (divergent) plate boundaries

This is where two plates move away from each other (Figure 8.8). Any gap that appears between these plates fills with molten magma and the lava that comes out forms/constructs new crust. One of the biggest and longest constructive margins is the **Mid-Atlantic Ridge,** which stretches down the middle of the North and South Atlantic oceans where the North and South American plates are moving away from the Eurasian and African plates. There are many undersea volcanoes along the ridge and minor earthquakes occur along it as the plates move. Both volcanic activity and earthquakes on these margins tends to be relatively gentle in comparison to the other margins.

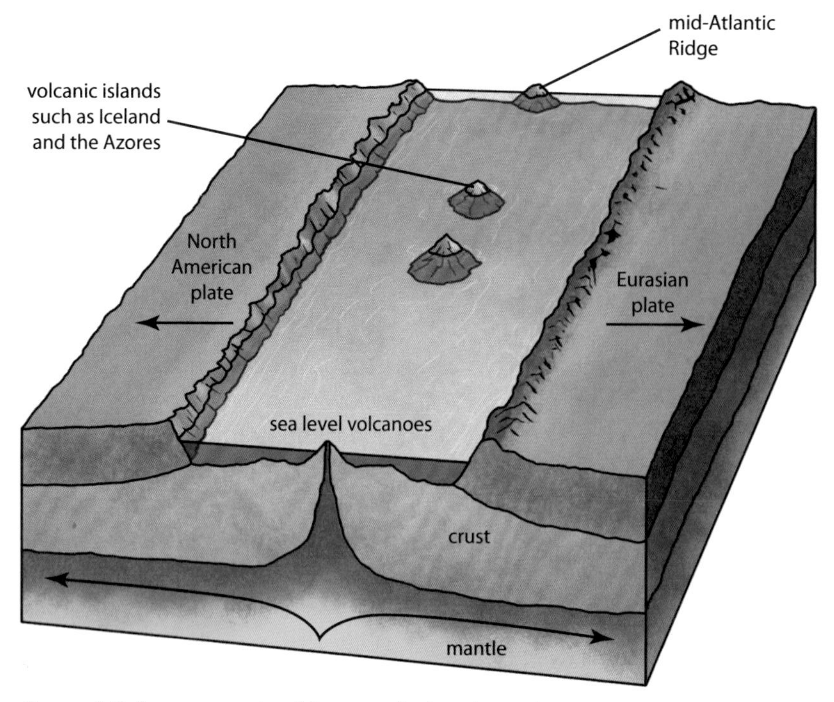

Figure 8.8 A constructive (divergent) plate boundary

Destructive (convergent) plate boundaries

These are found where plates made of heavier oceanic crust move towards plates made of lighter continental crust (Figure 8.9). Where they meet, the heavier oceanic crust is forced down under the lighter continental crust and forms a subduction zone. As the oceanic crust sinks deeper, it melts and forms magma. This may rise to the surface and emerge as lava. It forms very explosive volcanoes, as in the Andes Mountains in South America, where the Nazca Plate is moving under the South American Plate.

Where this process takes place in an ocean, the volcanoes may form a line of volcanic islands called island arcs, as in the islands of the West Indies in the Caribbean and in the Aleutian Islands, south west of Alaska. Destructive margins often suffer from very powerful earthquakes, such as the Chile earthquake of 1960, which was the largest ever recorded at 9.5 on the Richter scale, and the Japanese earthquake of March 2011, which was 8.9 and the fifth largest ever recorded.

TIP

If you are asked about the formation of volcanoes in island arcs, as well as mentioning the presence of plate boundaries, also explain that the subduction zone extends beneath these volcanoes and refer to the processes of subduction and melting taking place.

TERMS

subduction zone: the zone where one tectonic plate sinks (subducts) under another

island arc: a chain of volcanic islands located above a subduction zone at a tectonic plate boundary

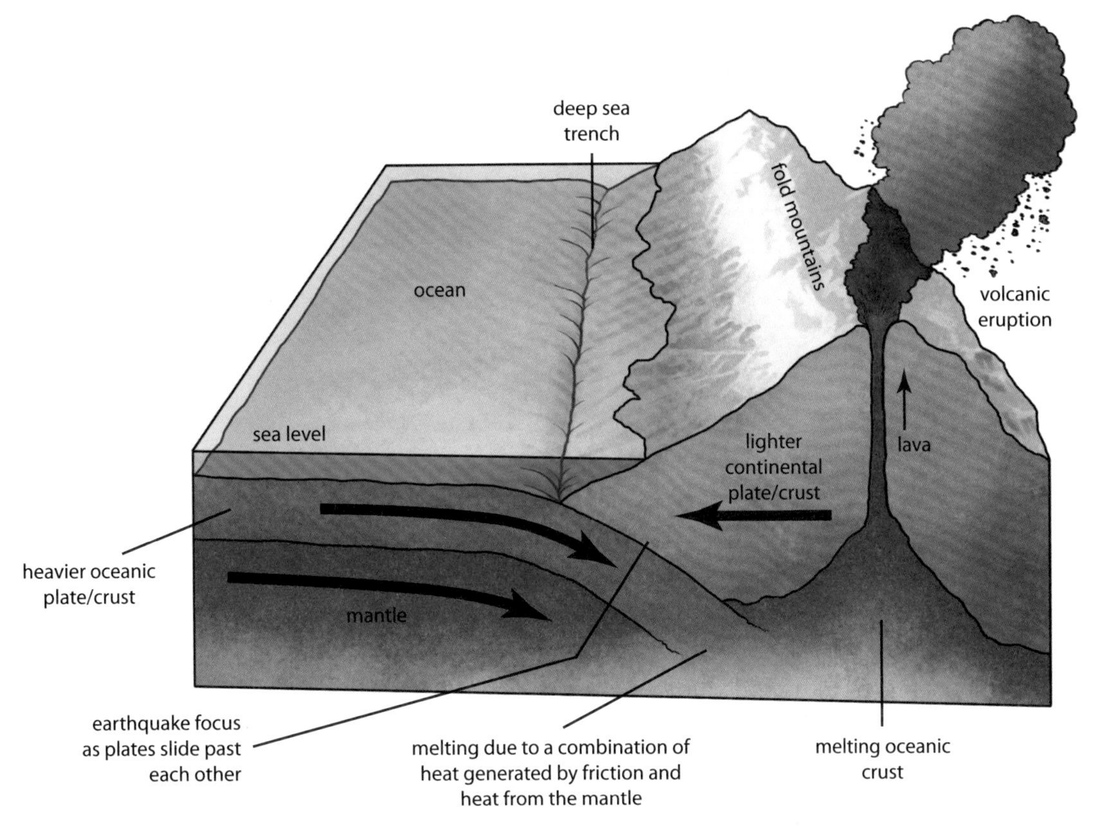

Figure 8.9 A destructive (convergent) plate boundary

Conservative plate boundaries

These occur where plates slide past each other (Figure 8.10). Sometimes they become locked together and pressure builds up until they tear apart, along a fault line. These movements can produce very powerful earthquakes, such as in Haiti in January 2010, but they do not produce volcanic eruptions and land is not created or destroyed. The San Andreas Fault in California is another famous example.

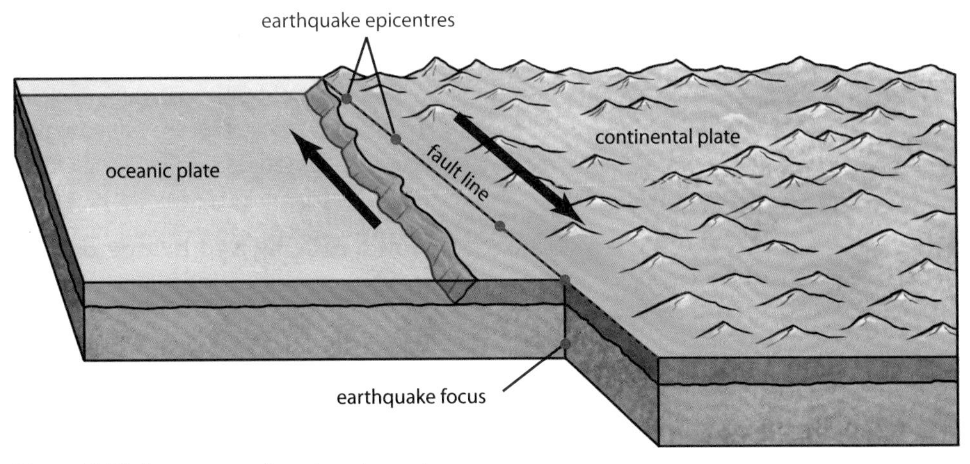

Figure 8.10 A conservative plate boundary

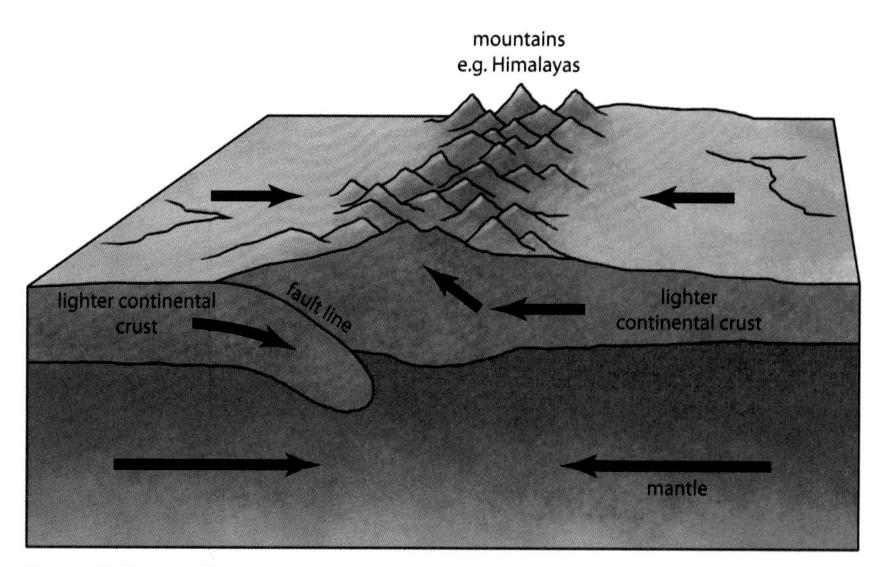

Figure 8.11 A collision plate margin

Collision plate boundaries

This is where two continental plates converge/move towards each other (Figure 8.11). These lighter plates are not dense enough to sink into the mantle. As the plates collide, they fold up and form fold mountains, such as the Himalayas. These collisions can also produce powerful earthquakes, as witnessed in Nepal in 2015, but they do not produce volcanic eruptions.

> **Self-test questions 8.1**
>
> 1 Describe the global distribution of volcanoes and earthquakes.
>
> 2 Explain how their distribution is related to movements at plate boundaries.

8.4 Effects of earthquakes and volcanoes on people and the environment

The effects of earthquakes and volcanos are often classified as being either primary or secondary. **Primary effects** occur as a direct result of the earthquake or volcanic eruption, such as buildings collapsing due to the ground shaking or loss of life due to pyroclastic flows.

Secondary effects occur as a result of the primary effects, such as the impact of tsunamis, fires caused by ruptured gas mains or higher insurance premiums.

Impacts on people

The impacts of earthquakes and volcanoes will vary according to a combination of factors at any particular place, but earthquakes and volcanic eruptions both have a similar range of hazardous results:

- The **loss of life** from collapsing buildings, bridges and elevated roads, fire and disease. Children and other dependents can be orphaned or left without any family support. Survivors may face serious psychological and emotional problems for many years afterwards.
- The **cutting of basic necessities** such as power, water, sewage treatment due to damage to power lines, water and sewage pipes.
- The **collapse of buildings** or the destabilisation of the base of buildings which may lead to collapse in aftershocks and future earthquakes and general damage to houses and buildings. People may be homeless for weeks, months or years.
- **Road, railway and bridge damage** which can make access difficult or impossible to earthquake-affected areas.
- The **loss of crops** as they are covered in ash, **loss of trees** in the areas near the volcano.
- **Fish can be killed** in rivers, lakes and hatcheries by ash getting into streams and rivers.
- The **spread of disease** due to polluted water and lack of medicine when water and sewage pipes are broken and water becomes polluted with sewage.

- The **loss of jobs and businesses** when factories, offices and places of work are destroyed.
- **Higher insurance premiums** for those people in earthquake-affected areas where insurance companies are not prepared to take the potential risk.

Most people agree that **human loss of life** is the most significant impact of earthquakes and volcanic eruptions. LICs and many MICs, such as Haiti, Afghanistan and Pakistan, do not have the resources to adequately deal with such disasters compared to HICs, such as the USA, Japan and Italy.

Earthquakes

The amount of damage caused by earthquakes will depend on a combination of factors:

- **strength** of the initial earthquake and the aftershocks that may follow it – the Haiti earthquake in January 2010 measured 7.0 on the Richter scale which was powerful enough to destroy buildings
- **depth** of the earthquake – many earthquakes take place deep in the crust, below 150 km, so much of their energy is absorbed by the crust above them
- **distance** from the epicentre – as the shock waves spread away from the epicentre, they become weaker
- **geology** of the rocks in the area – loose sedimentary rocks may liquefy and cause buildings and structures to sink into the ground; more solid, harder rocks will normally provide the safest foundations for buildings
- **building construction materials and designs** – steel-framed buildings are better able to absorb movement than concrete-framed buildings
- **space between buildings** – as buildings sway, they may hit each other if they are built too close to each other and become damaged
- **number of storeys** – in a tall, high-rise building, shock waves become amplified as they move up the building, which can cause them to sway and collapse
- **density of population** – a densely populated urban area is likely to suffer many more casualties and damage than a low-density rural area
- **time of the day** when the earthquake occurs – at night in residential areas most people will be inside their homes and asleep
  - In older residential areas, where buildings were not made resilient to earthquake movements – as in the Kyoto earthquake in 1995 – this can cause large loss of life and injuries.
  - During working hours, people in cities and towns will be working in office blocks or inside school. Such buildings will normally be earthquake proof.
- **secondary hazards** – these can include tsunamis on the coast, landslides and rock falls in mountainous areas, fires in urban areas – many caused by broken gas pipes and electricity lines
- **contamination** – after earthquakes, water supplies may become contaminated as they mix with sewage, and hunger may be a problem where food and water cannot reach affected communities; in winter, people without shelter and winter clothing may suffer from hypothermia.

Volcanoes

As seen earlier in this section, volcanoes produce a wide variety of hazards that can injure and kill people, and destroy property. However, unlike earthquakes, large, explosive eruptions can endanger people and property hundreds of kilometres away and even affect global climate. An explosive volcanic eruption can blast solid and molten rock fragments (called tephra) and volcanic gases into the air with tremendous force. The largest rock fragments (volcanic bombs) usually fall back to the ground within 2–3 km of the vent. Small fragments (less than about 1 mm across) of ash rise high into the air, forming a huge, billowing eruption column. These can grow rapidly and reach more than 20 km above a volcano in less than 30 minutes, forming an eruption cloud. The volcanic ash in the cloud can pose a serious hazard to planes. During the past 15 years, about 80 commercial jets have been damaged by inadvertently flying into ash clouds, and several have nearly crashed because of engine failure.

TERMS

tephra: rock fragments and particles ejected by a volcanic eruption

fumarole: an opening in or near a volcano, through which gases, such as sulphur dioxide, are emitted

Large eruption clouds can extend hundreds of kilometres downwind, resulting in ash fall over enormous areas. Heavy ash falls can collapse buildings, and even minor ash fall can damage crops, electronics and machinery.

Volcanoes can emit large volumes of gases during eruptions. Even when a volcano is not erupting, cracks in the ground allow gases to reach the surface through small openings called fumaroles. More than 90% of all gas emitted by volcanoes is water vapour (steam), most of which is heated ground water (underground water from rainfall and streams). Carbon dioxide is heavier than air and can be trapped in low-lying areas in concentrations that are deadly to both people and animals. Fluorine gas, which in high concentrations is toxic, can be adsorbed onto volcanic ash particles that later fall to the ground. The fluorine on the particles can poison livestock grazing on ash-coated grass and also contaminate domestic water supplies.

Lava flows can spread out in broad sheets several kilometres wide. Deaths caused directly by lava flows are uncommon because most move slowly enough that people can move out the way easily. Lava flows, however, can bury homes and agricultural land under tens of metres of hardened rock.

High-speed avalanches of hot ash, rock fragments and gas can move down the sides of a volcano during explosive eruptions or when the steep side of a growing lava dome collapses and breaks apart. These pyroclastic flows (see Section 8.1) tend to follow the direction of valleys on the side of a volcano and are capable of knocking down and burning everything in their paths. Lahars (flows of mud, rock and water) can rush down valleys on the sides of volcanoes at speeds of 30–100 km per hour and can travel more than 300 km, covering large areas of land. Some lahars contain so much rock debris (60–90% by weight) that they look like fast-moving rivers of wet concrete. These flows are powerful enough to rip up and carry trees, houses and huge boulders miles downstream.

Volcanoes can offer opportunities for people

Despite the hazards, many people live in areas where they are likely to encounter volcanic eruptions. They do this for a number of reasons.

- **Volcanic soils are often very fertile** and **yields of crops are high:** in the case of Mount Etna in Italy, the fertile volcanic soils support extensive agriculture with vineyards and orchards spread across the lower slopes of the mountain and the broad Plain of Catania to the south.
- **People can obtain hot water for heating** and also **generate electricity** from the volcano using the hot steam to produce geothermal power.
- **Volcanoes provide raw materials** such as sulphur, zinc, gold and diamonds which can be mined and sold.
- **Volcanoes can attract tourists** and local people can get jobs as tour guides. The people who live in towns near Mount Etna in Italy such as Messina and Catania, can earn money from renting accommodation to tourists.
- Governments, as in Italy, can set up **volcanic and earthquake prediction equipment** and then local people may feel more secure in living in high-risk areas.
- Many people have lived near volcanoes and earthquake zones all their lives. They are close to their family and friends, they work in the area and many cannot afford to move away.

8.5 Reducing the impacts of earthquakes and volcanoes

Responses to earthquakes and volcanic eruptions are both **short term**, such as emergency aid and disaster relief, and **long term**, such as risk assessment, rebuilding, improving hazard prediction and preparation for future hazards.

There are several things that can be done to help people live more safely in earthquake and active volcanic zones:

- **Improved technology** and **use of historical data** can improve the prediction and forecasting of earthquakes and volcanoes to provide warnings.
- **Mapping of high-risk areas** where buildings are at risk from liquefaction, mudslides, landslips, lahars, lava flows and pyroclastic flows can ensure that such areas are not used for building.
- **Improved building designs and materials** can mean that buildings do not collapse during earthquakes. Buildings with deep foundations made of special rubber can absorb shock waves more easily. Steel-framed buildings, with steel cross braces can move more than rigid, concrete buildings and stay intact. Automatic shutters that close on windows can stop glass falling on people below the buildings. Shatter-proof glass can also be used in windows.

Flexible pipes can be used for water and gas, as well as systems that shut off water and gas flows automatically when an earthquake occurs.

- **Ensuring that adequate emergency drills and procedures are in place,** both at home and for people at work and in schools and hospitals. The stockpiling of emergency supplies of medicines, drinking water, tinned food, tents and blankets can mean that there can be a rapid response to any earthquake or volcanic eruption.

Predicting volcanic eruptions

There are a number of ways scientists help predict and then prepare people for volcanic eruptions:

- **Seismometers** record the increasing number of earthquakes that are caused by huge amounts of magma pushing up under a volcano.
- **Tilt meters** measure the changes in the volcanoes' shape as they start to bulge with the upward movement of magma.
- **Thermometers** measure increases in temperature.
- **Gas sensors** monitor the release of gases from the volcano.
- **Satellites** can monitor changes in the shape of a volcano and increases in temperature.

Once lava has erupted, it may be possible to halt or divert lava flows by building concrete dams or spraying water on their fronts to solidify the lava and help dam and divert it.

Self-test questions 8.2

1. Explain why earthquakes of the same strength may cause different numbers of deaths.
2. Explain why many volcanic areas are popular places to live but can create problems for the people living in those areas.

Case studies

An earthquake – Sichuan province, China, May 2008

The Great Sichuan earthquake in China in May 2008 measured 7.9 on the Richter scale. The earthquake was caused by the movement of a large fault line between the Eurasian and Indo-Australian plates. The movement took place along a 240 km section of the fault down to a depth of 20 kms. In two minutes the land moved 9 metres along this fault causing extensive damage and loss of life on the surface. It occurred in the mid-afternoon when children were at school and people were at work. Over 7000 schoolrooms collapsed, mostly in rural areas, leading to the death of over 5000 students and the injury of around 15 000 more. The large number of school collapses was because many were not built to agreed adequate standards through corruption and mismanagement. Many other buildings collapsed, as they were poorly constructed and were not earthquake proof. The earthquake affected a wide area which contained large settlements and the region was densely populated. There were many gas leaks and fires as gas pipelines were ruptured by the movement. Roads were blocked and badly damaged which delayed emergency services. The rupturing of sewage and freshwater pipes meant that there was contamination of water and a lack of drinking water. In total, the earthquake killed almost 69 200 people and left over 18 200 missing.

There were many **long-term effects** which resulted from this earthquake:

- damage to workplaces, so that people were unable to earn a living
- a negative impact on the economy and industrial production
- damage to infrastructure – roads, railways, power lines, water, gas and sewage pipelines
- the economic cost of rebuilding the infrastructure and housing
- damage to schools and the disruption of education
- the psychological trauma of experiencing an earthquake and the loss of family and friends
- homelessness.

There are several ways to **protect people** in locations such as Sichuan from future earthquakes:

- build houses and schools which are more stable and designed to sway not collapse
- make sure that planning regulations are followed
- build deeper foundations on buildings
- use flexible water and gas pipes
- make sure there is an automatic gas switch-off when an earthquake occurs
- have frequent earthquake drills and make sure there is education of safety procedures and that people and the government stockpile supplies of drinking water, food, clothing/blankets and tents, etc
- build on solid ground
- ensure the emergency services are better prepared
- restrict building heights and improve their design.

Reasons why people live in high-risk areas such as Sichuan

There are many reasons why so many people live in areas such as Sichuan, even though they are at risk from earthquakes:

- They have lived there all their lives, are close to their friends and family, and have a sentimental attachment to the area.
- There is lots of work in the area and it has fertile farmland. This means that they can make a good living.
- There are good services/schools/hospitals in the area.
- There is pressure on living space in an area that is very mountainous.
- They are confident about their safety and are willing to take the risk as an earthquake has not occurred for many years.

A volcano – Mount Ontake, Japan, 2014

Japan has 110 active volcanoes, 50 of which have round-the-clock monitoring. Japan has about 10% of all the active volcanoes in the world. Its volcanoes form part of the Pacific Ring of Fire. The eruption of Mount Ontake was caused by the subduction of the oceanic Philippine plate beneath the continental Eurasian plate. The melted magma from the subducting oceanic Philippine plate rises through the continental crust and produces spectacularly violent eruptions.

The government has set up anti-disaster councils formed from the police, firefighters and volcano experts, who have created hazard maps showing the extent of expected damage, based on a multiple scenario of eruptions and they have devised evacuation plans for both residents and tourists.

Mount Ontake is a massive strato-volcano and the second highest in Japan. Although the volcano is under constant surveillance for changes in its shape – tilt meters on its sides, seismometers and a camera monitoring its state – an eruption in 2014 killed 57 people, highlighting the problems faced in predicting volcanic eruptions, even in HICs. There were no significant earthquakes that might have warned authorities in the lead up to this phreatic eruption. When it erupted, over 300 walkers and climbers were on the volcano and had little chance of escaping the effects of the volcanic bombs, ash and hydrogen sulphide gas.

The **short-term impact** was the effect of the eruption and its volcanic bombs, ash and gas on the 300 walkers and climbers. The **short-term response** included the very quick deployment of hundreds of trained and equipped emergency responders, including helicopters. The **long-term response** was a recognition that this was already a monitored volcano but the instruments on the volcano had not recorded any measurable ground deformation or any seismic activity. The **long-term impact** was therefore to look at ways of improving the monitoring of the volcano by looking at ways in which small-scale phreatic eruptions like this can be monitored as this eruption was neither expected nor its impacts prevented.

TERM

phreatic eruption: an eruption caused by groundwater in the sides of the volcano being turned into large volumes of steam and erupting along with large amounts of ash and volcanic bombs

Sample question and answer

1 a Suggest different ways to protect people from earthquakes. [4]

b Explain why many people continue to live in areas where they are at risk from earthquakes. [5]

c Explain the causes of an earthquake which has occurred in a named area that you have studied. [6]

TIP

In the first two parts of the question, you should list ways to protect people and reasons why people still live in earthquake zones. In the third part, use a detailed case study of an area affected by an earthquake with an explanation of the plate boundary that has caused the earthquake.

a There are many different ways to protect people from earthquakes. Firstly, their houses can be built to resist the effects of being shaken by the earthquake **[1]** so that they do not collapse. A steel frame can be used which moves and bends, rather than collapses unlike a concrete building **[1]**. They can also be built on deeper foundations **[1]**. People should make sure that they follow building rules and regulations **[1]** and authorities do not allow people to build with the wrong materials and that they are built on solid ground **[1]**. Authorities should make sure that people are aware of what to do when an earthquake hits and that they have emergency supplies ready of water and medicines **[1]**. (A maximum of 4 marks will be awarded.)

b Many people have lived for generations in earthquake areas so they are surrounded by their friends and family **[1]** and they have built their homes and developed their farms in that area **[1]**. Many people are just not rich enough to be able to move away and set up a new farm or buy a new house **[1]** and they may not get much money for their land or be able to sell their house because it is in a dangerous area **[1]** or find new employment to support themselves and their families **[1]**. When they balance the advantages and disadvantages of moving, they may decide to stay and take the risk **[1]**, especially if an earthquake has not happened for many years and they feel safe because of that **[1]**. (A maximum of 5 marks will be awarded.)

c California is in an earthquake zone and has had lots of earthquakes. Many of them are along the San Andreas Fault line which is a line of weakness between two plates called the North American and Pacific plates **[1]**. These plates are sliding past each other along this fault line **[1]** along what is called a conservative boundary **[1]**. Sometimes the two plates get stuck and pressure builds up **[1]** until they suddenly slip a few metres **[1]**. This big movement sends out shock waves which can destroy buildings and it has done this in both Los Angeles and San Francisco **[1]**.
(A maximum of 6 marks will be awarded).

Exam-style questions

1 What are volcanoes? [2]

2 Explain why some volcanoes are different in size and shape from each other. [4]

3 Why do mudflows occur during volcanic eruptions and what problems do they cause? [2]

4 In what ways do earthquakes cause damage to people and the environment? [4]

Rivers

Learning summary

By the end of this chapter, you should be able to:

- explain the main hydrological characteristics and processes which operate within rivers and drainage basins
- demonstrate an understanding of the work of a river in eroding, transporting and depositing
- describe and explain the formation of the landforms associated with these processes
- demonstrate an understanding that rivers present hazards and offer opportunities for people
- explain what can be done to manage the impacts of river flooding
- demonstrate knowledge of a case study of the opportunities presented by a river, the hazards associated with it and their management.

9.1 Rivers and drainage basins

Drainage basin characteristics

The area drained by a river and its tributaries is called a drainage basin, also known as a river basin or river catchment. All the precipitation that falls within a drainage basin will attempt to make its way towards the river channel, both on the surface or underground, and eventually to the river mouth. The dividing line separating drainage basins is called the watershed. Smaller river channels leading to the main river channel are called tributaries. Where two river tributaries/channels meet is called a confluence. The amount of water flowing down a river, the river's discharge, is measured by the number of cubic metres of water flowing past a measuring or gauging station every second. This figure is often abbreviated to the term cumecs. All these features are shown in Figure 9.1.

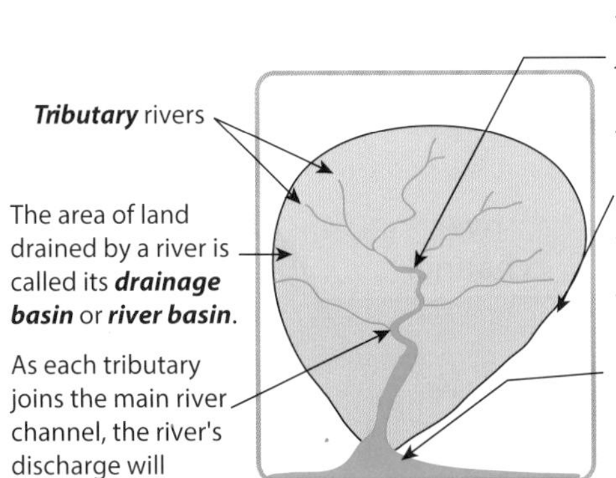
Where two tributaries join together, it is a ***confluence.***
Tributary rivers
The line that divides one river basin from another is called the ***watershed.***
The area of land drained by a river is called its ***drainage basin*** or ***river basin***.
The amount of water in a river is measured in ***cubic metres per second*** (often abbreviated to cumecs). It is called the rivers ***discharge*** and is measured at a ***gauging station***.
As each tributary joins the main river channel, the river's discharge will increase.

Figure 9.1 Features of a typical drainage basin

TERMS

drainage basin: the area of land drained by a river and its tributaries (streams). This is also known as the river basin or river catchment area

river mouth: where the river ends when it meets a body of water such as the sea or a lake

watershed: the boundary of a drainage basin

tributary: a stream or small river that joins a larger one

confluence: the point where two or more rivers meet

discharge: the volume of water passing a point or location along the river channel in a given time. It is usually measured in cubic metres per second (cumecs) at a gauging station in a river

Processes operating in a drainage basin

Within a drainage basin, water can be added as an input or lost as an output and the water can flow / be transferred through the basin in a number of ways (Figure 9.2).

The **input** is the water that is added to a drainage basin in the form of precipitation. This can occur in a number of forms, such as rain, snow and hail. It can also occur at different times of the year and in different amounts, intensities and frequencies throughout the year. While in the drainage basin, water may be held in a variety of stores for any length of time from seconds (on a leaf) to hours and days in various sized depressions, such as puddles, ponds, lakes and reservoirs, in the soil, or in gaps such as cracks, joints, bedding planes, fault lines and pore spaces in underground rocks in an aquifer.

TERMS

store: surface or underground locations in a drainage basin that hold and contain water

aquifer: a body of underground rock that, through its rock type or structures, holds water. This water can be abstracted from the rock by drilling wells and boreholes

The **outputs** from a drainage basin are the losses of water from the drainage basin in terms of evaporation, transpiration and river/channel flow:

- Evaporation is the loss of water as it is exposed to the air on the surface of the land, vegetation and bodies of water as it transfers from a liquid to a gas (water vapour). The rate at which evaporation takes place will depend on the temperature – the higher the temperature, the higher the rate of evaporation.
- Transpiration is the loss of water vapour from the stomata (very small openings) in the leaves of plants and trees. Rates of transpiration will depend on the amount and type of vegetation. A tropical forest ecosystem has very high rates of transpiration.
- The term evapotranspiration is sometimes used and it refers to the combined loss of water from both evaporation and transpiration.
- Finally, at the river's mouth, water is discharged into the sea or a lake.

Water can **flow / be transferred** within the drainage basin either above or below the ground. Above the ground, some rainfall may be **intercepted** by leaves and this water may then flow off the leaves and drip to the ground as **dripflow** and down the leaves and the stems of trees and plants as **stemflow**. The interception of rainfall by trees and other vegetation means that the soil may be protected as the intercepted water is released slowly to the land surface, allowing it to **infiltrate** (be absorbed) into the soil.

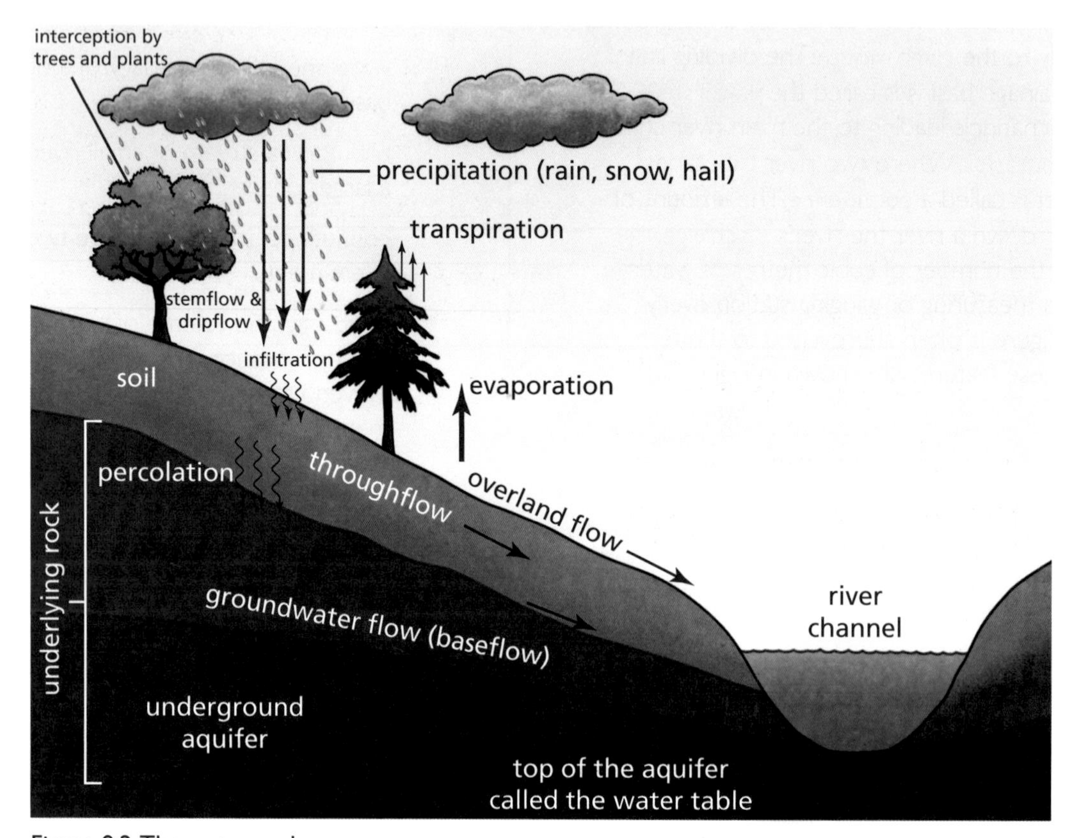

Figure 9.2 The water cycle

In areas that experience very high, intense periods of rainfall, there may be little or no interception of the rain and water may flow over the land surface. This may take place when either the water from rainfall arrives too quickly on the land surface and does not have enough time to be infiltrate the soil, or when rainwater falls onto an impermeable land surface such as a clay soil, or when rainwater accumulates on the surface because the spaces in the underlying soil and rock are filled with water, i.e. the soil and rock are **saturated** with water. This means that water is forced to flow over the land surface – a process known as **overland flow**, possibly causing soil erosion as the flowing water removes and transports away soil particles. Without a protective cover of trees and vegetation, soil erosion is often an inevitable result.

TERMS

impermeable: these rocks do not allow water to pass through – they are watertight, e.g. clay

infiltration capacity: the speed/rate at which water enters/infiltrates the land surface

porous: describes rocks that contain many small air spaces e.g. chalk – most porous rocks are permeable

permeable: describes rocks (e.g. limestone) that allow water to pass through them via cracks, faults/fault lines, joints and bedding planes (Figure 9.3)

The process of **infiltration** takes place when water enters small openings and pores in the soil and rocks in the ground surface. Every land surface has its own individual infiltration capacity. Areas with a low infiltration capacity can be very susceptible to flooding after heavy rain as water is unable to enter the ground fast enough and starts to accumulate on the surface.

The process of **percolation** is often confused with infiltration but this takes place after water has infiltrated the soil and then flows/percolates down through the soil and underlying rock as it is pulled down by gravity. The rate at which the water percolates will depend on how porous the soil or rock is and how permeable the rock is.

Water is stored underground in pore spaces, bedding planes, cracks, faults and joints, forming an aquifer. The top of an **aquifer** is called the **water table**. The level of the water table may rise and fall according to the input

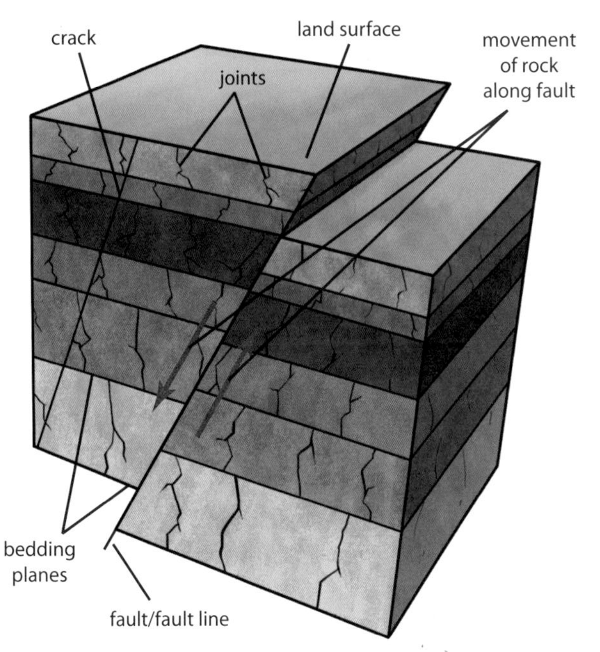

Figure 9.3 Geological structures

of water into the aquifer from rainfall and surrounding groundwater. In many areas, it is likely to fall in the hot summer months when less rainwater falls or water does not have time to infiltrate into the land surface or percolate underground before it is evaporated on the land surface and, therefore, does not reach the aquifer.

When water has entered the soil, it may move through it laterally/sideways as **throughflow**. It may then move through the soil down a slope towards a river channel. **Groundwater flow**, sometimes called **baseflow**, is water that has infiltrated and percolated into the rock below the soil and then moves laterally, in a downslope direction, to appear on the surface of the ground as a spring, or flow into a river channel. Groundwater flow will normally increase where the soil and rock are very porous and permeable, during periods of steady rainfall.

9.2 River erosion, transportation and deposition

Processes of erosion

TIP

The terms weathering and erosion are often confused with each other. **Weathering** refers to the actual breakdown of a rock 'in situ' (where it is found) by physical, chemical and biological processes, while **erosion** is the removal of this weathered rock material.

There are four processes of river erosion:

1 **hydraulic action:** where the weight and force of the water flowing in the river removes particles of rocks from the river channel's bed and sides.

2 **abrasion** (sometimes called **corrasion**): where the river's bedload (boulders, pebbles, gravel, sand and silt) as it rolls, bounces and collides with the channel bed and sides, removes particles of rock from the channel bed and sides.

3 **solution** (sometimes called **corrosion**): where some minerals (mainly the carbonate minerals found in rocks like limestone and chalk) are put into solution by the weak acids found in river water. These weak acids include carbonic acid which is formed as carbon dioxide gas in the atmosphere joins with rainwater and this has a great impact on the carbonate rocks and limestone in particular.

4 **attrition:** a process which does not erode the river channel bed and sides, but breaks up the river's bedload. It takes place when the rocks on the bed of the river are rolled along and they collide with each other and become smaller and rounder as a result. This means that the average size of rock particle (sediment) becomes smaller as it moves down towards the river mouth.

TERMS

weathering: the breakdown of rocks 'in situ'

erosion: the removal of weathered material from the land by water, ice or wind

bedload: fragments of rock which have come into contact with the bed of the river channel during their transportation

floodplain: an area next to a river that would be affected by flooding if the river overflowed its banks

delta: a landform, often triangular in shape, which develops where a river meets a slow body of moving water such as a lake or ocean. Sediment builds up above the water level forcing the river to split into distributaries to form a delta

gradient: how steep a slope, river channel or valley is

Processes of transportation

There are four processes of river transportation (Figure 9.4), which carry the load of the river:

1 **traction:** where the larger, *heavier* material that make up the river's bedload (boulders, pebbles and gravel either through rolling or bouncing are in actual contact with the river bed) is rolled along the river bed

2 **saltation:** where the *lighter* material that makes up the river's bedload (gravel, sand and silt) is bounced along the river bed

3 **suspension:** where the smaller, lighter material that makes up the river's *suspended* load (clay size – a particle that is less than 0.02 mm in diameter) is carried/suspended by the river

4 **solution:** where dissolved material that makes up the river's *solute* load is moved by the river in solution.

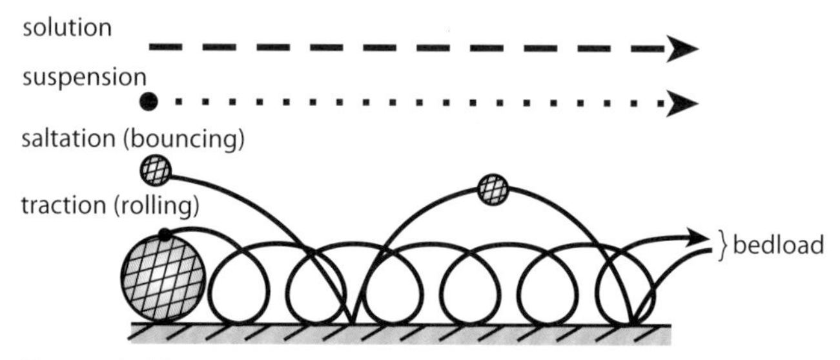

Figure 9.4 Types of river transportation

Processes of deposition

Where the river loses energy, it will start to drop/deposit its bedload and suspended load. This may take place in the river channel, on its floodplain or on its delta. The heavier material will be deposited first and as the river continues to slow down and become shallower, the smaller material will be deposited.

The river's velocity/speed will slow when:

- the gradient of the river channel decreases/becomes less
- the river channel bed becomes rougher and shallower
- the river meets a large, static body of water such as the sea or a lake and is slowed/halted.

9.3 The landforms associated with river processes

Forms of river valleys

Any analysis of rivers valleys makes use of a long profile and a cross section. The long profile is drawn from the source to the mouth of a river and shows the changes in the steepness/angle of the river channel from the source to the mouth. In a typical river channel, the angle will often be steeper in the upper course of a river than in the lower course.

The cross section is drawn across the river valley and will also change as you move down the long profile. In the upper course, it will often be steep-sided and narrow, and in the lower course, it will normally be gentle-sided and wide.

TERMS

long profile: the shape and gradient of the river channel from source to mouth

cross section: the shape across the river valley, from one side to the other

source: the starting point of a river, often a spring, melting glacier, lake or wetland

v-shaped valley: a valley with a V-shaped cross profile

vertical erosion: where a stream or river erodes the river channel bed at a faster rate than the channel or valley sides, often leading to a narrow, v-shaped, valley

lateral erosion: where a stream or river erodes away the sides and banks on each side of the river channel

Each section of the river has specific characteristics (these are also summarised in Figure 9.5):

- The **upper course** is a narrow, v-shaped valley and a narrow river channel; the river channel will often completely fill the valley floor; the river channel will be full of large, angular boulders and stones which makes the water turbulent. The river will often be flowing quite slowly as the channel is very rough; the river will have a smaller discharge than lower down its course. Vertical erosion dominates, where the river bed is eroded and lowered.
- The **lower course** is a very wide valley and larger channel; often a floodplain will be found beside the river channel; the river channel will contain much smaller, rounder rocks, often sand-sized or smaller. The river will normally be faster-flowing as the channel will be smoother. Lateral erosion dominates and deposition takes place.
- As the river moves between the two sections, there is a **middle course** where there is a combination of features as the river changes its characteristics.
- The **discharge** of the river will increase from the source to the mouth as smaller tributary channels add more water to the main channel.
- The **velocity** will also increase from the source to the mouth because as the discharge increases there is less friction with the channel bed and sides, which means that the river will flow faster.

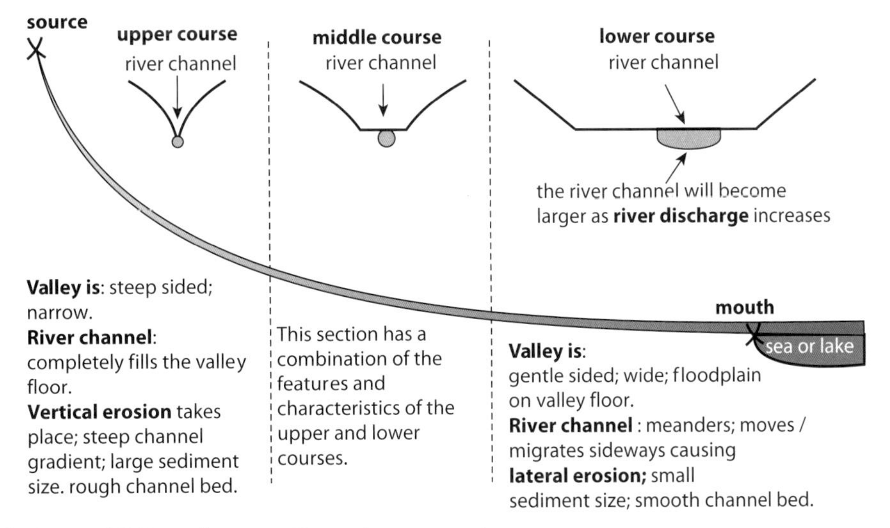

Figure 9.5 River long profile and cross sections

Waterfalls

Waterfalls are vertical breaks in the smooth long profile of a river channel and they vary enormously in size. There are three common ways for a waterfall to form in a river channel:

1 By **differential erosion** – where a band or layer of more resistant rock runs across the river channel. The softer, less resistant rock is eroded at a faster rate, causing a drop in the river bed – a waterfall (Figure 9.6).

2 By a **drop in sea level**, leaving the mouth of the river suspended above the lower sea level – this type of waterfall is sometimes called a knick point.

3 By an **earth movement** – often triggered by an earthquake, causing a drop in the river bed along a fault line.

TIP

If you are asked to describe the formation of potholes, diagrams such as Figure 9.7 can be used to illustrate your answer.

You may be asked to describe a waterfall formed by differential erosion. With the use of a diagram or a set of diagrams, as in Figure 9.6, you can explain how this happens.

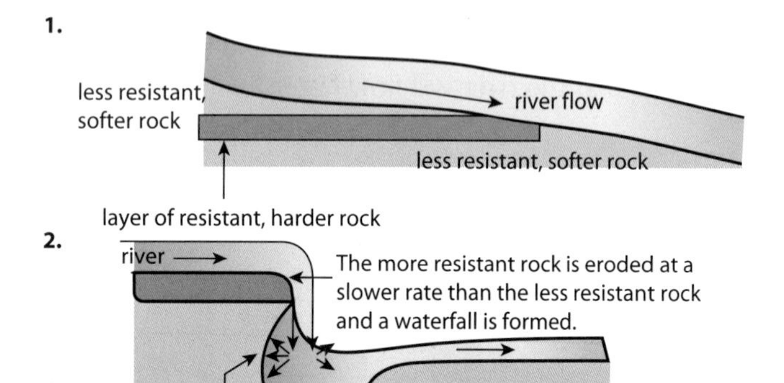
1.
less resistant, softer rock
river flow
less resistant, softer rock
layer of resistant, harder rock
2.
river
The more resistant rock is eroded at a slower rate than the less resistant rock and a waterfall is formed.
plunge pool
The less resistant rock is eroded by ***hydraulic action***. The harder rocks are undercut; they becomes ***unsupported*** and will eventually ***collapse***. The whole process will be continually ***repeated***.

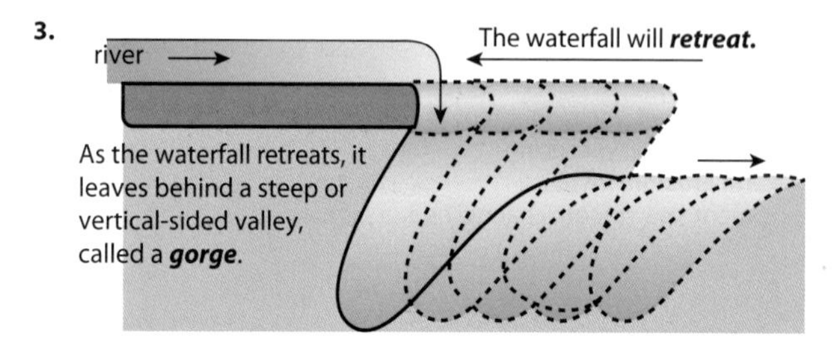
3.
river
The waterfall will ***retreat.***
As the waterfall retreats, it leaves behind a steep or vertical-sided valley, called a ***gorge***.

Figure 9.6 The formation of a waterfall through differential erosion

Potholes

Potholes are found in the bed of a river that is flowing over solid rock. They usually start to form when a weakness or crack or fault line is exposed in the rock in the river bed and differential erosion takes place (Figure 9.7):

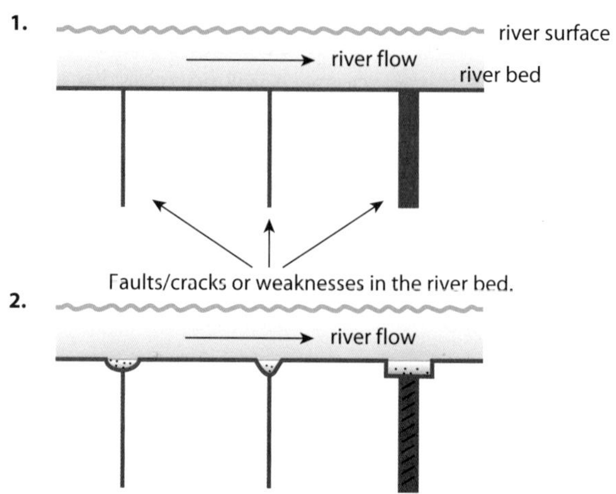
1.
river surface
river flow
river bed
Faults/cracks or weaknesses in the river bed.
2.
river flow
As they are weaker than the rocks on either side of them, the faults/cracks or weaknesses are eroded by ***hydraulic action*** and ***abrasion*** at a faster rate and is called ***differential erosion***.

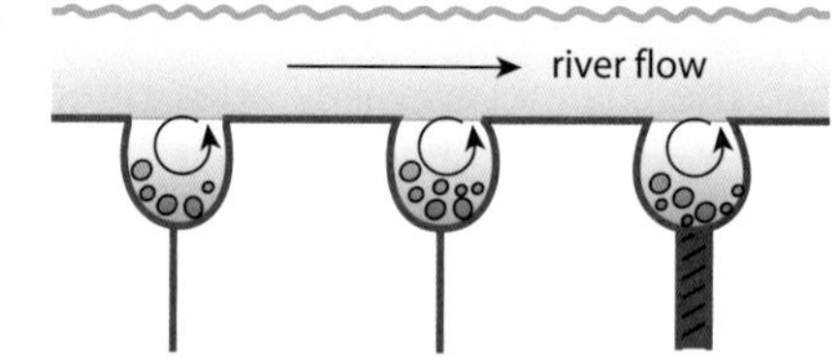
3.
river flow
The potholes are enlarged as the flow of the river spins rocks around the hole. ***Abrasion*** by the rocks enlarges the potholes. They may join up and so vertical erosion takes place.

Figure 9.7 The formation of potholes in a river bed

Meanders and oxbow lakes

In their natural state, unaffected and unaltered by people, rivers rarely flow straight. When rivers bend, they are said to meander and this term is used to describe any bend in a river. Meanders have a very distinctive shape and this is illustrated in the diagram taking a cross section, A to B, from Figure 9.8.

As shown in Figure 9.8, the *outside* bank of the meander, where the full force of the river is felt, is called a river cliff or bluff – this side of the channel is deep, fast flowing, is often undercut by the river, and has a large sediment size.

The *inside* bank is called the slip-off slope or point bar – this side is shallow, slow flowing, and has a small sediment size.

TERMS

meander: a bend in a river

river cliff/bluff: a steep section of the river bank caused by fast-flowing water eroding the outside bank of a meander

slip-off slope/point bar: a gentle slope on the inside of a meander formed by deposition where sediment is deposited on the inside of a meander where the river flows more slowly and with less energy

oxbow lake: a lake, often semi-circular in shape in a river floodplain, where a meander has been cut off from the river channel

bankfull discharge: the river discharge when the river channel full to the top of its banks and is just about to spill onto its floodplain

alluvium: sediments which are deposited by rivers

levée: a raised bank of sediment along the sides of a river channel

The cross section of the river channel A–B has a very distinctive shape and characteristics. These are shown in the cross section diagram A–B here.

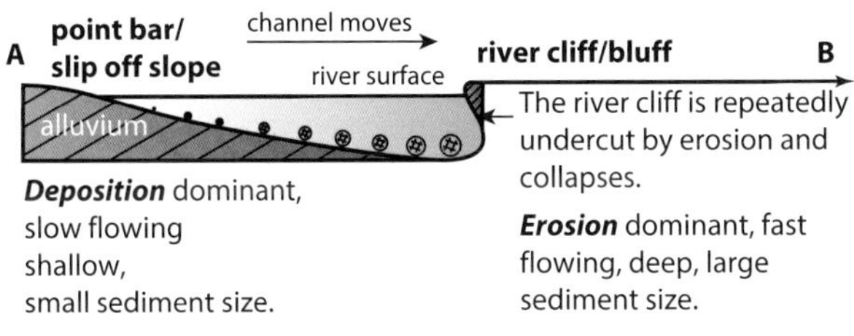

Figure 9.8 A cross section A–B taken through a typical meander

TIP

Make sure that you correctly identify which side of a meander is A and which is B – this is a typical task.

As a meander develops, its 'neck' becomes very narrow (Figure 9.9). With time, the river may break through the neck – often during high flow conditions, as in a flood event. This may then result in a section of the river channel being isolated from the main river channel to form a feature called an oxbow lake. With time, these often dry out and become oxbow scars.

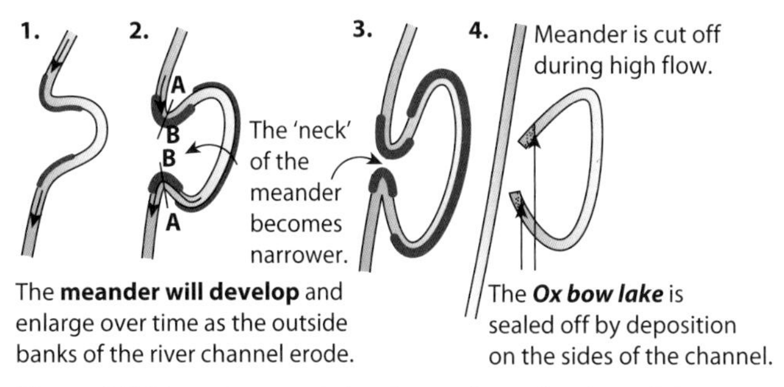

Figure 9.9 Meanders and the formation of oxbow lakes

Levées and floodplains

The middle and lower course of a river are characterised by having wide, flat areas beside them made out of material that has been deposited by the river during flood events (Figure 9.10). These features are called floodplains.

A river will flood when water in the river channel reaches the top of the river banks and flows over them. When a river reaches this level, it has reached what is called bankfull discharge. As the river water flows away from the river channel, it carries with it the bedload and suspended load that it has been transporting down the river. As it flows across the flat floodplain, it quickly loses velocity/speed and energy, and so deposits/drops the material it is carrying. The name given to all this deposited material dropped by a river is alluvium.

The larger, heavier material (pebbles and gravel) is dropped first, which means that the parts of the floodplain nearest the river channel are higher than the rest of the floodplain. These higher areas parallel to the river channel are called levées.

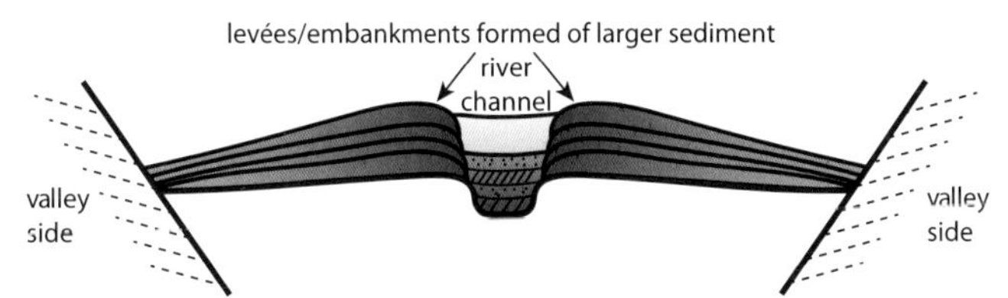

Figure 9.10 A cross section through a typical floodplain

Deltas

Where a river enters a sea or a lake, the remaining bedload and suspended load that it has been transporting will be deposited at its mouth. If there are strong tidal currents, as found on many coastlines, the material will be transported away and it often ends up on the beaches along the coast. However, if there are no tidal currents, the material will build up to form a delta. The deposition of the very small, fine sediment is helped by the fact that the salt particles in the sea cause the fine clay particles to 'stick' together (a process called flocculation) and as they become heavier, they drop to the seabed.

A typical delta is made up of three sets of deposits (Figure 9.11):

- The heavier bedload material is dropped first to produce **fore-set beds** at the bottom of the delta.
- The lighter, suspended material is carried further away from the coast and forms **bottom-set beds** (above the fore-set beds).
- On top of the fore-set beds as they build up, water will flood across them and deposit more layers of alluvium; these are called **top-set beds**.

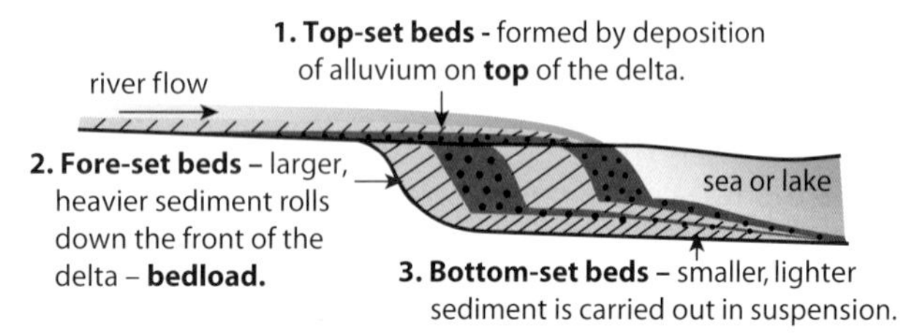

Figure 9.11 A cross section through a delta

On top of the delta, the river channel may split up into several smaller river channels, called **distributaries**. These then spread across the delta. The growth of vegetation on the delta will help trap more sediment and raise the delta above sea level.

Deltas may form a variety of shapes (Figure 9.12). The following are the two most common shapes.

1 **Arcuate** deltas, for example the Nile delta, which flows into the Mediterranean Sea. These have a triangular fan shape.
2 **Digital** deltas which have a shape that looks like the fingers on a hand. In some parts of the world, they are called **bird's foot** deltas. For example, the Mississippi delta near New Orleans, USA, which flows into the Gulf of Mexico.

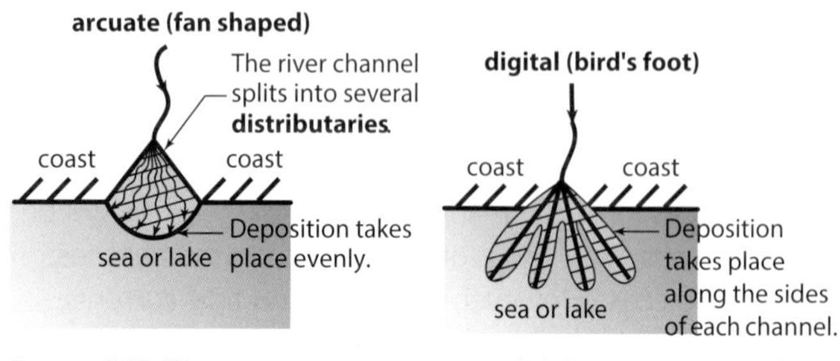

Figure 9.12 The two common types of delta – arcuate and digital

Self-test questions 9.1

1 Describe and explain how processes interact to form either waterfalls or oxbow lakes.
2 Why and where in a river's course does deposition take place?

9.4 River hazards and opportunities

Causes of river hazards

Rivers flood when the water in their channel reaches the top of their banks (bankfull discharge) and then flows across its floodplain. This usually happens in either of the following circumstances.

- When there has been a short period of heavy, torrential rain which the land cannot absorb quickly enough and overland flow may take place (when the water cannot infiltrate fast enough) – these are often called flash floods, when too much water arrives too quickly for the river to transport away.
- When there has been a long period of steady rain and the soil and rocks below the ground are full (saturated) with water so that any more rain that falls cannot infiltrate the soil and so it is forced to stay on the surface of the ground and then flows quickly towards the river channel and fills the river channel.

Sometimes both these situations may occur at the same time and produce floods that can be extremely serious and dangerous.

Rivers present a number of potential hazards through flooding and erosion:

- flooding which can destroy crops and buildings, and kill people and animals
- flood water which can be a source of water-related diseases as they provide a breeding ground for certain animals that spread disease, such as the mosquito, which can spread malaria and dengue fever, and the bilharzia snail; polluted water can also spread diseases such as cholera and diarrhoea
- floodplains and deltas often become very densely populated and so flood events can become extremely hazardous for large numbers of people.

- the alluvium that makes up a floodplain and delta is not very stable for building so foundations need to be carefully constructed.
- it is often very difficult and expensive to cross or bridge a large river as support structures may have to be built in the river channel, which is often both difficult and expensive. The structures are exposed to the full force of the river during flood events. Unfortunately, many are damaged and destroyed, causing bridges to collapse.
- fast-flowing, full rivers erode their banks so that undercutting and bank collapse may take place, placing property, housing, farm land, roads, railways, bridges and so on under threat.

The opportunities of living on a floodplain, a delta or near a river

Despite the hazards, there are also a number of opportunities presented by living on a floodplain, delta or near a river:

- The alluvium that has been deposited on them during floods provides extremely fertile soils, so they are often very important regions for agriculture – such as the river Nile in Egypt.
- The water from the river can be used for irrigation, allowing land that lacks water to be used for agriculture.
- Larger rivers are very important route ways for transport and communications (by the rivers themselves and by roads and railways built on the flat land on the floodplain).
- Many rivers are an important source of food (fish).
- They are important sources of fresh water.
- They provide large areas of flat land that can be used for building houses and industry.

9.5 Managing the impacts of river flooding

Flood hydrographs

Predicting floods can save lives and property. Flood hydrographs are one of the methods used to help in flood prediction. A flood hydrograph plots river discharge (in cubic metres per second – cumecs) and rainfall (measured in millimetres) over time.

TERM

flood hydrograph: a graph which shows the pattern of a river's discharge

channel engineering: planned human intervention in the river channel, altering its characteristics, or flow with the intention of producing some defined benefit, such as flood control

There are several important parts of a typical hydrograph (Figure 9.13):

- The rainfall peak is when the rainfall is at its heaviest.
- The **rising limb** on the graph represents the rising river discharge.
- When the discharge reaches its peak, it is called the **flood peak**.
- The falling limb shows the discharge as it drops.
- The **lag time** is the gap of time between the rainfall peak and the flood peak. This lag time is a very important time period as it gives people time to prepare for the flood. It can be several hours or several weeks on large rivers.

Every river has its own unique flood hydrograph and lag time. Unfortunately, some rivers have very short lag times and high flood peaks which make them very dangerous for people living beside the river. Prediction of floods is therefore increasingly important as more people live on floodplains and deltas than ever before.

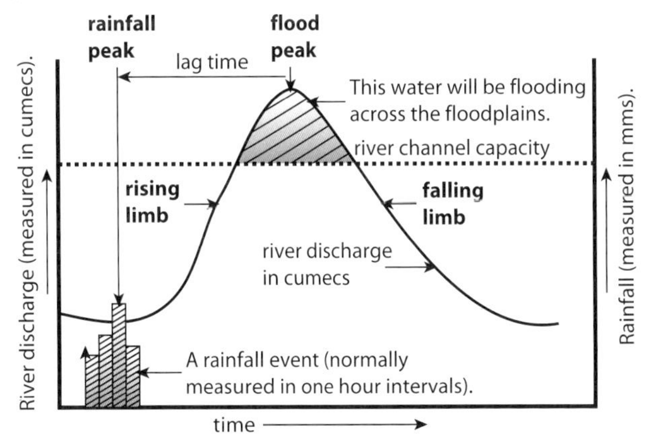

Figure 9.13 The features of a typical flood hydrograph

Flood prevention

There are two main methods of flood prevention. The first set of methods involves intercepting rainwater before it reaches and fills a river channel. The second set involves increasing the size/capacity of a river channel so that it can carry more flood water – this is called channel engineering.

Forests and vegetation naturally intercept rainfall but, on large areas of the Earth, deforestation and farming have removed the trees and vegetation so flooding now occurs where it never existed before or it is much worse than before.

To solve this problem, there are several methods which involve keeping water on the valley sides, allowing it to infiltrate into the soil and not allowing it to flow quickly over the surface of the land into the river channel. By preventing this surface runoff/overland flow, it also prevents the loss of soil – soil erosion – from farmers' fields. The methods include:

- planting trees – called **afforestation**
- leaving **permanent crops** such as grass on the valley sides
- **tiered or layer cropping**, where several layers of trees and crops are grown on the valley side by farmers to intercept the rain
- **contour ploughing** – ploughing across a valley side rather than up and down the valley side; each small bank of soil stops rainwater flowing down the valley side and the rainwater infiltrates into the ground and is available to the growing crops and is not lost as it flows away to the river channel
- **terracing** – where large, level steps are built into the side of a valley which provide areas of flat land for crops to be grown
- **tied ridging** – creating a grid of raised soil embankments, taking the shape of squares, from which rainfall cannot flow away down the valley side; again, the water is kept in the soil on the valley sides and the crops can make use of it
- leaving **crop stubble** in the fields which both stabilises and protects the soil and stops water flowing to the river channel
- **creating natural wetlands** – a natural floodplain stores a large amount of floodwater which reduces the threat of flooding further down the river; as many floodplains have now been drained and protected from flooding, the water that they would have stored is now moved further down the river putting flood defences under much greater pressure downstream and many are now not big enough to hold this extra water; in some rivers, like the Mississippi, floodplains that were drained and protected from flooding are now being allowed to flood after heavy rains to store floodwater
- **preventing infrastructure development** on threatened floodplains
- **floodplain retreat** – some government agencies, along the Mississippi, for example, are actually buying floodplain land from landowners and allowing these to flood during flood events as they reduce flood peaks further down the river.

FACT

Apart from keeping rainwater on the valley side and in the soil, all these methods prevent soil erosion where valuable, fertile top soil is carried into the river channel.

Much of this soil in the river channel is not transported away by the river and it is deposited and accumulates on the channel bed. The result of this is that the river channel becomes smaller – its channel capacity is reduced. This means that less water is needed to cause the river to flood. As a result, floods may happen more often and are bigger than they were in the past.

The second set of methods (channel engineering) increases channel capacity so that the river channel can hold more floodwater. The following methods are used:

- **dredging:** where the bedload is dug out by diggers and there is greater channel capacity as a result – this sometimes means that the soft channel sides may be eroded and collapse into the river channel so the channel sides are strengthened either by concrete or stone
- **wing dykes:** these are walls built out from the side of the river made out of concrete and stone on one side only of a river channel – the aim is to force the river into a smaller area which means it flows faster and carries away the bedload and alluvium and therefore does the job of dredging without the use of machines; these are used on the Mississippi river
- **two-stage channels:** this is where the top of one side of the channel is cut away to increase the amount of room in the channel for floodwater
- **building embankments or artificial levées:** these are high banks of soil, clay, sand and gravel built beside the river channel to increase the volume of the channel – sometimes these are built further away from the river channel to make the capacity even bigger

Advantages	Disadvantages
• They regulate river flow and can be used to prevent flooding below them. • The generation of cheap, renewable, clean electricity – HEP. The Three Gorges Dam on the Yangtze River in China is now the world's biggest power station. • This cheap electricity may attract industrial development. • Provide water for many uses – for people's domestic use, industry and for farming, including the irrigation of farm land. • Create employment during the construction phase of the dam. • Can become important fisheries if stocked with fish. • Make transport easier in rivers and can improve communications. • Provide tourism and leisure facilities, such as sailing, fishing, water skiing, swimming.	• Lakes and reservoirs formed behind them can flood large areas of fertile farmland. • The drowning of settlements and people's homes which can destroy communities and cause thousands of people to be forced to leave their land and homes. The Three Gorges Dam has forced 1.1 million people to move. • The loss of a natural river and its valley. • Reservoirs and lakes drown important ecosystems and can have an impact on rare plant and animal species. • The loss of a large recreational area. • A large concrete dam can have a big visual impact in a beautiful mountain area. • Can be very expensive to build. • Clear water erosion – as they trap the rivers load behind them as water comes out of the dam, it no longer has to spend most of its energy carrying the load. As a result, it is more powerful than when it was carrying its load, which means it has more energy for erosion. This means that it erodes its banks faster, and so they need protecting from erosion.

Table 9.1 Advantages and disadvantages of building a river dam

- **straightening the river channel:** this involves cutting out meanders so that floodwater can flow away much more quickly
- **holding dams:** these dams are built in the upper sections of rivers and hold back floodwater, from melting snow or heavy rainfall from monsoons, which they can then release after the flood threat is over – The Mississippi river in the USA has over 200 of these
- **check dams:** these trap the rivers bedload as it is being transported down the river in a flood so that the bedload does not fill up the river channel further down the river and reduce its capacity to carry water – They can be excavated when they fill up
- **overflow channels** or **spillways:** these allow water to flow away from the main river channel.

In addition to these methods, and often as a last resort, buildings and homes can be flood-proofed by sealing doors and other openings or raising them up above flood level.

The building of dams on rivers

The building of dams for flood control, water storage and the generation of hydro-electric power (HEP) normally produces both advantages and disadvantages to people and the natural environment (Table 9.1).

Self-test questions 9.2

1. Why do rivers flood?
2. Describe three likely impacts of flooding in an area.
3. What methods can be used to reduce the impact of flooding?

Case study

A river: the opportunities, associated hazards and their management – the Indus River Basin, Pakistan

The Indus River flows through one of the largest river drainage basins in Asia, crossing the borders of four South Asian countries – Afghanistan, China, India and Pakistan. Over 200 million people live within the basin.

The Indus drainage basin covers about 70% of the area of Pakistan and the importance of the Indus River to the country and people of Pakistan cannot be underestimated. More than 138 million people live and depend on the Indus river basin in Pakistan (out of a total country population of 193 million in 2016).

A large part of the Indus basin has a semi-arid to arid climate and its rainfall falls short of the human demand for water. The Indus is the key water resource for Pakistan's national economy, accounts for most of the nation's agricultural production and it also supports many heavy industries and provides the main supply of drinking water in Pakistan.

In order to exploit and use the water of the Indus River, the natural physical characteristics of the Indus River basin and its many channels in Pakistan have been changed enormously by human modifications over time. There are two mega dams (Mangla and Tarbela) and 320 barrages (many of considerable size), which provide the irrigation water to the floodplain, as well as being used to control the annual monsoon floods. Unfortunately, all these dams and barrages are not able to cope with the larger flood discharges that are generated by heavier monsoon rainfall events.

TERMS

mega dams: dams that are 150 or higher in height; there are currently more than 300 mega dams in the world

barrages: gated hydraulic structures built across the river or other water course to control, regulate and divert flows to the irrigation canals or to help the navigation of the river by boats

The factors of rising population pressure, climate change and continuous degradation of the basin's ecosystems have resulted in increased flood risks, made worse by inadequate flood planning and management on its floodplain.

With so many millions of people relying on the Indus River, any flooding by the river can have devastating effects. From 1950 to 2012, there were 22 major floods in Pakistan's section of the Indus River basin, killing over 9300 people, affecting over 110 000 villages and causing a cumulative direct economic loss of about $20 billion. The major floods in 2010 were Pakistan's most damaging on record. They affected all the provinces and regions of Pakistan, killed 1600 people, caused damage totalling over $10 billion, flooded / inundated an area of about 38 600 km^2, damaged nearly 2 million houses and forced over 20 million people to leave their homes and businesses. It is estimated that the Pakistan's GDP growth rate of 4% prior to the floods fell to minus 2–5% and has been followed by several years of below-trend growth.

Figure 9.14 shows the location of the Indus River, the area of flooding and the major barrages/dams on the river.

What causes the flooding on the Indus River?

Flooding in Pakistan has generally been caused by the heavy concentrated rainfall during the monsoon season, in the months of July and August. However, the 2010 floods were a combination of natural and human factors that produced flood peaks that were far in excess of the river channel's capacity to hold them.

Natural factors

- The monsoon rainfall can be exceptional – some areas in northern Pakistan can receive more than three times their annual rainfall, a total of 280 mm, in a matter of 36 hours.
- Steep valley sides that are capable of producing rapid overland flow/surface runoff.
- Degraded ecosystems and a lack of protective vegetation cover mean that the interception of heavy rainfall is greatly reduced and soil erosion is accelerated.

Human factors

- The Indus River basin lacks an appropriate flood policy.
- It lacks comprehensive laws.

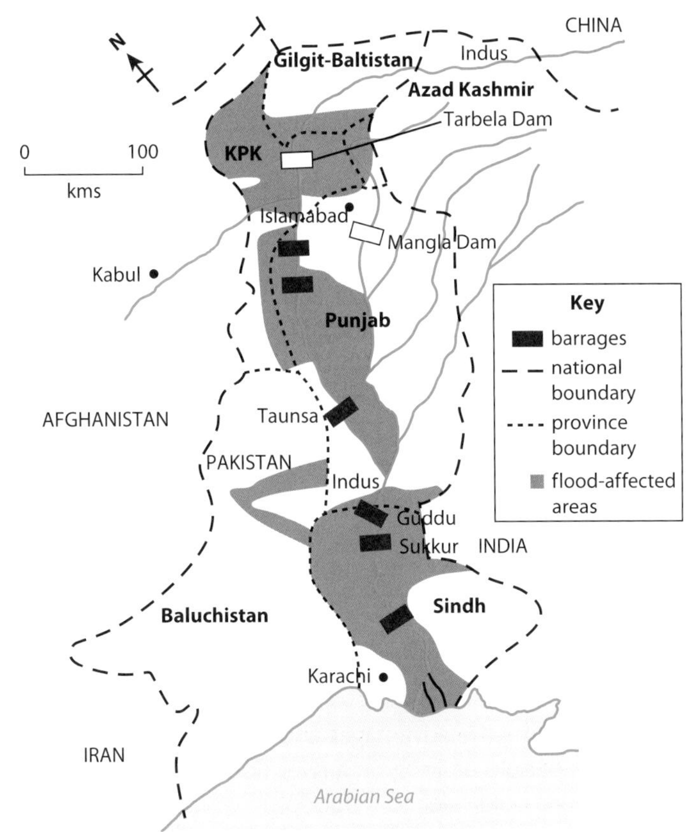

Figure 9.14 Indus River, Pakistan, showing the flood-affected areas

- There is inadequate flood-control infrastructure.
- Poor anticipation of the scale of potential floods.
- Protective levées/embankments could not bear such sustained flood pressure for several days and large numbers and sections of them failed and collapsed resulting in terrible damage.

What is being done to manage the flooding on the Indus River?

The Pakistan government has been relying on a traditional flood control approach based on structural measures, through the building of dams and building of artificial levées, but the 2010 flood exposed the weaknesses of this approach.

Structural/hard engineering flood-protection measures

- **Levées** – 6000 km of artificial levées/embankments provide most of the flood protection and over 1400 spurs (these are levées or stone walls constructed to divert the flow of the river at critical locations; on the Mississippi river in the USA similar structures, called wing dykes, perform a similar function). These are used to protect the main towns and important infrastructure, such as major roads and railways. The levées now cover most of the critical stretches along the river channels. The height of these levées is fixed at 1.8 metres, though some areas probably need higher levées and many were not properly constructed or poorly maintained.
- The **river channel shape and position** has been altered at key locations to control actively meandering channels and to save erodible beds and banks from erosion.
- **Reservoirs** – the 2010 flood demonstrated the effectiveness of the country's two main reservoirs at Mangla and Tarbela in holding back floodwater. However, sedimentation in both has significantly reduced their storage capacities. The building of new reservoirs along the Indus is really important, not only for future flood protection, but to supply the rapidly increasing need for water and energy.

Non-structural flood-protection measures

- flood risk assessments
- **floodplain zoning** and **land-use planning** and much stricter enforcement
- early **flood warnings** to communities at risk from flooding, and greater flood **preparedness** on the part of these communities – however, these have not been fully incorporated into the country's flood management planning
- building **earth mounds or elevated platforms** to serve as temporary refuges for people during floods.
- better **flood forecasting**, with all neighbouring countries involved.

Sample question and answer

Using an example which you have studied, explain how and why a delta is formed at a particular location. [5]

TIP

Students often fail to develop their answers to 5-mark, case study questions. Make sure that you include place-specific references in your answer, but only ones that are relevant to the actual question.

The River Nile in Egypt has formed a very large delta where it reaches the Mediterranean Sea. As the river reaches its mouth, it starts to slow down **[1]** and it no longer has the energy to carry its load **[1]**. The river will then deposit the sediment (alluvium) **[1]** it is carrying at the mouth of the river. As there are no major sea currents to carry away the sediment, it has built up to form a delta **[1]**. The salt in the seawater has helped by causing the small clay particles to join together and become larger and heavier so that they have sunk to the seabed **[1]**, a process called flocculation **[1]**. The growth of vegetation on the Nile delta has helped trap more sediment and raise the delta above sea level **[1]**. Over time, the sediment has accumulated and increased the size of the delta **[1]**.
(A maximum of 5 marks will be awarded.)

Exam-style questions

1 a What is the name given to the line dividing two river basins? [1]

b What is a confluence in a river? [1]

c Explain how the processes of hydraulic action and abrasion may erode the bed and banks of a river channel. [4]

d Name and explain two processes by which a river transports its load. [1]

2 Describe the advantages and difficulties for people of living on a river floodplain. You should refer to a floodplain which you have studied. [7]

Coasts

Learning Summary

By the end of this chapter, you should be able to:

- demonstrate an understanding of the work of the sea and wind in eroding, transporting and depositing
- describe and explain the formation of the landforms associated with these processes
- describe coral reefs and mangrove swamps and the conditions required for their development
- demonstrate an understanding that coasts present hazards, including coastal erosion and tropical storms, and offer opportunities for people
- explain what can be done to manage the impacts of coastal erosion
- demonstrate knowledge of a case study of the opportunities presented by an area of coastline, the hazards associated with it and their management.

10.1 Coastal erosion, transportation and deposition

Processes of erosion

As with river erosion, there are four processes of coastal erosion:

- **hydraulic action**: where the weight (this can be up to 20 tonnes per square metre) and force of a wave crashing against a cliff removes particles of rocks from the cliff – it also includes the process where air is trapped by a wave in a crack in the cliff and the enormous hydraulic pressure this creates opens up the crack further, which weakens the cliff
- **abrasion** (sometimes called **corrasion**): where boulders, pebbles, shingle and sand are picked up by a wave and thrown against the cliff – This constant collision removes particles of rock from the cliff face
- **solution** (sometimes called **corrosion**): some minerals (mainly the carbonate minerals found in rocks, such as limestone and chalk) are put into solution by the weak acids found in seawater
- **attrition**: this process does not actually erode the cliff but it is the process that breaks up the boulders, pebbles, shingle and sand on the beach – it takes place when the rocks on the beach are rolled up and down the beach by swash and backwash as this happens, they collide with each other and become smaller and rounder as a result.

TERMS

swash: the movement of a sea wave up a beach after the breaking of a wave

backwash: the movement of a sea wave down the beach after the breaking of a wave

Processes of transportation

There are two main processes of transportation:

- Longshore drift (Figure 10.1) takes place when sand and shingle is moved by wave action. For this process to take place on a beach, the waves must break **across** the beach at an **oblique** angle:
  - When the wave breaks, its **swash** transports sand and shingle up and across the beach.
  - When the swash runs out of energy, it returns back down the beach, carrying the sand and shingle with it, as **backwash**.
  - As this process is continually **repeated**, it means that the sand and shingle will be moved along the coastline.
  - Where there is a break in the coastline, this results in the formation of spits, bars and tombolos. To prevent the process, groynes can be built to trap the sand and shingle.

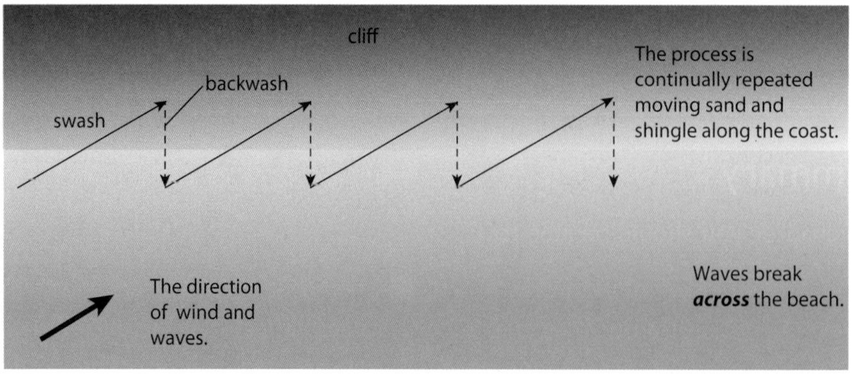

Figure 10.1 The process of longshore drift

TERMS

longshore drift: the movement of material along a beach transported by wave action

spits: ridges of sand or shingle attached to the land, but ending in open sea

bars: ridges of sand or shingle across the entrance of a bay or river mouth

tombolos: spits connecting an island to the mainland

groynes: wooden, stone or concrete barriers built perpendicular (at right angles) to the coast in order to break waves and reduce the movement of sediment along the beach

TIP

Processes can often be explained with simple, labelled diagrams. Practise drawing the process of longshore drift. You can also use it when answering questions on the formation of spits, bars and tombolos.

- **Wind action:** Smaller grains of sand can be moved by the wind and can form sand dunes (Figure 10.2) at the back of a beach.
  - For this to happen there needs to be:
    - a **large**, **sand beach** to supply the sand
    - a **strong**, **onshore wind** to first dry out the sand and then transport it inland
    - an **obstruction** to trap the sand, such as seaweed, at the top of the beach – the **strand line**.
  - The sand will **accumulate** (build up) into a small dune, about one metre high, called an **embryo dune.**
  - **Pioneer species** of plants, such as marram grass, will colonise the small dune. The roots and stems of these plants will trap more sand and speed up the process of deposition so that the sand builds up in to bigger mobile or yellow dunes.
  - As this process continues, the sand dune will increase in size and height to become **fixed** or **grey dunes**.

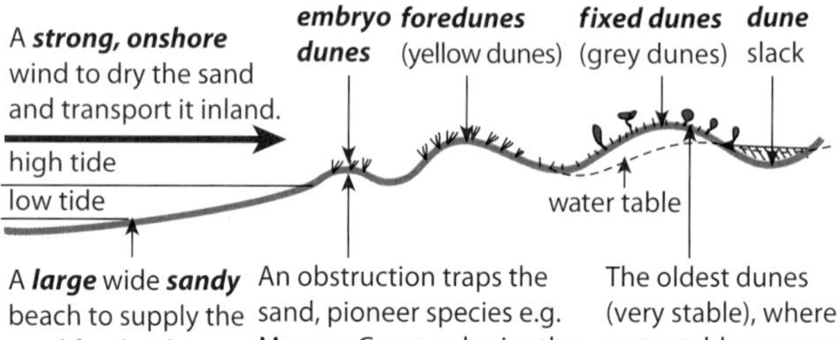

Figure 10.2 The formation of sand dunes

Constructive and destructive wave action

A **constructive wave** has its swash **stronger** than its backwash, which will result in sand and shingle being moved **onshore**, therefore building up the beach as it **brings in** more sand and shingle than it takes away. Constructive waves tend to form over a long distance (this distance is called the **fetch** of a wave). They are usually small, flat waves with a long wavelength, up to 100 metres, and they have a low frequency, about 6–8 every minute. These waves steepen slowly when they reach a beach and gently spill forward.

A **destructive wave**, in contrast, has its swash **weaker** than its backwash, which will mean that sand and shingle will be moved **offshore**, therefore 'destroying' the beach as it **takes away** more sand and shingle than it brings in. Destructive waves tend to form over a short distance – a short fetch. They are usually large, steep waves with a short wavelength, 20 metres, and a high frequency, about 10–14 every minute.

Processes of deposition

When a wave loses its **energy** it will **drop/deposit** sand and shingle. This results in a number of features of deposition being formed along a coastline. These include beaches, spits, bars, tombolos and saltmarshes.

10.2 Landforms of coastal erosion

Cliffs and wave-cut platforms

Cliffs are formed when waves erode a coastline (Figure 10.3). Cliffs go through a repeated cycle:

- They are **undercut** by hydraulic action, abrasion/corrasion and solution/corrosion.
- This will form a cliff notch at the bottom/base of the cliff.
- This leaves the rocks in the cliff above **unsupported**, so that they will eventually **collapse**.
- The collapsed material will be **broken up by attrition** and **removed** by wave action.
- The process will then be **continually repeated**.

As the cliff retreats, it will leave behind a gently sloping platform of rock called a wave-cut platform. Many of these are covered by the collapsed rocks from the cliffs, which will form beaches.

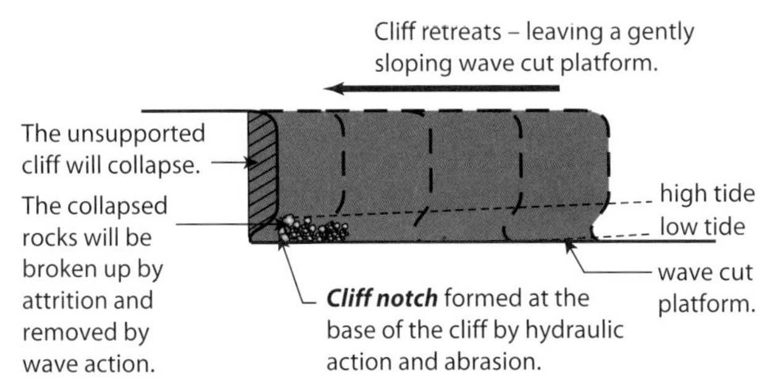

Cliff retreats – leaving a gently sloping wave cut platform.
The unsupported cliff will collapse.
The collapsed rocks will be broken up by attrition and removed by wave action.
high tide
low tide
wave cut platform.
Cliff notch formed at the base of the cliff by hydraulic action and abrasion.

Figure 10.3 The formation of a cliff and wave-cut platform

Caves, arches, stacks and stumps

These features form in narrow, rocky headlands. The sequence starts when a vertical crack, or fault, in the headland is eroded by hydraulic action, abrasion/corrasion and solution/corrosion to form a cave. Sometimes the crack or fault at the back of the cave may be eroded through to the surface of the headland so that when a large wave enters the cave, water is pushed out of the top of the cave to form a **blowhole** (gloup).

As the cave enlarges, it may erode all the way through a headland to form a natural arch. Over time, the roof of the arch may become weakened and will collapse. This leaves an isolated pinnacle of rock, called a stack. Over time, this will become eroded to form a smaller stump (Figure 10.4).

TERMS

saltmarsh: an area of mudflats formed when sediment is deposited in the low wave energy area behind a spit

cliff notch: a small overhang formed at the base of a cliff by wave action (erosion)

wave-cut platform: a wide, flat area of rock at the bottom of cliffs seen at low tide

headlands: areas of more resistant rock jutting out from the coast into the sea

cave: a hollow at the base of a cliff which has been eroded by waves

arch: a rock bridge formed at a headland that has been partly broken through by the sea

stack: an isolated column of rock sticking out of the sea just off the coast that was once attached to the land

stump: a short piece of rock at the end of a headland, formed after a stack has collapsed

bay: a broad coastal inlet, often curved and with a beach, between two headlands

FACT

Where a cave forms in a cliff and enlarges, its roof may gradually weaken and collapse, leaving a narrow, steep-sided inlet in the cliff, called a **geo**.

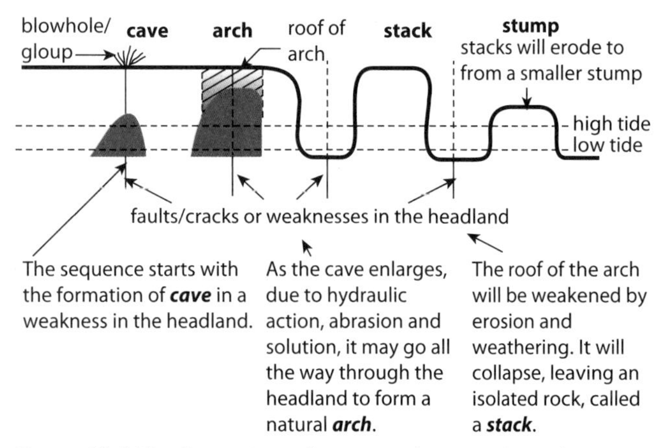

Figure 10.4 The formation of caves, arches, stacks and stumps

TIP When asked to **explain** the formation of landforms of erosion, always name the **processes** of erosion and explain how they operate to create the landform.

Bay and headland coastlines

Where a coastline is made out of one type of rock and has no weaknesses, it will erode back at a constant rate. However, if there are weaknesses in the coastline such as sections of softer, less resistant rock, **differential erosion** will take place (Figure 10.5). As the softer, less resistant rock is eroded at a faster rate it will form a bay, leaving the harder, more resistant rocks projecting out to sea as **headlands**. Such a coastline is called a **discordant coastline**.

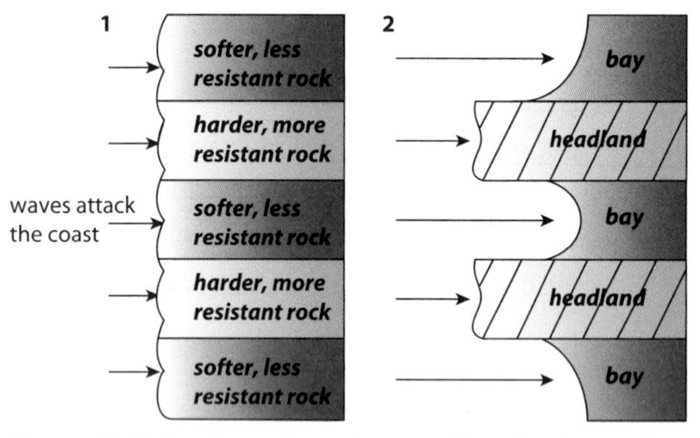

Figure 10.5 The formation of bay and headland coastlines

10.3 Features of coastal deposition

Beaches, spits and bars

Beaches are accumulations of sand and shingle. They form where cliffs have been eroded and the collapsed rock has been broken up by the process of attrition into sand and shingle, and/or where longshore drift has deposited sand and shingle. Beaches are important coastal features as they protect many coastlines from erosion by wave action.

Spits are formed when longshore drift is operating, moving sand and shingle along a coastline. If there is a break in a coastline, such as an estuary or river mouth, longshore drift will continue to deposit sand and shingle and so the beach will start to grow away from the coast, forming a **spit**. This will be long and narrow and have one end attached to the coast and one end in open water.

If a spit continues to grow, it may stretch completely across a river mouth or estuary until it reaches the other side. It then becomes a **bar**. The water trapped behind the bar becomes a **lagoon**. Over time, this will often become filled with alluvium brought down by a river and may disappear. Should a spit develop and grow until it reaches an island offshore, it will form a feature called a **tombolo** (Figure 10.6).

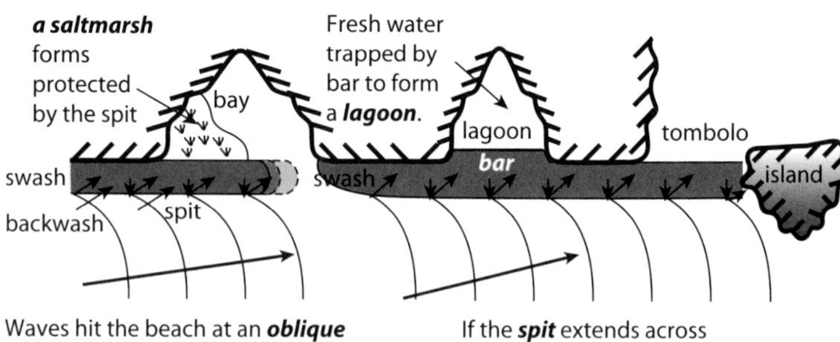

Figure 10.6 The formation of spits, bars and tombolos

Saltmarshes

Saltmarshes form in the coastal intertidal zone, in sheltered bays and estuaries or in the protected area behind a spit or tombolo, which are regularly flooded by the tides. In such sheltered places, there is little or no wave action or longshore drift and so very fine sand, silt and clay will be deposited. Saltmarshes develop a very distinctive ecosystem. **Pioneer species** of salt-tolerant plants called halophytes (herbs, grasses or low shrubs) will colonise the deposited material, which will trap more sediment and bind it together. Each high tide will bring in more sediment which will continue to build up the feature to form the saltmarsh.

Self-test questions 10.1

1 Name and describe three processes of coastal erosion.

2 Describe and explain the formation of a cave, arch and stack.

10.4 Coral reefs and mangrove swamps

Coral reefs

Although corals are found throughout the world's oceans, large coral reefs are only found between 30° north and south of the Equator (Figure 10.7).

There are three types of coral reef:

- **barrier reefs** – these are usually the largest reefs and are separated from the land by a deeper water lagoon
- **fringing reefs** – these are attached to the land and are found around many Caribbean and South Pacific islands
- **atolls** – these are ring-shaped and usually rise out of very deep water, well away from the land. They enclose a lagoon and there may be small reefs within the lagoon.

TERM

coral reef: a ridge of rock near the surface of the sea, formed by the growth and deposit of coral (calcium carbonate)

FACT

The Great Barrier Reef in Queensland, Australia, is the largest reef in the world. It covers an area of 348 000 km^2 and is over 2000 km long.

The formation and location of coral reefs is controlled by seven limiting factors (conditions):

- **temperature** – the mean annual temperature has to be over 18 °C. The optimal temperatures for them is between 23–25 °C
- **depth of water** – coral reefs can only grow in depths of water less than 25 m
- **light** – the shallow water allows light for tiny photosynthesising algae, called zooxanthellae; in return for the corals providing the algae with a place to live, these tiny algae provide the corals with up to 98% of their food

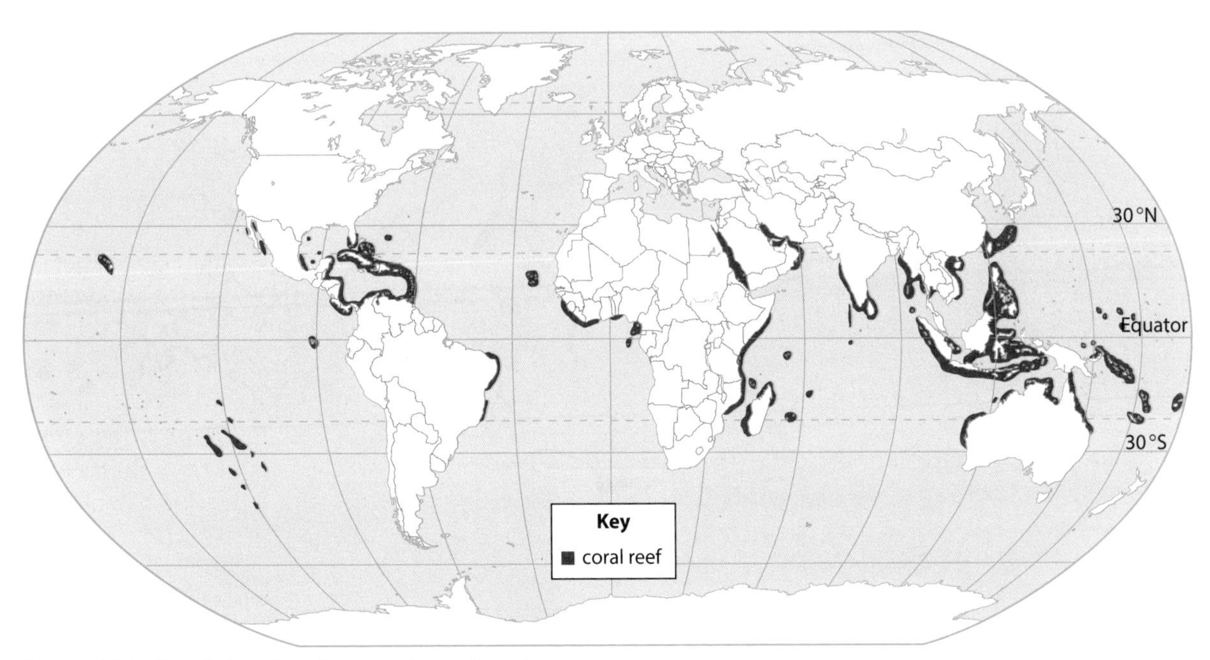

Figure 10.7 The global distribution of coral reefs

- **salinity** – corals can only live in seawater, but they cannot tolerate seawater of high salinity
- **sediment** – sediment clogs up the feeding structures and cleaning systems of corals – cloudy water also reduces light penetration in the water, reducing the light needed for photosynthesis
- **wave action** – coral reefs prefer areas of high energy wave action – this ensures freshly oxygenated water
- **exposure to air** – corals die if they are exposed to air for too long – they can only survive and grow, therefore, at the level of the lowest tides.

Mangrove swamps

Mangrove swamps are found along coastlines that experience tropical and subtropical climates, mainly between 30° north and 30° south of the Equator (Figure 10.8). Mangrove swamps should naturally cover 60–75% of tropical and subtropical coastlines. Globally, they cover an area of about 138 000 km², in 118 countries.

Mangroves are tropical trees and shrubs which grow in the intertidal zone, in low-energy, relatively sheltered, shallow coastal and estuarine environments, where fine sediments (often with a high organic content) collect in areas of the coast that are protected from high-energy wave action.

TERM

mangrove swamps: tidal swamps that are dominated by mangroves (shrubs or small trees with numerous tangled roots that grow above ground)

FACT

Mangrove trees are salt-tolerant (**halophytes**). They contain a complex salt filtration system and complex root system to cope with being immersed in saltwater.

Mangroves are **adapted** to life in harsh coastal conditions:

- **anaerobic (low oxygen) conditions** – some species, like the red mangroves, prop themselves above water level with **stilt roots** and can then absorb air through pores in their bark, called **lenticels** – others, like the black mangroves, live on slightly higher ground and have **pneumatophores** (specialised root-like structures which stick up out of the soil like straws for breathing), which are also covered in lenticels

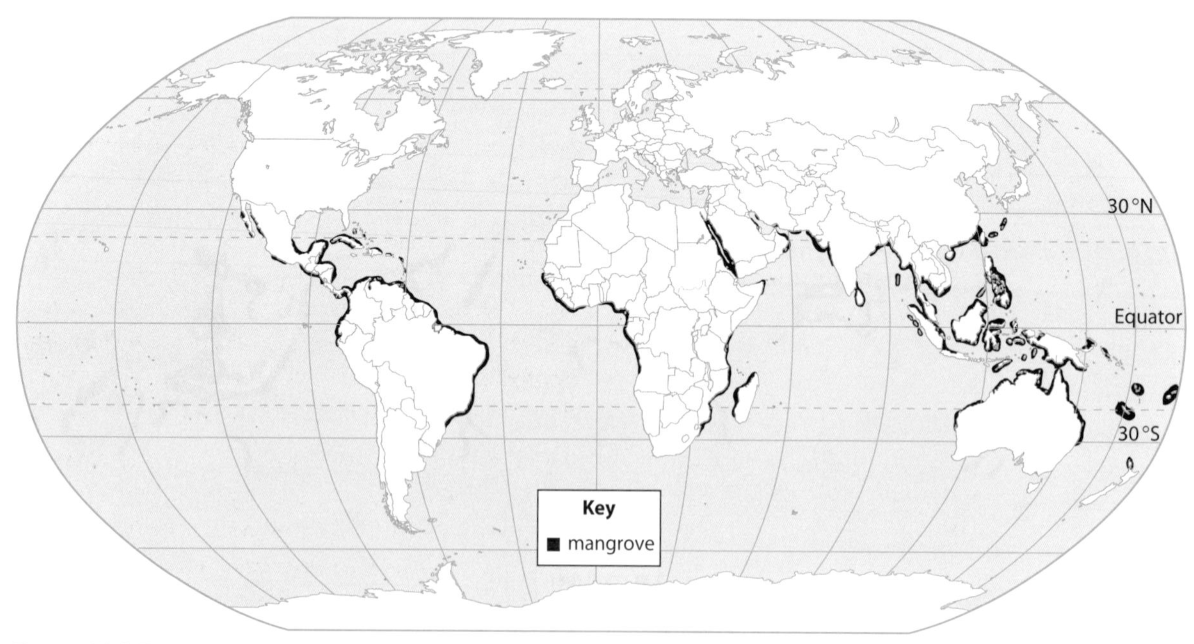

Figure 10.8 The global distribution of mangrove swamps

- **salt intake** – some mangroves have an impermeable/waterproof membrane in their roots which acts as a filter to keep salts in their roots to limit the amount that enters the rest of the tree
- **water loss** – freshwater is limited in the salty intertidal soils in which mangroves live and so mangroves try to limit the amount of water they lose through their leaves – they can restrict the opening of their **stomata** (pores on the under leaf surfaces, which exchange carbon dioxide gas and water vapour during photosynthesis) – they also vary the direction/orientation of their leaves to avoid the harsh midday sun and so reduce evaporation from the leaves.
- **nutrients** – mangroves have difficulty obtaining nutrients as the soil they live in is always waterlogged so little free oxygen is available – however, their aerial roots, the pneumatophores, allow the mangroves to absorb gases directly from the atmosphere, and other nutrients such as iron, from the soil/mud
- **survival of their offspring/seeds** – mangrove seeds float and are therefore able to disperse and float away from the parent tree:
  - Also, many mangroves are **viviparous** – their seeds germinate while still attached to the parent tree and once germinated, the seedling can produce its own food via photosynthesis.
  - It then drops into the water, and can be transported great distances.
  - They can survive drying out / dessication, and remain dormant for over a year before arriving in a suitable environment.
  - When a floating seedling is ready to root, it can adjust its density to float vertically. In this position, it is more likely to lodge in the mud and root. If it does not root, it can alter its density and drift again in search of more favourable conditions.

FACT

Mangrove swamps are one of the most productive and biologically important ecosystems of the world because they provide unique products and services to both people and coastal and marine systems.

Mangrove swamps have three major roles:

- **coastal protection** – they help **stabilise** and **protect shorelines** and reduce the devastating impact of natural disasters such as tsunamis and tropical storms, hurricanes/cyclones/typhoons
- **a breeding ground and nursery** – they provide breeding and nursing grounds for a wide range of marine and fish species, such as fish, crabs, oysters, and other invertebrates and wildlife such as birds and reptiles
- **a source of food, medicine and raw materials** – they provide food, medicine, fuel and building materials for local communities.

10.5 Coastal hazards and opportunities

The world's coasts present hazards, including coastal erosion and tropical storms, as well as offering opportunities for people.

Hazards

Human activities along the coastline are often responsible for causing major disruption to natural coastal systems. These activities may be **direct**, such as fishing, locating houses and industry, and tourist and recreation use, and **indirect** through pollution, global warming and contributing to the rise in sea level.

The continual increase in human population and their activities on the coast means that when natural processes and events take place, such as **coastal erosion** and **storm activities** (hurricanes/cyclones/typhoons and tsunamis), increasingly large numbers of people are affected.

Opportunities for people

The dynamic processes that occur within the worlds different coastal zones produce **diverse and productive ecosystems**, which are of great importance for the global population in the following ways:

- Coastal areas equate to only 8% of the world's surface area but they provide 25% of global productivity. Approximately 70% of the world's population, almost 5 billion people, is within a day's walk of the coast. Two-thirds of the world's cities are found on the coast.

- The coast contains **resources**, such as fish, minerals and industrial, living and recreational/tourist locations, which are considered to be common property. They provide opportunities for **development** and they are in high demand from the people on the coast for a wide variety of reasons, such as subsistence use, recreation/tourist and economic development.

TERMS

tropical storms: areas of very low pressure in low latitudes, with strong winds and heavy rain

Integrated Coastal Zone Management (ICZM): the management of the coast using an integrated approach, taking into account all aspects of the coastal zone, including geographical and political boundaries

littoral cell: a length/section of the coast that is isolated from adjacent sections of coast and has its own sources of sediment

10.6 Managing the impacts of coastal erosion

Coastlines are in need of very careful **management**. Due to the complex nature of all the human activity on the coast, a **holistic approach** is required to obtain a **sustainable** outcome for the world's coasts. Some form of Integrated Coastal Zone Management (ICZM) is needed.

People will normally defend a coastline when something of economic value is threatened by coastal erosion, such as a settlement, an industrial area, a port or an important transport link. People can then do one of two things – either they can **build coastal defences** or they can **do nothing and abandon the coast**. The **advantages** and opportunities need to be weighed up alongside the **disadvantages** of living and locating near the coast.

Decisions about how to defend a section of a coast can be taken using various types of assessments:

- **cost-benefit analysis** considers the social and economic aspects of a strategy. The benefits of a scheme (new businesses or jobs and savings in lives and property) are divided by the costs of building and maintaining it.
- **Environmental Impact Assessments (EIA)** try to assess the effects any strategy will have upon an area. It is especially important along coastlines as attractive scenery and ecosystems are valuable tourist assets.
- **Shoreline Management Plans (SMP)** try to decide upon the most appropriate scheme for each part of a littoral cell.

When a decision has been made to protect a coastline, rather than abandon it to natural processes, there are a variety of sea defences that can be used and they are usually split into two methods of defence – **hard engineering** and **soft engineering**.

Hard engineering

Hard engineering involves building permanent, rigid structures, such as sea walls and revetments. These are the traditional forms of coastal defence which are used to **reflect**, **deflect** and **absorb wave energy**. They are often very expensive, not sustainable and sometimes cause damage to other parts of the coastline through beach erosion/scour and preventing the movement of sediment by longshore drift.

- **Sea walls** have been the traditional, high-cost solution to protect valuable stretches of coast and property. Sea walls can cost up to $15 000 per metre of length and this means that careful consideration has to be given to the design, size and materials. A **vertical sea wall** may **reflect** waves and cause the beach to be scoured/eroded, resulting in the sand being carried offshore. Dropping the beach in front of the sea wall results in waves spending much more time in contact with the sea wall and hitting it with greater amounts of energy. This may result in erosion of the base/foundations of the sea wall and can lead to the eventual collapse of the wall. Many sea walls are now built at an inclined angle to **deflect** wave energy to stop beach erosion and they may also cause the beach in front of them to build up and further protect the wall and coast.
- **Revetments/wave ramps** are used in lower-energy environments than those that need a sea wall. Waves are forced to run up the inclined ramp and lose energy as they do so. Some have a small wave return wall at their top to stop any waves over-topping the ramp.
- **Offshore breakwaters** are made out of resistant **rock armour** or **concrete**. They are built up from the sea bed up to high tide level to absorb the energy of incoming waves and so protect the coastline behind them.
- **Gabions** are **wire cages** of varying size, from half a metre to a metre square, filled with resistant

rocks. They can be joined together to extend their length and height and function as a sea wall. They can only be used in very low-energy situations as that are relatively easily removed by more powerful waves.

TIP

In some textbooks, rock armour/rip-rap and groynes are identified as hard engineering strategies, in other textbooks they are included in the soft engineering group of defences.

Soft engineering

Soft engineering involves building less rigid, often more sustainable structures such as replacing and building up a beach. These techniques are often much cheaper and can be more environmentally friendly.

- **Rock armour/rip rap** uses large boulders of hard, resistant rock placed at the base of a cliff. The boulders force the waves to break, dissipating their energy and protecting the cliff from erosion. Where resistant natural rock is not available or too expensive, moulded concrete shapes are used.
- Some coastal defence structures, such as **groynes**, are a combination of both hard and soft engineering. Groynes are made out of wood, concrete or large boulders and built at right angles to the coast to trap sand and shingle being transported/moved along the coast by longshore drift. This helps build up the beach in height and width. The larger beach then protects the coast as it absorbs the energy of the waves and prevents them reaching the cliff or coastline.
- **Beach nourishment** is where sand and shingle is added to a beach to build it up in height and width to better protect a coastline from wave action. Care needs to be taken to use the same type and size of sand, otherwise it may be removed by wave action if it is too small and will need to be replaced, at further expense.
- **Cliff stabilisation** is carried out by putting drains in cliffs to release groundwater and stop the cliffs collapsing if they absorb the increased weight of the water after heavy rainfall. Some softer cliffs can be **terraced** – cut into giant steps – to make them more stable.

Managed retreat/coastal realignment

Several national governments have decided as a result of the huge costs of repairing and replacing old coastal defences, such as around the UK coastline, to allow some areas of coast to erode. These areas are where the cost of building a defence is greater than the value of the land being defended. This approach to managing the coast is called **managed coastal retreat or managed realignment**.

Self-test questions 10.2

1. Describe the conditions required for the development of coral reefs.
2. Explain why the methods used to protect a coastline have to be carefully planned.

Case study

An area of coastline – Unawatuna, Sri Lanka

Unawatuna, a small fishing settlement on the south coast of Sri Lanka, was once named as having 'the best beach in the world'. The publicity it gained from this exposure led to the small settlement developing into a **tourist resort** based on its picturesque sheltered bay and coconut palm-fringed beach. To cater for the increasing number of visitors and tourists, the whole beach front was developed. Much of the development has been in an unplanned and inappropriate manner, and includes hotels, guesthouses, restaurants, cafes and shops.

After the **Asian tsunami** devastated large sections of the eastern and southern coastline of Sri Lanka on **26 December 2004**, a great deal of new development took place, catering for an ever increasing number of tourists. In several coastal locations, decisions were made to provide **long-term protection** from possible future tsunamis and tropical storms. In the small Unawatuna bay, it was decide that a **breakwater** would provide the best solution. Unfortunately, the building of the breakwater and possible changes to the offshore coastal relief/topography during the tsunami has caused the beach to begin re-aligning its plan shape. Between 2008 and 2014, sand began moving from the eastern end of the beach to accumulate at the western end. As a result, the eastern end saw the almost complete removal of the beach in front of several properties, mainly hotels, restaurants and cafés (Figure 10.9). Waves were able to reach these exposed properties and started to erode and undermine the foundations of the buildings.

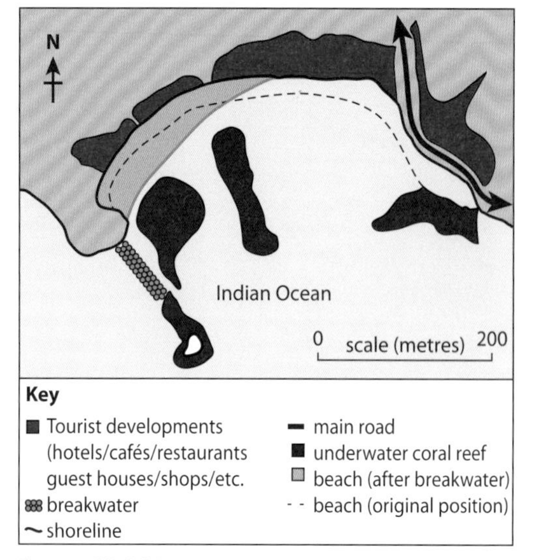

Figure 10.9 Unawatuna bay and beach, before and after the building of the breakwater

A variety of schemes and methods were then used by individual property owners and the government in an attempt to defend these properties – a mixture of **hard engineering** and **soft engineering** methods.

The **hard engineering** involved building permanent, rigid structures – small, four-metre high, vertical **sea walls**. Where vertical sea walls were built in front of some properties, they **reflected** the waves and this caused the beach to be further **scoured/eroded**, resulting in the sand being transported offshore by the reflected waves. With no beach, the sea walls started to be undercut. The sea walls should have been inclined, not vertical, to deflect rather than reflect waves. There should also have been a **curved wave return wall**, at the top of the wall, to further **deflect** the waves.

Other hard engineering methods involving **rock armour** and **gabions** also failed. The rock armour was made up of randomly deposited rocks which were too small to resist the levels of wave energy (which had not been measured prior to commencement). **Geotextile bags** were filled with sand but, again, they were placed vertically and reflected the waves, further increasing beach scour. The gabions were broken up and destroyed by increasingly large waves caused by the almost complete erosion and disappearance of the beach.

The **soft engineering** involved a new artificial beach being added to further protect the wall – the process of **beach nourishment**. Several hundred tonnes of sand were brought by truck from harbour dredging at another location on the coast. Unfortunately, the sand was too fine for the level of wave energy on the beach and was quickly removed by wave action. Some of it has gone on to the underwater **coral reef** and has killed the sensitive corals.

In 2015, an **international coastal management company** was engaged to pump several hundred thousand tonnes of sand from offshore sources onto Unawatuna beach and built the beach up both in height and in width. This solution, however, though currently protecting the hotels and other properties, may not be sustainable as the breakwater remains in place and the reason for the original movement of the beach has not been ascertained. By 2016, much of the new sand had been moved by wave action. As is the case globally, expensive coastal protection methods are often used with little knowledge or study of the coastal processes operating at specific locations.

Sample question and answer

Describe how a spit and a bar may be formed on a coastline. [6]

TIP

This question can be answered in a written form (as below), but it may be easier and better to explain the formation with the use of a simple annotated diagram or diagrams.

Spits and bars are formed when the process of longshore drift **[1]** is operating along a coastline. This process moves sand and shingle along a coastline **[1]**. It takes place when the waves hit the beach at an oblique angle and break across the beach rather than straight up and down the beach **[1]**. When the waves break, their swash transports sand and shingle up and across the beach, this is called swash **[1]**. When the swash runs out of energy, it returns back down the beach, carrying the sand and shingle with it, this is called backwash **[1]**. As this process is continually repeated **[1]**, it moves sand and shingle along the coastline. If there is a break in a coastline, such as a river mouth, longshore drift will cause sand and shingle to build up so that the beach will start to grow away from the coast, forming a spit **[1]**. If the spit continues to grow, it may eventually reach right across a river mouth until it reaches the other side, when it becomes a **bar [1]**. The water trapped behind the bar is known as a **lagoon [1]**. (A maximum of 6 marks will be awarded.)

Exam-style questions

1 a What is a headland? [1]

b Name two other natural features which are formed by coastal erosion. [2]

c Explain how bays and headlands may be formed along a coast. [4]

2 For a named area which you have studied, explain how and why coastal sand dunes have formed in that area. You may use a labelled diagram to help in your answer. [7]

Chapter 11

Weather

Learning summary

By the end of this chapter, you should be able to:

- describe how weather data is collected and explain the characteristics, siting and use made of weather instruments, including observations of types and amounts of cloud
- make calculations using information from weather instruments
- use and interpret graphs and other diagrams showing weather and climate data.

11.1 How weather data is collected

Measuring the weather is an activity that takes place all the time globally. The terms 'weather' and 'climate' are often confused with each other. There are thousands of weather stations worldwide, each with several instruments, measuring the different weather variables at any particular moment in time. The more important ones take both remote digital and manual readings every hour of the day and observe these variables constantly. The measurements that are taken are recorded. This recorded data is then used to describe the climate. A period of at least 30 years of weather readings is needed to be able to give accurate averages of all the weather variables, but also to give the extremes of these variables which enables us to describe the climate, and any changes in it.

TERMS

weather: the state of the atmosphere at any particular moment in time

climate: a description of the averages and extremes of weather variables of an area over an extended period of time

Stevenson screen: a container in which weather/meteorological instruments are placed

isohyet: a line on a map joining places of equal rainfall

Stevenson screen

Weather stations have a number of characteristic features and instruments. A typical weather station will contain several standard instruments. Some of them will be kept inside a raised white box called a Stevenson screen. This box protects the instruments from precipitation and direct heat radiation from outside sources, while still allowing air to circulate freely around them. It allows the shade temperatures to be taken – which are the temperatures shown on weather reports and forecasts. It therefore forms an important part of a standard weather station.

FACT

The Stevenson screen was designed by Thomas Stevenson (1818–1887), a British civil engineer and father of the author Robert Louis Stevenson.

The Stevenson screen will contain several instruments, including thermometers (ordinary, maximum/minimum), a wet and dry bulb thermometer (hygrometer), and, sometimes, a barometer. The screen provides a standardised environment in which to measure temperature, humidity and atmospheric pressure. The traditional Stevenson screen has several characteristic, standardised features:

- **double louvered sides**, which allow air to circulate freely around the instruments but not blow directly on them
- a **standard height for the instruments** in the screen, which is about 1.25 m above the ground
- a **double roof** to provide a layer of air between the two roofs which helps insulate the screen from the heat of the sun
- **a white painted exterior** to reflect the sun's radiation

- positioned **in an open space** to minimise the effects of buildings and trees; siting of the screen is very important
- in the Northern Hemisphere, **the door of the screen should always face north** so as to prevent direct sunlight on the thermometers when it is opened. The **opposite** applies in the Southern Hemisphere.

You may be asked to label or annotate the standard features of a Stevenson screen on a diagram you are given, so make sure you revise this list.

The use of a standardised environment allows temperatures to be compared accurately with those measured in earlier years and at different places around the Earth.

Rain gauges

Traditional rain gauges are made out of non-corroding copper or plastic. They have a number of characteristic features and their location is very important to ensure that they give accurate readings:

- **located away from buildings and trees** which might affect their collection of rainfall
- the top of the rain gauge needs to be over **30 cm high** to avoid surface water running into it and the spray from rain splash entering it as raindrops hit the ground
- the **base of the rain gauge is sunk into the ground** so that it is not easily blown or knocked over
- normally **located on either grass or gravel**, which absorbs the impact of falling raindrops and stops rain splash
- a **collecting funnel** which has a standardised diameter which will be linked to the size of the measuring cylinder used to measure the rain that is collected
- the funnel has a **very narrow opening** to reduce any possible loss of the water that is collected in the collecting cylinder by the process of evaporation
- a **measuring cylinder**, to measure the collected water
- **a flat surface** for the measuring cylinder to make sure the reading can be taken from the **bottom of the meniscus** formed by the water in the measuring cylinder (Figure 11.1).

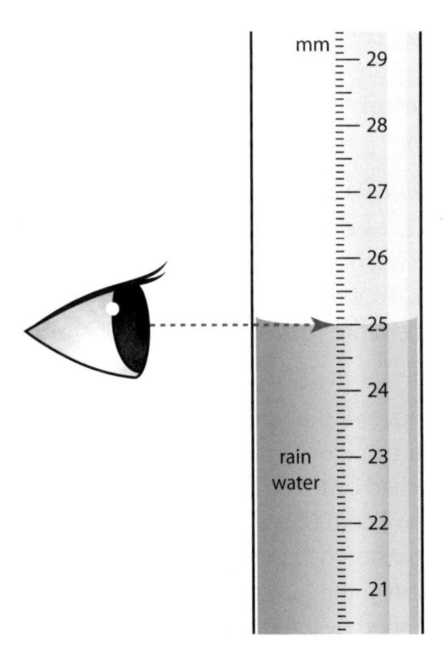

Figure 11.1 Reading the water level in a measuring cylinder; the bottom of the meniscus gives a reading of 25 mm

Once measurements have been taken, the water is then poured away and the rain gauge put back together again. The narrow opening on the funnel is inspected to make sure no obstructions have fallen into it, like leaves or insects. Readings may be taken every hour in some weather stations or, more often, once every day at the same time. Increasingly, remote automatic rain gauges are used to give continuous readings, which are more accurate than hourly or daily readings.

Annual rainfall is normally mapped using isolines – for rainfall, these lines are called isohyets.

Maximum and minimum thermometers

Temperature is measured using a thermometer, containing either coloured alcohol or mercury, which expands and contracts in a glass tube as temperatures rise and fall. A maximum and minimum thermometer (sometimes called a Six's thermometer) is used to measure the hottest and coldest temperatures recorded during a period of time. This is usually one day: a 24-hour period.

They have a number of characteristic features:

- They are made up of a **U-shaped glass capillary tube** with two separate temperature readings: one for the maximum temperature and one for the minimum temperature.

- There are **bulbs at the top of each arm** of the U-shaped tube. The one at the top of the minimum tube contains **alcohol**; the other, at the maximum end, contains a vacuum.
- In the bend of the U is a **section of mercury** which is pushed around the tube by the expansion and contraction of the alcohol in the minimum bulb. It is the alcohol that measures the temperature; the mercury indicates the temperature reading on both the minimum and maximum scales.
- As the mercury moves, it pushes **two small steel markers** which are inside the tube. They record the furthest point reached by the mercury in each side of the tube. When temperature changes, the mercury is moved by the expansion or contraction of the alcohol. The markers remain in the tube at the furthest position they have been pushed by the mercury.
- They can be **reset by using a small magnet**, which is used to pull the steel marker back down to the mercury.
- They record the extremes of temperature experienced by the thermometer since it was last reset.

The instrument is used to measure the diurnal range of temperature. These can then be used to work out the daily, or mean, temperature.

TERMS

diurnal range: the daily range of temperature

relative humidity: the amount of water vapour (moisture) in the air, expressed as a percentage of the maximum amount of water vapour that the air could hold at a given temperature

Wet and dry bulb thermometers (hygrometers)

A wet and dry bulb thermometer (hygrometer) is used to measure the relative humidity of the air. The instrument is made up of two thermometers: a **dry bulb thermometer**, which measures the temperature of the air, and a **wet bulb thermometer**, which measures the temperature of the air if there was 100% humidity – called **absolute humidity**. This bulb is kept wet by being covered by an absorbent muslin wick which is kept wet by its end being kept in a container of distilled water.

FACT

The Six's thermometer was invented by James Six in 1782, and named after him.

Relative humidity can indicate the likelihood of rainfall.

If the humidity of the air is 100%, the readings on both thermometers will be the same – it will normally be raining or there will be mist or fog surrounding the instrument. If it is less than 100%, the wet bulb reading will be lower than the dry bulb reading. By recording the temperatures of both thermometers, it is possible to work out the relative humidity, as a percentage of the air surrounding the instrument. This is done by:

- recording the temperatures of both the wet and dry bulb thermometers
- recording the difference between the two readings
- looking up the results in a relative humidity table.

Modern digital electronic devices have often replaced this instrument and they use the temperature of condensation, changes in electrical resistance and changes in electrical capacitance to measure humidity changes.

Sunshine recorder

A sunshine recorder is used to measure the length of time that the sun has shone during the day in between periods of cloud cover, fog or when the sky has been obscured by smoke or some other form of air pollution. The **Campbell–Stokes sunshine recorder** is one of the most common instruments used to do this. A mounted glass sphere is set up to concentrate the sun's rays so that when the sun shines through the glass, it burns a line on a piece of heat-sensitive paper, which is replaced daily. The strip of paper is calibrated in hours and is held in place in the crescent shaped holder under the glass sphere.

Due to the rotation of the Earth, the sun moves across the sky and the point of burning on the heat-sensitive paper moves with it. This leaves a burnt line along the paper. When the sun goes behind a cloud, or below the horizon when it sets at night, it stops burning a line, so the length of daily sunshine hours can be calculated.

Barometers

Barometers measure atmospheric pressure. Worldwide atmospheric pressure is commonly measured in **millibars (mb)**. The average world atmospheric pressure is 1013.25 mb. Above this, we are said to be in an area of **high pressure**, below it we are in **low pressure**.

TERMS

atmospheric pressure: the pressure exerted on the surface of the Earth by the weight of the atmosphere above it; the force per unit area

knots: units of speed equal to 1 nautical mile per hour, or exactly 1.852 km/hour or approximately 1.151 mph

wind rose: a diagram showing the relative frequency of wind directions at a place

Two types of barometer have been used for measuring atmospheric pressure – a **mercury, or Fortin, barometer** and an **aneroid barometer** (most commonly used). A **mercury barometer** is created by placing the open end of a metre-long glass capillary tube, containing a partial vacuum, in a reservoir of mercury. Mercury is forced up the tube by atmospheric pressure. As atmospheric pressure changes, so will the height of the mercury in the tube – it rises when atmospheric pressure increases and vice versa. An **aneroid barometer** has a small, flexible container inside it which contains a partial vacuum. Small changes in external atmospheric pressure cause the container to either expand or contract. This expansion and contraction moves small mechanical levers and springs so that the tiny movements of the container are amplified and displayed by a needle on the front of barometer. The movements of the needle can then be observed on a scale in either millibars or inches, which corresponds to the height of mercury in the Fortin or mercury barometer. Most aneroid barometers include an extra external needle which is used to mark the current measurement so any changes can be seen.

One form of aneroid barometer that is seen in some weather stations is a **barograph**. This instrument records the changes in atmospheric pressure by means of an ink pen on a small lever attached to the aneroid barometer. It traces a line on graph paper attached to a revolving metal cylinder.

Anemometers

Cup anemometers are usually used in weather stations to **measure the speed of the wind**. (though digital anemometers are available). They must be placed in the open, away from buildings or trees that can alter the wind speed. They will normally be placed on a tall mast on top of a building, or 10 m above ground or building level. They normally consist of three, sometimes four, hemispherical cups attached to horizontal arms, which are mounted on a vertical shaft. The air flow rotates the cups and this can be calibrated on a speedometer. The readings can be in per second (m/s), km or miles per hour (kmph/mph), or, more commonly in meteorology, in knots.

Wind vanes

These instruments are used to **indicate wind direction**. Like the anemometers, they need to be in an open space unaffected by buildings and trees, so that they are normally placed on 10-metre-tall masts on the top of buildings. They consist of a horizontal, freely rotating arm placed on top of a fixed vertical pivot. Attached to the shaft of the fixed pivot are four fixed pointers showing the four points of the compass. This means that wind direction can be read visually. Alternatively, the direction is transferred electronically to an instrument which can give a more accurate reading either in the points of the compass or as a degree bearing. This information can then be transferred to a circular graph to produce a wind rose.

FACT

An enormous range of digital instruments are available to display and record weather data. Some are fixed in place and read temperatures inside and outside, atmospheric pressure, wind speed and direction, and rainfall.

Cloud cover

The amount of cloud covering the sky is measured in **oktas**. One okta is one-eighth cloud cover, that is one-eighth of the sky is covered/obscured by cloud. If half the sky is covered in cloud, there would be four oktas cloud cover; if all the sky is covered in cloud, there will be eight oktas. Figure 11.2 shows how this information is recorded.

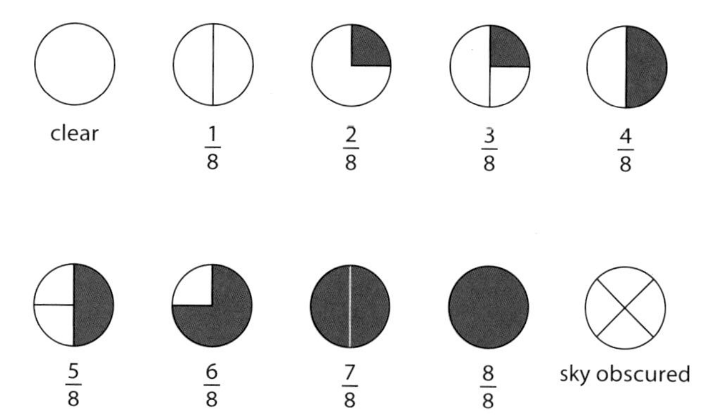

Figure 11.2 Cloud cover, measured in oktas

Types of cloud

The type of cloud and its height can also be recorded with the use of cloud identification charts. The most common types of cloud that are observed are named in Figure 11.3. Note , with their abbreviations to make recording in the field easier.

	name of cloud	abbreviation	height level
high level:	Cirrus – wispy, hair like	Ci	above 5 000 metres
	Cirrostratus – fine layer, possible halo	Cs	
	Cirrocumulus – lumpy layer	Cc	
mid level:	Altostratus – thin layer	As	2 000–5 000 metres
	Altocumulus – lumpy layer	Ac	
low level:	Stratus – uniform layer	St	surface–2 000 metres
	Stratocumulus – undulating or lumpy layer	Sc	
	Nimbostratus – thick rain bearing layer	Ns	
	Cumulus	Cu	
	Cumulonimbus	Cb	from sea level to 12 kms

Figure 11.3 A cloud identification chart

Self-test questions 11.1

1 What is meant by the terms 'weather' and 'climate'?

2 Describe and explain the factors you would take into account in locating a rain gauge and a cup anemometer.

11.2 Interpreting weather data

Today, weather data collected from the instruments in a weather station is usually transmitted directly to a computer in a meteorological office and a software package creates graphs and analyses the data automatically. This digital recording means that more readings can be taken on a 24-hour basis in many more locations. The data is recorded on an internationally agreed standard **weather station model** (Figure 11.4), using standard **symbols** (Figure 11.5) so that information can be shared easily and quickly with other stations and meteorologists worldwide. The individual measurements/values are always in the same position on the station model.

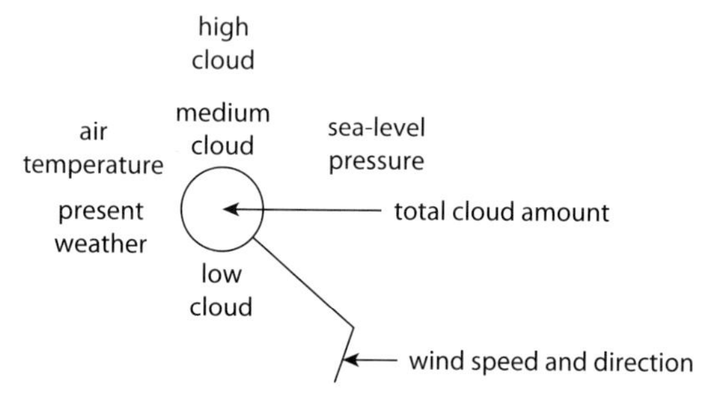

Figure 11.4 Weather station model

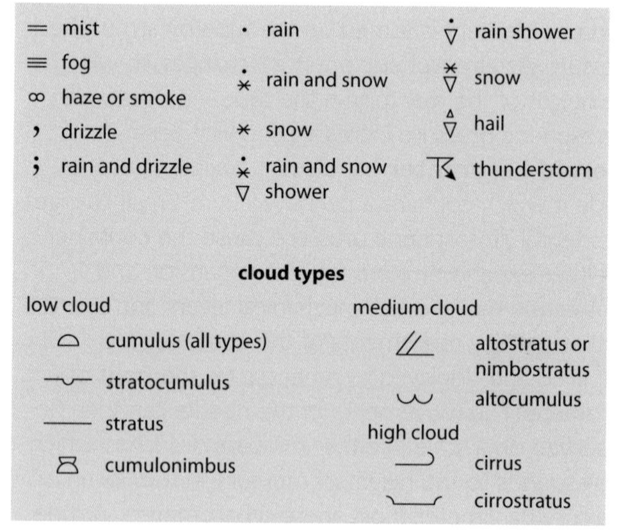

Figure 11.5 Some standard weather symbols for types of precipitation and cloud types

Wind direction and speed

On a weather map, wind direction is shown by a short line, which points in the direction *from* which the wind is coming. In Figure 11.6, the line attached to the circle indicates that the wind is coming *from* the north-west.

Figure 11.6 Showing the wind direction – from the north-west

Wind speed is shown using 'feathers' attached to the end of the line. Each feather is 10 knots of wind speed, half a feather is 5 knots (Figure 11.7).

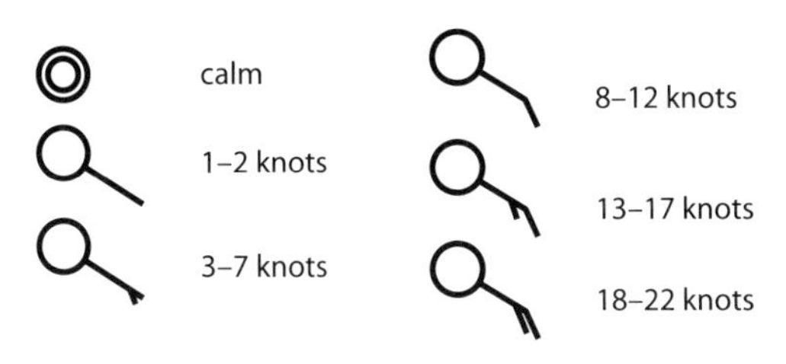

Further half feathers are added up to $4\frac{1}{2}$, which show a wind speed of 45 knots. Then:

Figure 11.7 Showing wind speed

Past and present weather

Past weather, from the previous weather recording, and present weather is also shown on the weather station model. Figure 11.8 shows that it is now raining (present weather), but previously it had been drizzling (past weather).

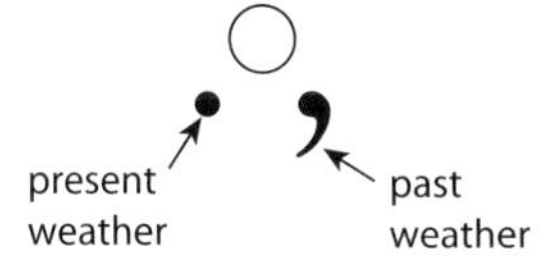

Figure 11.8 Showing past and present weather – drizzling and now raining

Atmospheric pressure

Figure 11.9 shows the symbols that are used to show pressure trends. Two pressure readings are added above and below this symbol (Figure 11.10). The figure above the symbol is the present pressure in tenths of millibars (with the 10 or 9 omitted from in front of it), and the figure below the symbol indicates how much the pressure has changed, in tenths of millibars, since the last reading.

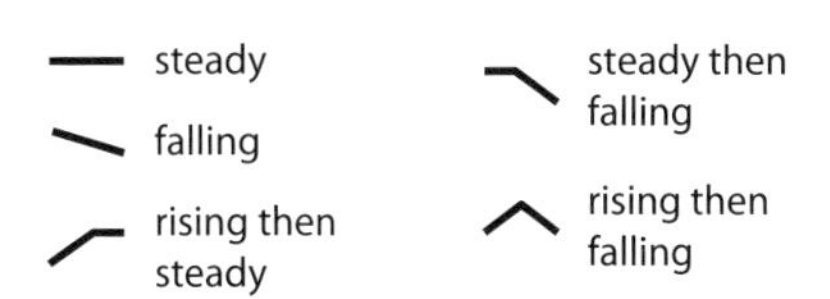

Figure 11.9 Pressure trend

An example of a weather station model

The use of a standard weather station model means that weather data can be shared globally and possible forecasts/predictions can be made. The data can also be stored to build up a picture of the **climate** at this location. When this information is established, any changes in climate over time can be highlighted.

Figure 11.10 is an annotated example of how the weather would be described on a weather station model.

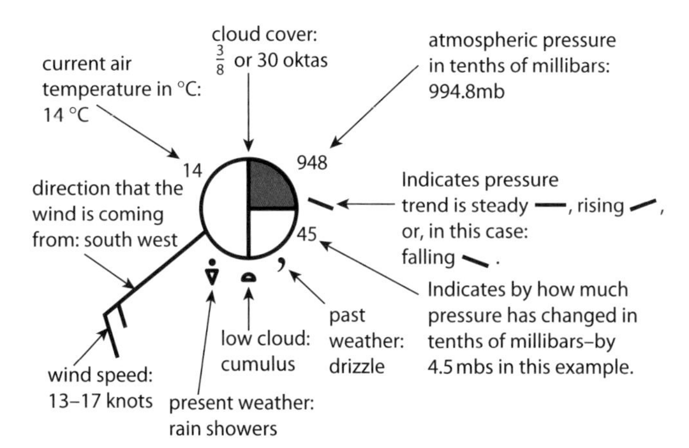

Figure 11.10 An example of a weather station model

Self-test questions 11.2

1 On Figure 11.4, what is the wind direction and wind speed?

2 Make up a station model using the following information:

a 4 oktas cloud cover

b wind direction south

c wind speed 3–7 knots

d snow showers

e current air temperature 16 °C

f cirrus cloud

Sample question and answer

1 a Name the two instruments used to measure wind speed and relative humidity. [2]

b What is meant by the term 'diurnal range of temperature' and how is it calculated? [2]

c Give two reasons why the maximum and minimum thermometer should be placed inside a Stevenson screen. [2]

TIP Ensure that you know the names of the weather instruments, their functions and where they are sited in a weather station.

a Cup anemometer [1] and a wet and dry bulb thermometer (hygrometer) [1].

b The daily range of temperature [1]; calculated by subtracting the minimum temperature reading from the maximum temperature reading [1].

c So that direct sunlight does not fall onto the glass and alter the temperature of the mercury or alcohol inside the thermometer [1] and so that the wind does not blow directly on to the instrument, causing the temperature to be cooled [1].
(A maximum of 6 marks will be awarded).

Exam-style questions

1 Study Figure 11.11, which shows two instruments that are used to measure characteristics of the weather.

a Name the two instruments. [2]

b Give the present and past reading for the instrument on the left of Figure 11.11. [2]

c Explain how you would read and reset each instrument. [6]

2 a Why is the instrument on the right in Figure 11.11 kept in a Stevenson screen? [2]

b Explain the factors that are important in locating a Stevenson screen and the reasons behind its characteristic features. [6]

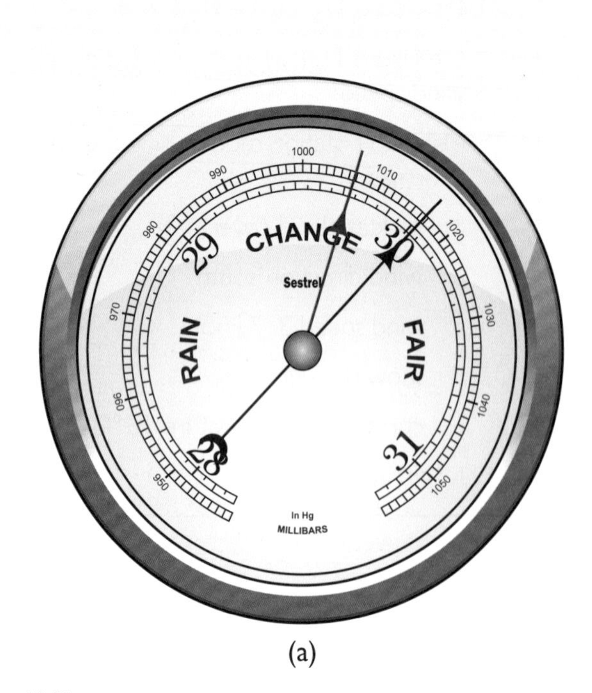

(a)

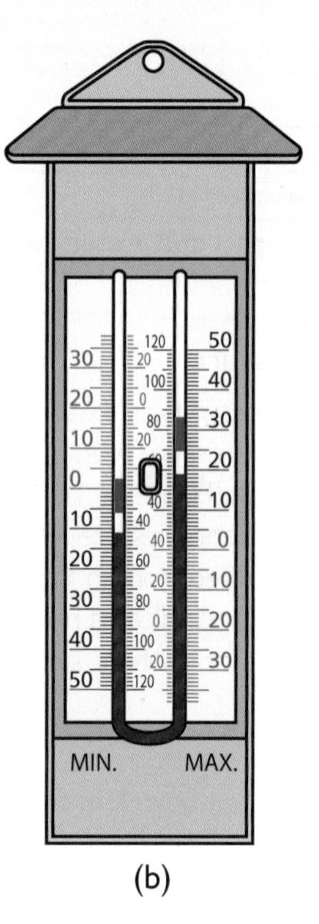

(b)

Figure 11.11

Climate and natural vegetation

Learning Summary

By the end of this chapter, you should be able to:

- describe and explain the characteristics of two climates – equatorial and hot desert – including the factors influencing the characteristics and climatic graphs showing them
- describe and explain the characteristics of tropical rainforest and hot desert ecosystems, including the relationship between natural vegetation, soil, wildlife and climate
- describe the causes and effects of deforestation of tropical rainforest, including the effects on the natural environment (both locally and globally) and on people
- demonstrate knowledge of case studies of an area of tropical rainforest and an area of hot desert.

12.1 Characteristics of equatorial and hot desert climates

The equatorial climate

The equatorial climate has distinct characteristics:

- a **low range of monthly average (mean) temperatures** – normally only a 2–3 °C difference between the hottest and coldest months.
- **constant high temperatures** – an average (mean) monthly temperature of about 26 °C with no month below 18 °C. Daytime highs of about 30 °C dropping to 23 °C at night, giving a daily/diurnal range of about 7 °C.
- a **high rainfall** – 1750 and 2500 mm annual rainfall, **evenly distributed throughout the year**, with the driest month above 60 mm.
- **high humidity** – usually over 80%.

Regions with an equatorial climate are located near the Equator (Figure 12.1). Several factors influence the characteristics of the equatorial climate:

- At these locations, the angle of the **midday sun is near vertical** all year round; the lowest angle the sun reaches at midday is only $66\frac{1}{2}$°. These regions, therefore, experience constant heating of the Earth's surface. The term insolation is often used to describe this heating. The more the insolation, the higher the temperature.
- This heat is transferred from the surface of the Earth to the air above it. The air expands and rises, causing a region of **low pressure** to form along the Equator, giving rise to prevailing winds that blow from the north east in the Northern Hemisphere and the south east in the Southern Hemisphere towards the Equator. These are known as the trade winds (Figure 12.1).
- As the air is warm, it is capable of storing considerable amounts of water vapour and so can produce daily, prolonged, heavy convectional rainfall and frequent thunderstorms in the Equatorial region.

TERMS

equatorial climate: the constantly hot and wet climate of regions near the Equator

insolation: a measure of the amount of solar energy received per square centimetre per minute at the Earth's surface

prevailing winds: the direction from which the wind blows into an area for most of the year

trade winds: the prevailing pattern of easterly surface winds found blowing between the tropics and the Equator

convectional rainfall: this occurs when land is heated up and the warm air above it rises, cools and condenses to give clouds and rain

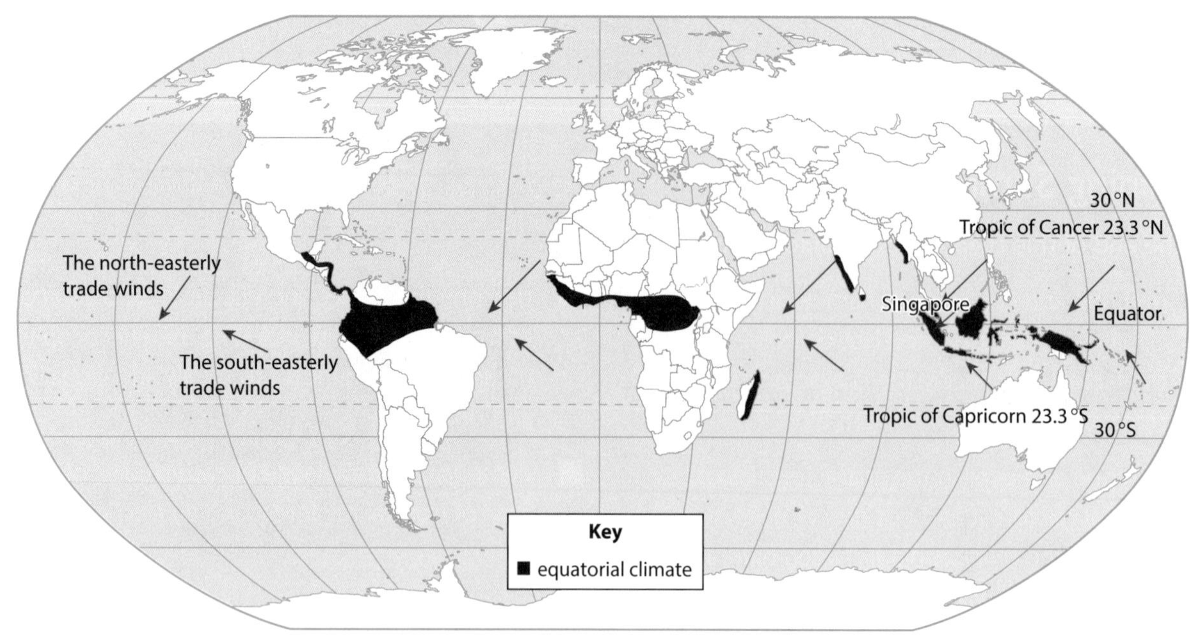

Figure 12.1 Equatorial climate regions and trade winds

- The high levels of **soil moisture** and rainwater lying on the dense vegetation cover leads to high levels of evaporation and transpiration. All this contributes to a repetitive climate pattern of hot, humid air, dry but misty mornings and late afternoon downpours and convectional storms.

The equatorial climate of Singapore

Figure 12.2 shows a climate graph for Singapore, showing average temperatures, rainfall and humidity.

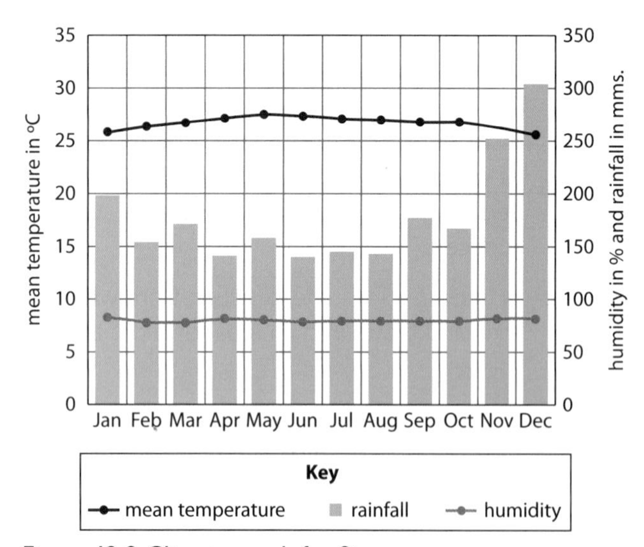

Figure 12.2 Climate graph for Singapore

Singapore lies just one degree or 135 km north of the Equator. Temperatures are at a constant high of around 27 °C all year. The annual range of temperature in Singapore is very small, only about 1.5 °C, which is much smaller than its daily/diurnal range of 6–7 °C. Rainfall is very high (the annual total is 2415 mm) and is evenly distributed throughout the year, with the lowest monthly total being 170 mm in July. Relative humidity is high all year, and the rainfall is often torrential, accompanied by thunder and lightning in late afternoon and evening thunderstorms..

The hot desert climate

The hot desert climate has several distinct characteristics:

- **average (mean) monthly temperatures** of over 29 °C in the hot summer season, but 10 °C in the cool winter season – hot deserts do not have a cold season
- **daytime temperatures** of over 38 °C – temperatures can reach over 50 °C in summer, but are capable of falling as low as 5 °C at night, so they have a large diurnal/daily range of temperature – however, the annual range of temperature is not very high, typically around 15 °C
- **low annual rainfall**, generally less than 250 mm – they are also areas where more water is lost by evapotranspiration than falls as precipitation

- **isolated, irregular rainfall events**, resulting in annual rainfall in some years being less than 80–100 mms – when rainfall does occur it is sporadic, unpredictable and can be torrential, causing potentially dangerous flash floods
- **low humidity**, 25–30%

TERMS

climate graph: a graph showing the average weather for a period of time for a named location. Temperature is shown as a line graph and precipitation as a bar graph

hot desert climate: the constantly hot and dry climate of regions between 15° to 30° north and south of the Equator

latitude: this is used to locate the north–south position of a point on the Earth's surface. Latitude ranges from 0° at the Equator to 90° (North or South) at the poles. Lines of latitude run east–west as circles parallel to the Equator

rain shadow: a region having little rainfall because it is sheltered from prevailing rain-bearing winds by a range of hills or mountains

ocean current: any permanent or continuous, directed movement of ocean water that flows in one of the Earth's oceans

Several factors influence the characteristics of the hot desert climate:

- Most hot deserts are located in latitudes from 15° to 30° north and south of the Equator (Figure 12.3). **Atmospheric pressure is high** as the air is usually descending and therefore warming. Rainfall cannot occur where air sinks.
- Many deserts are a **long distance from oceans**, seas or large lakes which means that they receive little rainfall.
- Other deserts have **prevailing winds** that blow over large areas of land (offshore) and so there is no source of moisture.
- Some deserts are in areas of rain shadow – where moist air is blocked by tall mountains.
- Many of the hot deserts occur on the western coast of continents with a cold ocean current flowing past them (Figure 12.3). Coastal hot desert areas such as the Namib in SW Africa and the Atacama in western South America have climates that are influenced by being located beside the sea and having these cold ocean currents flowing past them. Summer temperatures are cooler as a result – coastal areas of the Namib and Atacama deserts can have the hottest month at 19 °C, and have a low annual range of temperature, often only 5 °C.
- The **relief/topography** of the land can impact on the hot desert climate. Low areas on the inland/lee side of mountains can be extremely dry.

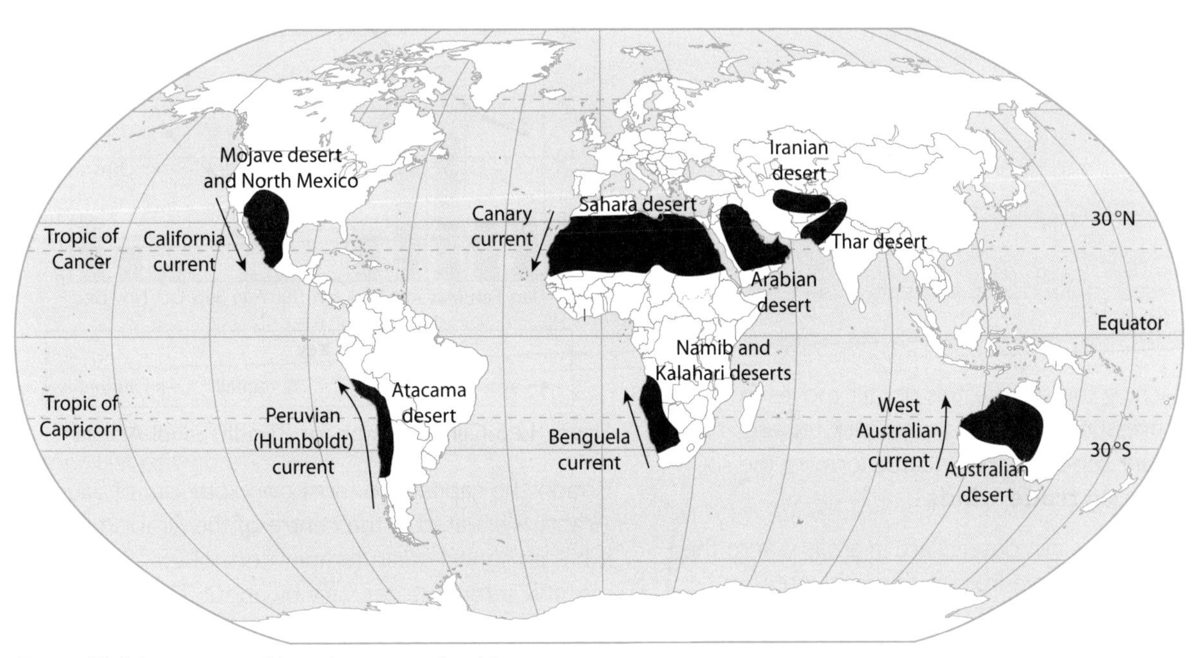

Figure 12.3 Locations of hot deserts and cold ocean currents

TERMS

altitude: the height above sea level

arid: a region is arid when it has a severe lack of available water, which hinders or prevents the growth and development of plant and animal life

Coriolis effect: this causes a deflection in global wind patterns. The anticlockwise rotation of the Earth deflects winds to the right in the Northern Hemisphere and to the left in the Southern Hemisphere

Reasons for the hot desert climate

There is a meteorological reason that hot deserts form where they do:

1 Once the air that rises along the Equator, creating the equatorial climate, reaches the tropopause (the boundary between the atmosphere and space), it then travels north and south towards each tropic. As the air moves northwards and southwards at high altitude, it cools.

2 At about latitudes 30°N and 30°S, the air begins to **descend/sink** towards the Earth, creating **high pressure** at the surface. This descending air is a major cause of the lack of rainfall and high aridity in these regions. Descending/sinking air becomes **compressed** and this compression causes **warming**, which results in a decrease in the relative humidity of the air.

FACT

For rain to occur, the air must be rising, as it does over the Equator, where the rising air cools and condenses to give thunder clouds and heavy rainfall.

Easterly trade winds blow out from the tropics towards the Equator and westerly winds blow out towards the poles.

3 After reaching the surface, this dry air moves from the high pressure, hot desert area back towards the low pressure region at the Equator, forming the so-called prevailing **trade winds**.

Many of the world's hot deserts are in areas where the **prevailing winds** move from continental interiors towards the coast, which means that they are dry and so precipitation is low. North-east Trade Winds blow in the Northern Hemisphere and south-east Trade Winds in the Southern Hemisphere. Their direction results from two controlling factors:

a Winds blow out of high pressure areas and into the low pressure areas (Figure 12.4).

b They are deflected as they do so by the Coriolis effect caused by the Earth's rotation.

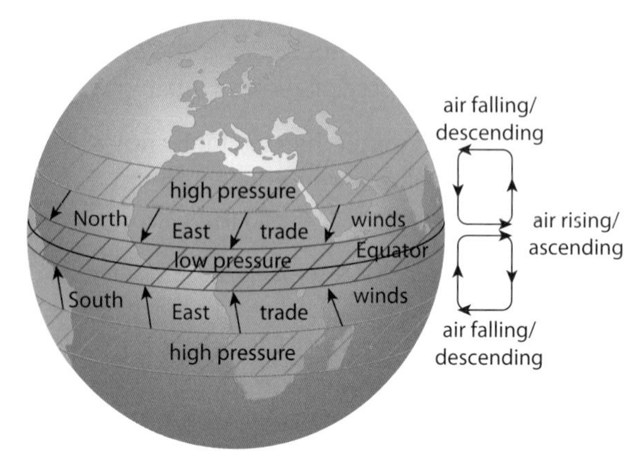

Figure 12.4 Equatorial and tropical air circulation, pressure and winds

The hot desert climate of Riyadh, Saudi Arabia

Figure 12.5 shows a climate graph for Riyadh in Saudi Arabia, showing average temperatures, rainfall and humidity.

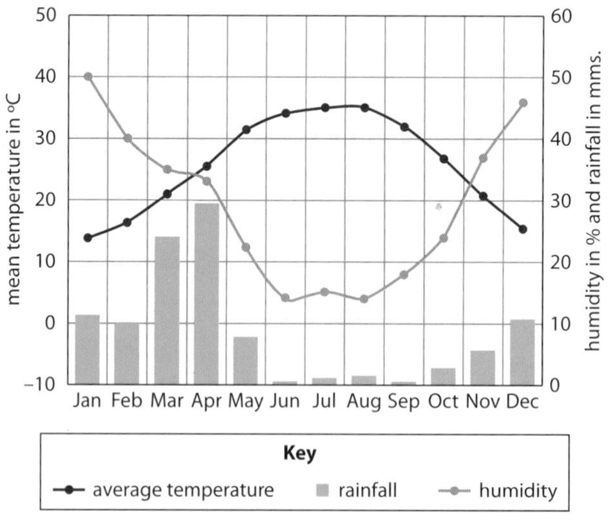

Figure 12.5 Climate graph for Riyadh, Saudi Arabia

Riyadh, the capital and most populous city of Saudi Arabia, is situated in the centre of the Arabian Peninsula. It has an annual mean temperature of 25.7 °C and mean monthly temperatures vary by about 21 °C. The mean maximum temperature in August is 45 °C. Its winters are warm with cool, windy nights. Riyadh experiences very little rainfall, especially in summer, but receives over half of its total annual rainfall of 101 mm in the two months of March and April. It also experiences dust storms from the surrounding desert, during which the dust can be so thick that visibility is under 10.

Self-test questions 12.1

1 Climate graphs may be included in an examination paper and to answer the questions, you may need to accurately read off information. Using Figure 12.5 showing the climate graph for Riyadh, Saudi Arabia, answer the following questions.

a In which months of the year does the hot season occur?

b Which month has the most rainfall and what is its value?

c Which months have the least rainfall?

d Which month has the warmest average temperature and what is its temperature?

e Which month has the coldest average temperature and what is its temperature?

f Describe two factors that influence the characteristics of the hot desert climate.

12.2 Characteristics of tropical rainforest ecosystems

Tropical rainforest (TRF) ecosystems are usually found between about 15° north and south of the Equator. **Climate** is the most important factor in determining where TRFs can be found – they are usually found in regions of **equatorial climate** (see Section 12.1). There are three main areas where they are found:

- **Amazonia** (which includes northern South America – mainly Brazil and Central America). Amazonia represents over half of the Earth's remaining rainforests, and it makes up the largest and most biodiverse tract of TRF in the world (covering 5 500 000 square kilometres), with an estimated 390 billion individual trees divided into 16 000 species.
- **Central Africa** (which includes both Central Africa and southern West Africa)
- **Indo-Malaysia** (which includes Indonesia, Malaysia and several other countries in SE Asia and northern Australia)

FACT

There are several other smaller areas outside these main three areas, such as Madagascar and many of the Pacific Islands.

TERMS

ecosystems: communities of living (biotic) and non-living things (abiotic) interacting with each other in an area

biodiverse: describes the variety of organisms present in different ecosystems

adaptations: changes and mechanisms that help organisms survive in their ecological habitat

TRFs are **home to more species than all other ecosystems combined** – 82% of the world's known plant and animal species are found in tropical rainforests. This includes over one-third of the world's tree species.

For the TRF ecosystem to develop, a number of natural/physical factors combine together with climatic factors. For example, the constant high temperatures and rainfall results in a **constant growing season**, which means that trees and plants grow all year round and therefore drop their leaves all year round. This **leaf litter rapidly decomposes** and is **recycled as nutrients** for the trees and plants to grow.

The structure of the tropical rainforest

The trees, plants and animals have many special adaptations to living in this wet tropical environment. Tropical rainforests have five distinct layers (Figure 12.6), each with its own characteristic features and adaptations.

TIP

You may be asked to describe the characteristics of the natural vegetation in a tropical rainforest, so make sure you know these.

Emergent layer (45–55 m)

This contains a small number of very tall, large trees which grow above the general canopy. Their trunks are straight and, in their lower sections, branchless as they attempt to get to the light above the canopy. The trees need to be able to withstand hot temperatures and strong winds. As a result, many have large **buttress roots** for support. The tall emergent trees often have many other plants growing on them like **lianas** (which are like vines and they use the tall trees as support to get up to the light above the canopy) and **epiphytes** (which grow on the trunks and branches of the trees and trap water and dead leaves falling from the tree or running down their trunks which become the nutrients for the plant).

FACT

Eagles, butterflies, bats and certain monkeys, such as the orangutan and the proboscis monkey in Borneo Malaysia, inhabit this layer. All are threatened due to habitat loss, especially through the clearance of forest for oil palm and human habitation.

Canopy layer (30 m)

This forms a roof over the three remaining layers. Most canopy trees have smooth, oval leaves that come to a point, called a **drip tip**, to get rid of the rain that falls on the leaf, which could encourage the growth of mould on the leaf. The canopy is a thick, dense layer of leaves and branches. Very little light gets through this layer to the layers below. The canopy, by some estimates, is home to 50% of the Earth's plant species. Many animals live in the canopy as food is abundant, including snakes, parrots, toucans, sloths and tree frogs. The canopy often supports a rich flora of epiphytes, including orchids, bromeliads, mosses and lichens.

Under/sub canopy layer (15 m)

Little light reaches this layer so the plants often have large leaves, up to 2 metres across, to catch the sunlight. There is usually a large concentration of insects here.

Shrub layer (3–4 m)

It is very dark in this layer – only about 5% of the sunlight shining on the rainforest reaches down to this layer. Tree ferns and small shrubs grow here in low light conditions. Many have large leaves to catch the little light that gets down to this layer. Many animals live in the shrub and under canopy layers, including jaguars, red-eyed tree frogs and leopards.

Floor/ground layer (0 m)

As less than 1% of the sunlight makes it down to this layer, almost no plants grow in this here. Due to the lack of sunlight reaching the forest floor and the warm, humid conditions, things begin to decay quickly. The forest floor is often completely carpeted by leaves and these are quickly recycled into nutrients and taken up by the trees and plants. The soil gives the impression of being extremely fertile but in fact almost all of its fertility comes from the thin layer of rapidly recycled leaf litter. Therefore, if the rainforest is removed by the process of deforestation, the soils stops receiving dead leaves and plants and it quickly, within a couple of years, turns into infertile soil on which little will grow. Giant anteaters live in this layer.

TERMS

deforestation: permanently removing forest so the land can be sued for something else

indigenous people: people native to an area

slash and burn: a form of agriculture where the natural vegetation is cut down and burnt as a method of clearing the land for cultivation

Indigenous peoples

The TRF is the home for many indigenous people, who are **hunter-gatherers** – people who hunt animals and gather food from the forest – fruit, berries, seeds and roots. These peoples live in, and totally depend on, the rainforests, so that complete cultures may be lost, as they are forced to migrate and start to live new lives.

Some of the indigenous peoples are subsistence farmers as well as hunter gatherers. They clear, and then burn, small patches of the forest to grow crops in. This is called slash and burn agriculture and was thought in the past to be sustainable (unlike the plantation agriculture which can only survive with addition of artificial chemical fertilisers). However, even this small-scale farming causes the soil to lose its fertility over time and it does not recover fully even when left to rest for several years.

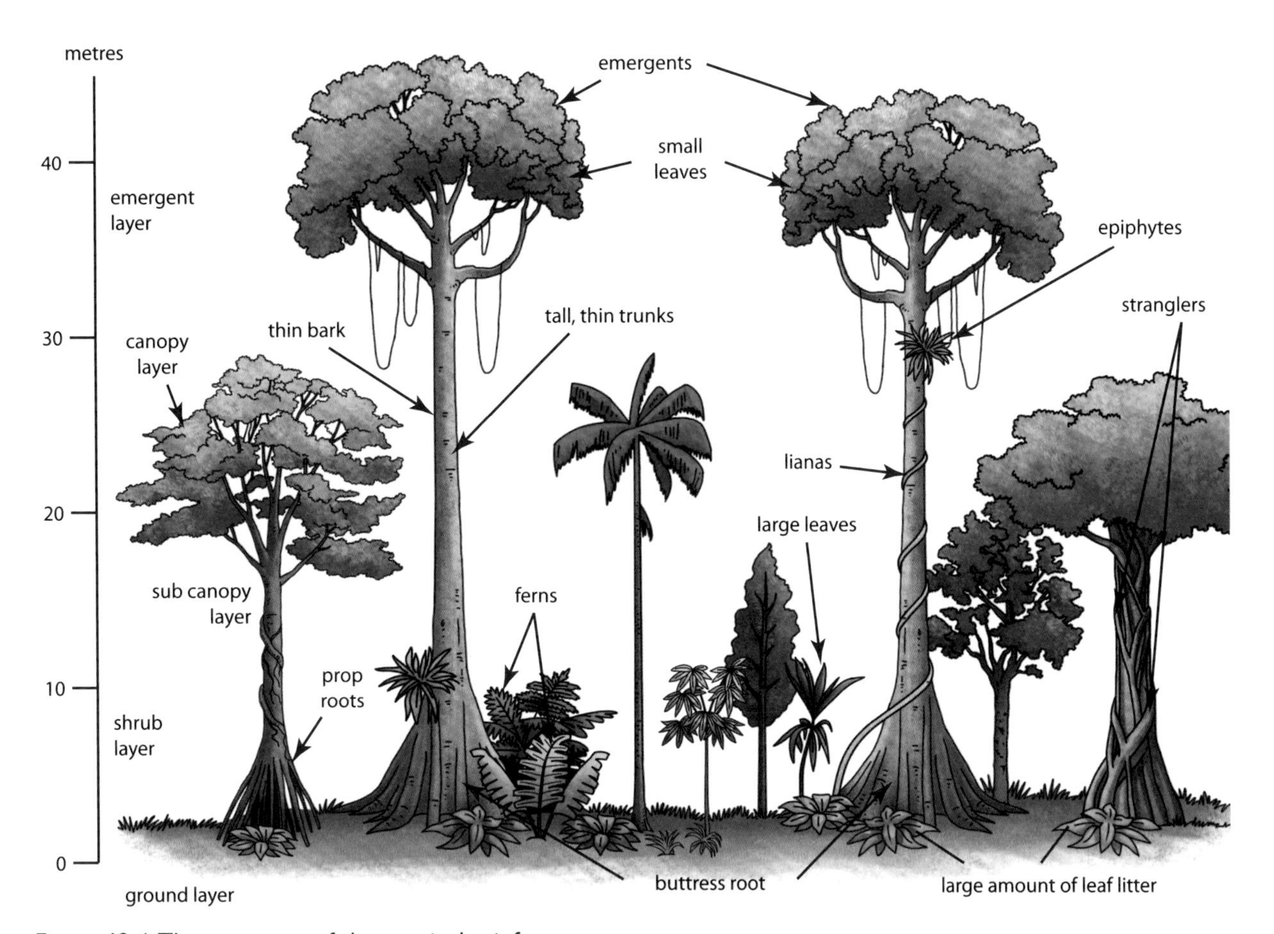

Figure 12.6 The structure of the tropical rainforest

12.3 Deforestation in the tropical rainforest

The three main areas of TRF – Amazonia, Central Africa and Indo-Malaysia – are almost all in countries which have experienced rapid population growth and development. This has meant that more land is needed for settlements, for farming to feed the increasing numbers of people, to provide employment and to exploit the resources needed for development. The area of clearance of the forest in Amazonia has been estimated at between 20 and 40% – at a rate of about 15 hectares a minute! In Malaysia, 70% of the Malay peninsula was covered in rainforest in 2000 but, at present rates of clearance, this could be as low as 25% by 2020.

Deforestation happens for many reasons, and each has an impact on the environment and people:

- Where the rainforest is **logged for its valuable timber**, such as mahogany and teak, it means that the habitat for thousands of species of plants and animals are destroyed. This destroys delicate food webs and food chains and the animals disappear, often to become extinct. It also takes away the habitat for the indigenous peoples.
- **Plantation agriculture** where the forest is cleared to create huge farms for the growing of plantation crops such as sugar cane and oil palms – both now in great demand as biofuels. Malaysia has cleared large areas of TRF and is now the world's biggest exporter of palm oil.
- **Cattle ranching** to meet the growing demand for beef and burgers from HICs and MICs in particular.
- **New settlement** to provide land for small-scale farmers. The Brazilian government has used just over 5% of its rainforest to provide land for some of the country's 25 million landless people. Alongside some stretches of the 12 000 km of new roads built through the rainforest, strips of land 10 km wide have been cleared to provide new settlers with land to farm.
- **Mining** – the TRFs often have important deposits of valuable minerals under them such as gold, coal, iron ore and bauxite (the ore from which aluminium is made). Many of these mines are some of the biggest

in the world and are open-cast – where the soil and forest are removed and the minerals are removed. The area is totally destroyed by the mining activity.

- **HEP, dams and reservoirs** – some of the world's largest rivers flow through TRFs and they are an important source of energy (HEP), which requires dams. Large areas of TRF have been deforested to make way for some of the world's biggest dams and reservoirs and this has had a major impact on the ecology of the areas they have flooded.
- **Cultivated foods and spices** – many of the fruits and nuts we eat and buy in supermarkets, such as coffee, chocolate, banana, mango, papaya, macadamia nuts, avocado, and sugar cane all originally came from TRFs. Many are grown on plantations in cleared areas of rainforest. However, without the forest, there are no supplies of soil nutrients, so many of these crops rely on the heavy use of artificial chemicals.
- **Climate change and carbon sinks** – TRFs play an important part in regulating climate and the gases that make up the atmosphere. Deforestation of trees means that there is less evapotranspiration so there is less water vapour in the atmosphere above rainforests, which means that there is a reduction in rainfall in the areas of TRF and this can increase the threat of droughts. As the world produces more carbon dioxide, the rainforests are responsible, through the process of photosynthesis, for turning the carbon dioxide into oxygen.

FACT

It is thought that tropical rainforests may provide one-third of the Earth's fresh oxygen supply.

- **Loss of indigenous people** – many of these activities have had a great impact on the traditional societies and people of the TRFs. In Amazonia, the indigenous population before European settlers arrived is estimated to have been 6 million. By 2003, this had fallen to 200 000.

Self-test questions 12.2

1 Describe two factors that influence the distribution of tropical rainforests.

2 Describe and explain the impacts of human activities on the natural environment of a named area of tropical rainforest that you have studied.

12.4 Characteristics of hot desert ecosystems

Deserts are classified in two main ways:

1 According to the **amount of precipitation and the rates of evaporation** (calculated as potential evaporation). In this classification, there are two types:
   - **extremely arid areas**, which have at least 12 consecutive months without rainfall, e.g. the Atacama Desert
   - **arid areas**, which have less than 250 mm of annual rainfall, e.g. the Namib desert in south west Africa ranges from 2 mm in its most arid areas to 200 mm in its wettest areas.

2 Whether they are **sandy** (**ergs**) or **rocky** (**regs**) – most people think of deserts as being vast areas of sand but, in fact, sand covers only about 20% of the Earth's deserts. Nearly all the reg deserts have had their fine sand blown away. This is a process called **wind deflation** (in some books it is called **aeolian deflation**).

TERMS

wadi: the bed or valley of a stream in regions of south-western Asia and northern Africa that is usually dry except during the rainy season

oasis: an isolated area of vegetation in a desert, typically surrounding a spring or a similar water source, such as a pond or small lake. They also provide a habitat for animals and people if the area is big enough

xerophytes: a species of plant that has adapted to survive in an environment with little liquid water

Deserts often contain other common features such as river deposits – mainly small dried up deltas called alluvial fans and dried up lakes called playas or river channels and beds, sometimes called wadis, and wetter areas where the water table comes to the surface, called oases.

Desert vegetation

Deserts typically have a very sparse vegetation cover but it is very diverse. More typical plants are small bushes or shrubs. Most desert plants have to be drought and/or salt-tolerant. These plants are called xerophytes. They adapt in four main ways to the lack of water:

1 Some store water in their leaves, roots, and stems – like the prickly pear cactus.

2 The trees that are found in hot deserts, such as the camel thorn tree, have long tap roots that may reach down as far as 60 metres to the water table.

3 Wide-spreading shallow roots just below the ground surface to absorb water when it rains before it quickly evaporates, from a larger area of the ground.

4 The development of small, spiny leaves which give off less moisture than leaves with greater surface areas.

The stems and leaves of these plants play another important role in that they reduce the speed of sand-carrying winds and therefore protect the ground from erosion.

The animals found in the hot deserts have to adapt to the lack of water and often migrate long distances to find available food. Many are grazing herbivores – and where there are herbivores, further up the food chain are the carnivores.

Human life in deserts

A desert is a hostile, potentially deadly environment for unprepared humans. In hot deserts, high temperatures cause rapid loss of water due to sweating, and the absence of water can result in dehydration and death within a few days. In addition, unprotected humans are also at risk from heatstroke. Humans may also have to adapt to **sandstorms**. Despite this, some cultures have made hot deserts their home for thousands of years, including the bushmen of the Kalahari, the Bedouin in the Arabian peninsula, the Tuareg and Pueblo people. Modern technologies, including advanced irrigation systems, desalinisation and air conditioning, have made deserts much more hospitable.

Most traditional human life in deserts is **nomadic**. Their life depends on finding water, surviving frequent droughts and on following infrequent rains to obtain grazing for their animals. Any permanent settlement in deserts requires permanent water and food sources and adequate shelter, or the technology and energy sources to provide it. Larger settlements are only usually found where these conditions exist, such as in the oil-rich economies of Kuwait, the Gulf States and Saudi Arabia.

Overcoming desertification

To help overcome desertification, trees are being reintroduced to the desert to stabilise the sand and act as wind breaks. The choice of trees and shrubs is extremely important – they need to be able to survive, be of low maintenance and have several other uses.

TERM

desertification: a type of land degradation in which a relatively dry area of land becomes increasingly arid, typically losing its sources of water as well as its vegetation and wildlife. It is caused by a variety of factors, such as through climate change and through the overexploitation of the land and its water resources

The **ghaf tree** is an indigenous species, specifically to the arid desert areas of the UAE, Oman and Saudi Arabia. It is a drought-tolerant, evergreen tree. The tree taps groundwater with its roots, penetrating as deep as 30 metres underground to access it. Because of its extensive root system, it stabilises shifting sand as well as acting as a windbreak. It also fixes atmospheric nitrogen and adds organic matter through leaf litter decomposition, rejuvenating poor soils. It provides much needed shade and shelter to the farmers working in the fields as well as to the camels, sheep, goats and wildlife during the summer months.

The flowers, fruits, leaves, bark, branches and roots of the ghaf tree, all provide resources and habitat for a variety of native fauna and flora, making the tree a keystone species – one that plays an integral part of the food chain in an ecosystem. For example, many birds build nests on the ghaf tree.

The ghaf tree also provides wood for use in construction and house-building. It is much valued as a fodder tree. The leaves are of high nutritive value and, on average, yields of 60 kg of green forage can be obtained from a full grown tree. It is one of the most important feed species for desert livestock, contributing a major proportion of the feed requirements of camels, cattle, sheep and goats. It has a seed pod yield of nearly 14 000 kg/km^2 and these too are eaten by the livestock.

Self-test questions 12.3

1 Describe two factors that influence the distribution of hot deserts.

2 With the use of a case study, describe the threats posed by people in a hot desert to its natural ecosystem.

Case studies

An area of tropical rainforest – Sinharaja, Sri Lanka

Sri Lanka highlights some of the major issues faced by the world's TRFs. It is estimated that only around 1.5% of the country's original forest remains.

The Sinharaja rainforest lies in the south west of the country and what remains is now protected in the Sinharaja Forest Reserve. It is also now a designated National Park and a biodiversity hotspot. The reserve is only 89 km^2 in area, but it has a rich mix of endemic species, including trees, insects, amphibians, reptiles, birds and mammals.

The threats to the reserve are typical of those faced by TRFs globally. There are four main threats:

1 **Habitat change and loss** as a result of the expansion of human settlements, agricultural land, gem mining and related infrastructure, such as road networks.

2 The **over exploitation of the flora and fauna species** by the killing of animals for bush meat, the killing of elephants to supply the ivory trade, the export of ornamental fauna and flora, timber felling/illegal logging and the illegal trading of economically valued species.

3 **Invasive alien species** – species whose introduction and/or spread outside their natural habitats threatens biological diversity. Many of these are non-native species that, in the absence of their natural predators found in their countries of origin, have multiplied dramatically, often at the expense and extinction of native species.

4 **Pollution and climate change** resulting in the extinction of endemic/indigenous animals and plants, an increasing potential for **forest dieback** – where trees and plants are put under increasing pressure/stress, become weaker and are more prone to disease and fungal attack.

Of all the threats to the protection of Sinharaja, it is the **socio-economic threats** relating to the people and organisations found in the immediate vicinity of the reserve that are perhaps among the most important. **Land being cleared** by local people for farming and cultivation is the biggest problem. These are closely followed by licenced timber contractors who open up routes and roads to facilitate their logging operations on the edge of the reserve which, unfortunately, makes the reserve more accessible to illegal timber operations. **Illegal gem mining** is also a serious problem, organised by wealthy gem dealers from outside the Sinharaja region.

The future protection of the rainforest

A **conservation plan** has been set up between the Sri Lankan government and the international non-government organisation (NGO) the International Union for Conservation of Nature (IUCN). To protect the reserve, a scheme of **zonation and management** has been set up for specified areas, both inside and outside the reserve, to provide essential forest products for sustained use and to meet the needs of the local people and eliminate their former dependence on resources within the reserve.

Other strategies include:

- Establishing a 3.2 km-wide **buffer zone** round the reserve, protecting the core area and using the buffer zone for various uses, such as sustainable subsistence farming and the harvesting/gathering of rainforest plant products such as fruit, leaves and nuts and seeds
- Relocating illegal settlements and villages to areas outside the reserve.

An area of hot desert – the Namib Desert, South West Africa

The Namib is a coastal desert in southern Africa. The name Namib means 'vast place'. It stretches for more than 2000 km along the Atlantic coasts of Angola, Namibia and South Africa, covering an area of almost 81 000 km^2. It is almost completely uninhabited by people except for several small settlements and nomadic pastoral groups. It contains some vast sand dune areas, with some of the tallest and most spectacular dunes in the world, along with rocky desert. The dry climate of the Namib reflects the almost complete lack of rivers or bodies of water on the surface.

The aridity of the Namib Desert is caused by the descent of dry air in the tropics, causing high pressure. It often receives less than 10 mm of rain annually. Rainfall is both scarce and unpredictable due to several factors:

- Winds coming from the Indian Ocean to the east lose part of their humidity when passing the Drakensberg mountains, and are essentially dry when they reach the desert.
- Winds coming from the Atlantic Ocean are prevented from travelling inland by winds moving towards the coast from the area of high pressure over the desert. Morning fogs coming from the ocean and pushing inwards into the desert are a regular phenomenon along the coast, and much of the life cycle of animals and plants in the Namib relies on these fogs as the main source of water.

A number of plant and animal species are found, highly adapted to the arid climate of the area. The Namib Desert beetle has an outer hard skin that causes humidity from the morning fogs to condense into droplets, which roll down the beetle's back into its mouth. Another beetle builds 'water-capturing' webs, while black-backed jackals lick humidity from stones. Gemsboks, a large antelope, can raise the temperature of their bodies to 40 °C in the hottest hours of the day.

Economically, the Namib Desert is an important location for the mining of tungsten, salt and diamonds, while the Namib-Naukluft National Park, covering 50 000 km^2 extends over a large part of the desert. It is the largest game reserve in Africa and one of the largest in the world and has several well-known visitor attractions. It now provides employment in tourist accommodation, guides and rangers.

Threats to the natural environment

- The **impact of off-road driving**, which can cause long-lasting damage to the delicate desert vegetation. Lichens are particularly sensitive to mechanical damage as they grow extremely slowly and cannot quickly recover. Most of the damage is done by vehicles from mining companies on prospecting expeditions.
- The **drop in the water table** caused primarily by the extraction of groundwater, which supplies the domestic consumption of the urban areas of Walvis Bay and Swakopmund and the enormous demands made by a uranium mine near Swakopmund. At present, they are trying to meet the requirements by prospecting for more underground water sources. However, if water were to be found, roads, pipelines, and power lines would have to be constructed through the most pristine dune desert in the world.
- **Pastoralists** who graze large herds of goats and small groups of donkeys. The livestock have **overgrazed** some areas and are competing for food with wild animals, such as gemsboks.

Sample questions and answers

1 What is meant by the term 'ecosystem'? [1]

2 Describe the distribution of tropical rainforests. [2]

3 For a named area of tropical rainforest which you have studied, describe and explain the characteristics of its climate. [7]

TIP

Make sure you know the meaning of the command words in questions. For 'describe', say what something is like or where it is. For 'explain', you need to write about why something occurs or happens.

1 An ecosystem is a community of living and non-living things interacting with each other in an area. [1]

2 Tropical rainforests are found close to the Equator, between about 15 degrees north and south of the Equator [1]. They are found in the Amazon in South America, in South East Asia and in West Africa [1].

3 A tropical rainforest like Sinharaja rainforest in Sri Lanka [1] has a very distinctive climate with high temperatures averaging about 30 °C for most of the year [1], and a small annual range of temperature. There is also very little difference, about 2–3 °C, in the monthly average temperatures [1]. This is because this area is very close to the Equator and the sun is overhead all year round [1]. It has a high rainfall of about 2000 mm [1], spread evenly through the year [1]. The heavy rainfall is due to the region having very hot temperatures during the day, producing strong convection currents which cause the formation of convectional rain clouds [1] and these result in heavy rain every day. (A maximum of 10 marks will be awarded).

Exam-style questions

1 Explain why either a named area of tropical rainforest or hot desert is under threat of clearance. [6]

2 Discuss how a named area of threatened tropical rainforest is now being managed more sustainably. [5]

3 Describe and explain two consequences of deforestation. [6]

THEME 3: ECONOMIC DEVELOPMENT

Development

Learning summary

By the end of this chapter, you should be able to:

- use a variety of indicators to assess the level of development of a country, including GNP per capita, literacy, life expectancy and composite indices
- identify and explain inequalities between and within countries
- classify production into different sectors (primary, secondary, tertiary and quaternary) and give illustrations of each
- describe and explain how the proportions employed in each sector vary according to the level of development, using development indicators and employment structure to compare countries at different levels of economic development and over time
- describe and explain the process of globalisation, the role of technology and transnational corporations, and the impacts of globalisation at a local, national and global scale
- demonstrate knowledge of a case study of a transnational corporation and its global links.

13.1 Indicators of development

In Geography, **development** can be defined as the process of change which allows all the basic needs of a country or region to be met, thereby achieving greater social justice and quality of life and encouraging people to fulfil their potential.

For an explanation of some of the terms used in this chapter, such as GDP and GNI as well as HIC, MIC, and LIC, see Chapter 1, Section 1.1.

How development is measured

Traditionally, the term development is linked with economic growth and it is seen as the process by which countries and societies advance and become richer. International organisations and national governments have looked at ways to encourage the economic growth of poorer nations to enable them to catch up and narrow the development gap with richer nations. Traditional measures of development focused on economic data, to show rates of economic growth and the acquisition of wealth, using indices such as **GNP** (Gross National Product), **GDP** (Gross Domestic Product) or **GNI per capita** (Gross National Income per person).

The difference between GNP and GDP is that, although both reflect the national output and income of an economy, GNP includes the value of all goods and services produced by nationals whether in the country or not, while the GDP is a measure of national income/national output *within* a particular country. GNI is the total value of goods and services produced within a country together with the balance of income and payments from or to other countries. GNI is increasingly becoming the preferred economic indicator for many international organisations. Figure 13.1 shows GNI per capita and highlights the stark income differences between HICs and LICs. All three indicators are frequently referred to as 'per capita' as the totals for GNP, GDP and GNI are divided by the population of a country or region. GNP is nearly always calculated on a per capita/per person basis so that differences in the population size of countries are neutralised.

Development, however, is a much more complex and wide-ranging process than pure economic outputs. It involves a combination of factors, including cultural, economic, environmental, political, social and technological change. It is also important to recognise that development has social and quality of life implications and measures such as life expectancy, education level and access to sanitation are important.

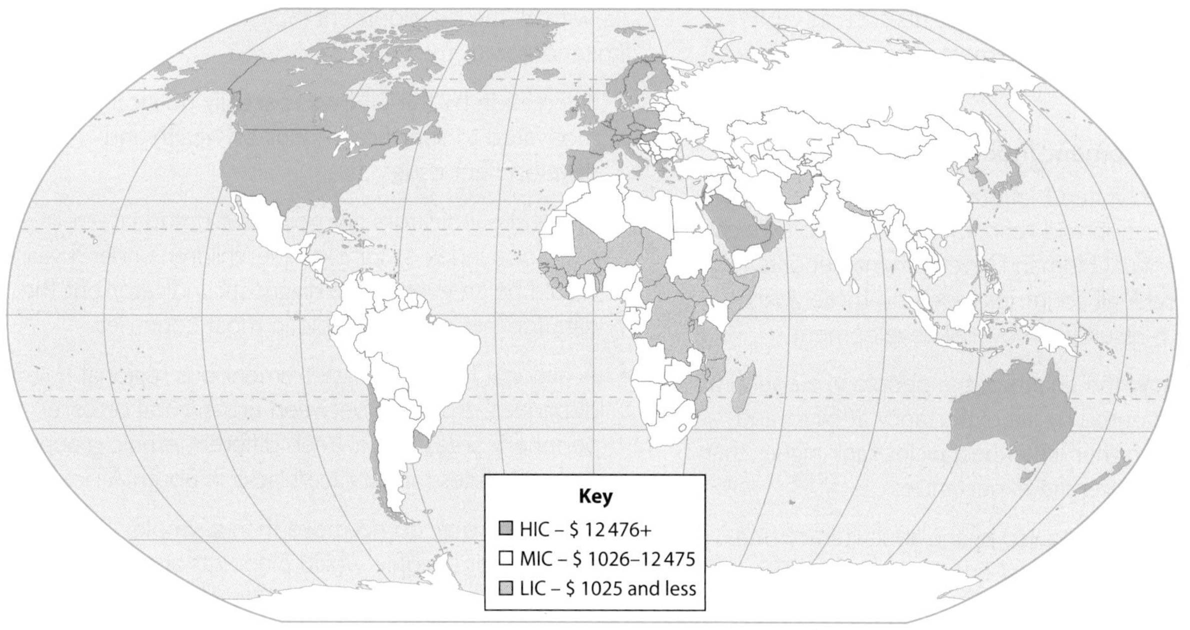

Figure 13.1 GNI per capita, 2017, in USD($)

In trying to actually measure development, we need to look at factors which affect the quality of life of a person, region or country. These include:

- **social indicators** such as health, literacy/education and well-being
- **political factors** such as people's personal freedom, freedom of speech, the right to vote, freedom from discrimination and the role of disadvantaged groups in society.

Such qualitative aspects of development become increasingly difficult to measure using numerical or quantitative measures.

It is therefore common for development to be assessed and analysed using a range of variables (sometimes called **multivariate analysis**). The measurement of a range of development variables, not just economic growth, allows further analysis of what has been achieved by a country. Comparisons can be made between countries to identify disparities at a continental or global scale. This has led to global concepts such as the North–South divide and the development gap which have attempted to highlight the differences between the rich 'North' and poor 'South' hemispheres.

TERMS

development gap: the differences between the most and least advanced countries – the HICs and the LICs

North–South divide: a socio-economic and political divide between the North – including the USA, Canada, Western Europe, developed parts of Asia, and Australia and New Zealand, which are not actually located in the geographical North but share similar economic and cultural characteristics – and the South (including Africa, South America, developing Asia and the Middle East)

index (plural indices): pointers or indicators

FACT

It is worth noting that, in more recent years, the picture has got more complicated, with the rise of BRIC (Brazil, Russia, India and China) and MINT (Mexico, Indonesia, Nigeria and Turkey) economies.

Recognising the complex nature of development is the reason development is often measured using an index, which combines a range of data. Indices are considered more accurate than single data points such as GNP/GDP per capita. One of the most common of these measures is the Human Development Index (HDI).

Human Development Index (HDI)

Since 1980, the United Nations (UN) has been working on the construction and refinement of the HDI which it uses in its annual Human Development Reports. HDI attempts to rank all countries based on three goals/outputs which result from overall development:

1. **Longevity:** the health of the people in a country. This is measured by life expectancy at birth, and those with higher life expectancies rank higher than those with lower life expectancies.
2. **Knowledge:** measured by the adult literacy rate combined with the gross enrolment ratios of students in primary school through to university level.
3. **Income:** adjusted to measure real per capita income, including purchasing power adjusted to local cost of living. This dimension is measured with the GDP per capita in purchasing power parity (PPP) terms.

In order to accurately calculate each of these dimensions for the HDI, a separate index is calculated for each of them based on the raw data gathered.

TERM

purchasing power parity (PPP): GDP per person adjusted for the local cost of living

trade blocs: groups of countries form economic agreements and unions

Using this information, it is then possible to rank countries (on a scale between 0 and 1) into groups and so carefully monitor changes in countries over time, either as they develop or deteriorate. Over half of the world's population live in countries with 'medium human development' (51%), while less than a fifth (18%) live in countries falling into the 'low human development' category. Countries with 'high' to 'very high' human development account for slightly less than a third of the world's total population (30%).

FACT

Norway, with an HDI of 0.944, has topped the global country list since 2015, while Niger is at the bottom, with an HDI of 0.348.

As with all the development indicators/indices, there are limitations with the HDI:

- Its creation by the UN was partially politically motivated to specifically focus on health and development issues.
- The three indicators are good but not ideal – a nutritional index of, for example, children under 5 years, would be an even more diagnostic indicator but the data for this is not available in most countries.
- A national HDI can mask tremendous regional disparities, especially between urban/rural or core/periphery areas or between different ethnic groups such as whites/mixed-race/blacks in South Africa.
- This index may do no more than a simple development profile which pinpoints anomalies between economic and social development in a more standardised way.
- The index is one of relative not absolute development, so that if all countries improve at the same rate, the poorest countries will not get credit for any progress.

There are other measures of development and quality of life. For example, the **Multidimensional Poverty Index (MPI)** developed by the UN in 2010. It uses different factors to determine levels of poverty rather than just economic factors. It can be used to create a picture of people living in poverty, and allows comparisons both across countries, regions and the world, and within countries by ethnic group, urban/rural location, as well as other key household and community characteristics.

13.2 Inequalities between and within countries

Inequality between countries

Globally, it is calculated that 2.8 billion people live on under $2 per day, in what is classed as 'moderate poverty'. Another 1.1 billion people live on less than $1.25 per day, in 'extreme poverty'. Over time, a greater proportion of wealth has concentrated in the hands of the richest 20% of people, compared to the poorest 20% of people.

Many HICs are much richer than the LICs, starkly highlighting this development gap. A number of factors normally combine to produce these differences between countries:

- **Geographical location and physical environment:** countries found in the interior of continents, such as Afghanistan in Asia and Niger in Africa, without access to the sea have generally developed more slowly than countries with access to the sea and the world's oceans, as it affects their ability to trade.
- **Size of country in area:** many small countries have developed more slowly than large countries. Size in area can affect the availability of natural resources; larger countries, such as the USA, Canada, Australia and Russia, often have a wider range of natural resources. There are exceptions – Singapore and Luxembourg are both top-end HICs.
- **Climate:** polar and tropical countries have developed more slowly than those countries with a more equable, temperate climate. Polar/sub-polar and tropical climates often produce more infertile soils which can be difficult to farm sustainably and productively. These climates can limit the range of species of plants and animals that can be farmed.
- **Economic policies:** many of the HICs have applied economic policies which have encouraged strong economic growth. The setting up of trade blocs, such as the European Union (EU) or the North American Free Trade Association (NAFTA), have seen their collective economies grow compared to many countries that have developed alone. Many LICs have often found it hard to market their products to these rich, large markets because of the regulations, laws, import taxes and tariffs placed on their products.

FACT

The EU forms the richest trading group and market in the world and has a population of over 500 million.

- **Stable governments:** many European countries, the USA, Canada, Australia and New Zealand, have democratic political systems which they believe encourages economic growth. Political unrest, such as in Syria, Afghanistan, Libya and Iraq, can impact on vital infrastructure development and can adversely affect economic and social development. China has a stable government which has put in place economic and population policies that have led to very rapid and sustained economic growth to become the world's second largest economy.
- **Population policies:** governments can encourage or discourage higher birth rates to affect population growth. A growth in population can provide economic benefits, such as a large and productive workforce and an increasingly larger and richer market within the country. However, some LIC economies have been unable to expand fast enough to keep up with their population growth, and have not put in place successful population policies, leading to high levels of unemployment, especially among younger people, and high levels of poverty.

Inequality within countries

The development gap normally relates to global inequality. However, it also relates to disparities between the rich and the poor **within** countries. This type of inequality has risen in many countries – seven out of ten people live in countries where the gap between the rich and the poor in their country has increased.

As the economies of LICs and MICs grow, people are being left behind, trapped in chronic poverty often dependent on jobs that are precarious and low paid. Meanwhile, in these same countries, the rich seem to be getting richer. India has seen a tenfold increase in the number of dollar billionaires in the last decade. During the same period, an estimated one in three children have been diagnosed as undernourished.

Every aspect of an individual country's government and economy has the potential to affect equality within a country – finance, education, housing, employment, transport and health policies, and how they are applied to different areas of a country, from urban to rural areas, from the capital to the rest of a country, and between the different cultures and peoples within a country. In Bolivia, babies born to women with no education have infant mortality rates greater than 100 per 1000 live births, while the infant mortality rate of babies born to mothers with at least secondary education is under 40 per 1000. Life expectancy at birth among indigenous Australians is substantially lower (59.4 for males and 64.8 for females) than that of non-indigenous Australians (76.6 and 82.0, respectively).

Self-test questions 13.1

1. Name two ways in which the development gap can be measured.
2. Describe and explain two factors that can cause differences in development between countries.

13.3 Classifying production

There are four main sectors which, together, make up the employment structure of a country or region:

1. **Primary sector:** industries that extract raw materials such as mining, quarrying, farming, fishing and forestry. These primary products may be sold directly to customers or moved on to secondary industries to be processed.
2. **Secondary sector:** industries that process and manufacture the products of the primary industry, the raw materials, such as iron and steel making or processing food, or assemble the component parts made by other secondary industries such as car assembly.
3. **Tertiary sector:** industries that provide a service or skill such as education, health care, retailing, office work, transport and entertainment.
4. **Quaternary sector:** industries that provide information and expertise such as the microelectronics industry.

Formal and informal sectors of employment

Jobs can be divided into two groups – formal and informal employment. Those in the formal sector are 'official' jobs, where the worker is usually registered with the government and may be taxed, but, at the same time, will be eligible for paid holidays and health care benefits. These jobs usually provide better security of employment and are often better paid, with workers getting a regular weekly or monthly salary. Jobs in the formal sector normally include those in government services, education and health care. Formal employment is most common in HICs.

These formal sector jobs contrast with those in the informal sector. These jobs are often part time, temporary, outside the tax system, lacking any benefits and job security. They are often low paid. They are found in street markets as market traders, food stall workers, and shoe shiners or as farm workers such as fruit pickers. Informal employment is most common in LICs.

TERMS

formal sector: encompassing all jobs with normal hours and regular wages, on which income tax must be paid

informal sector: encompassing all jobs which are not recognised as normal income sources and on which taxes are not paid

employment structure: the percentage of workers in the primary, secondary, tertiary and quaternary sectors in a country

13.4 Employment structure and development

The proportion of people employed in the four sectors in any country or region is called the employment structure. Employment structure varies considerably from country to country and within countries, and it changes over time as a country develops economically. Figure 13.2 shows theoretical employment structures for typical LICs, MICs and HICs. It is a useful indicator that can be used to measure a country's level of development and then compare it with other countries, or compare regional or rural/urban areas within a country.

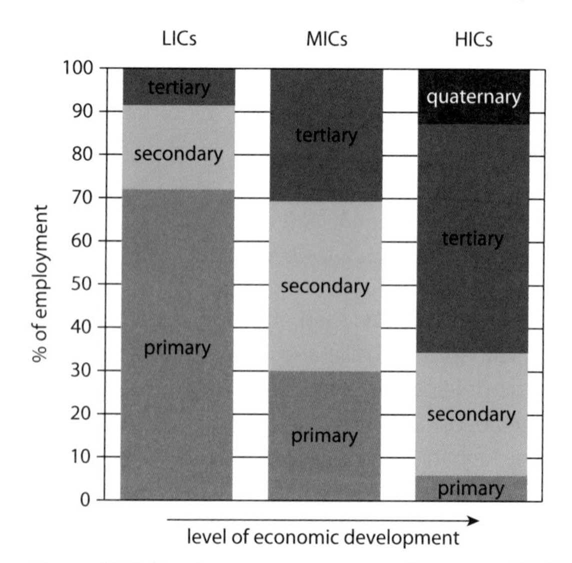

Figure 13.2 Employment structures for typical LICs, MICs and HICs

LICs frequently have a very high percentage of their workers in primary industry. A large proportion of the people work as subsistence farmers in the primary agricultural sector, as this is the only way they can make a living. In subsistence farming, people are often poor in economic terms because they eat/subsist on the crops and animals they grow and rear, and there is often very little surplus crops and animals to sell for cash in local markets. The same situation is common for coastal and lake-side fishing settlements.

There is also a small percentage of workers in secondary industry in LICs and it is often craft-based such as carpentry or the making of and repair of simple tools. Many LICs have a relatively small educated workforce and may lack adequate infrastructure, such as an efficient transport network of roads and railways, which is not attractive to foreign companies or multi/transnational corporations. Jobs in the tertiary/service sector often require a higher level of education which is not available to most people, who may only have been able to complete primary education, before being required to work on the family farm or business for little or no personal income.

TERM

multi/transnational corporations (MNCs/TNCs): global corporations, companies or businesses with their headquarters and research development activities in one country and their factories and/or production centres in other countries

In **HICs,** the proportions of people working in the different employment sectors has changed dramatically. In the 1800s, most people in HICs would have been employed in the primary sector. Many people worked on the land, and made their living from agriculture and related products. There are now far fewer people employed in these primary jobs in HICs, as agriculture is both extremely productive with improved crops and animals and the use of agrochemicals, but also heavily mechanised so there are very few actual jobs in the sector.

As HIC economies developed in the late 1800s and early 1900s, more people were employed in manufacturing industries, such as shipbuilding, iron and steel making and textile production. By 1900, over half of the workers in the UK and many other HICs in Europe and North America were employed in secondary-sector industries. However, the percentage of jobs in the secondary sector began to fall in many HICs as more manufacturing jobs moved to **MICs** where the labour costs were lower. Higher labour costs in the HICs, computer automation and the use of robot technology meant that only a few workers were needed for the control room and maintenance of the machines.

The tertiary sector has grown considerably in the HICs. The number of jobs in health care, education, financial services and retail industries has grown enormously in the last 20 years in most HICs. Many people left the rural areas in the search for jobs in the urban towns and cities. By the year 2017, 84% of the UK workforce were employed in tertiary industries and only a small number were employed in primary industries, with only 1.6% of the workforce employed in agriculture accounting for only 1% of the UKs GDP. This has all changed the work that people do, and also where they work.

In the UK, financial service industries, such as banking and insurance, now provide over 30% of the country's GDP and employ large numbers of highly-educated people. The quaternary sector has become very important in HICs. Research and development jobs of most multi/transnational companies along with their marketing and advertising jobs, are still based in the HICs while their manufacturing is based in MICs. The aerospace industries, such as Boeing in North America and Airbus in the EU, are highly concentrated in HICs where they can have access to a highly-educated workforce and the research facilities of leading universities.

13.5 Globalisation

The term **globalisation** refers to any process of change operating on a world/global scale and having global/worldwide effects.

There are three main forms of globalisation:

- **Economic:** largely caused by the growth of multinational/transnational companies/corporations (MNCs/TNCs)
- **Cultural:** the impact of Western culture, art, media, sport and leisure activities.
- **Political:** the growth of Western democracies and their influence on other countries.

The causes of **economic** globalisation include:

- the growth of TNCs/MNCs
- improvements in transport technology and transportation networks, allowing an increase in the fast, efficient and relatively cheap movement of commodities and people around the world
- accelerating advances in ICT – information and communication technologies; this allows management decisions and information to be instantly communicated worldwide
- the formation and growing importance of regional economic or trading blocs, such as the EU and NAFTA; the encouragement of free trade between the countries of these blocs further breaks down economic barriers between countries.

TIP

For more information on hi-tech industries (involved in ICT), see Chapter 15 *Industry*.

The impact of globalisation

The impact of globalisation is very obvious in the emergence of the MICs and in the growth of TNCs. TNCs have looked to LICs and MICs in order to cut their costs and increase their profits. This has led to a process of called global shift. This is because the TNCs are searching for the cheapest locations to manufacture and assemble components, so that the low-cost, labour-intensive parts of the manufacturing process are shifted to the MICs and LICs, where costs are substantially lower than in HICs.

TERM

global shift: where production processes are relocated from HICs (in the EU, USA and Japan) to MICs and LICs in Asia (for example, China, Vietnam, India, Cambodia, Laos and now Burma/Myanmar) and South America

The **role of the governments** in many MICs has been important. Their role has sometimes been **direct**, for example, in terms of their economic policy, their budget priorities, through offering financial incentives to TNCs and by restricting the activities of trades unions, and **indirect**, for example, in relation to investing in education and skills training, improving their energy supply networks and infrastructure and investment in their transport infrastructure.

Multinational or transnational corporations (MNCs/TNCs)

In the past, many companies were small to medium size and manufacturing in only one country. However, many companies decided to take control of all stages of their business, from the sourcing of their raw materials and component parts, through manufacturing and processing to transportation and marketing. This has often meant operating in several countries so that they became **multinational** or **transnational companies** (**TNCs**). These companies have become some of the biggest and most powerful companies in the world and many have operating budgets and profits exceeding that of many countries. They include the large car manufacturing companies such as Volkswagen of Germany, Toyota of Japan and General Motors of the USA, but also include large clothing manufacturers such as Nike of the USA and Adidas of Germany, pharmaceutical companies such as Glaxo Smith Kline of the UK and Pfizer of the USA, or food companies such as Nestlé of Switzerland and Del Monte of the USA.

FACT

Volkswagen, for example, accounts for almost one quarter of Slovakia's, in Eastern Europe, total exports.

Characteristically, the headquarters and research and development activities of TNCs are often in the HIC country of origin, while their manufacturing is often carried out in MICs and LICs. Globally, they directly employ around 45 million people, provide jobs indirectly for millions more workers and control over 75% of global trade.

There are advantages and disadvantages of TNCs locating in LICs and MICs (Table 13.1).

Advantages	Disadvantages
• more people in employment • higher wages than from their existing work • an improvement in skills • improvements in transport infrastructure – to roads and railways • improvements in service infrastructure, e.g. electricity, water supply, and sanitation • rise in the standard of living • improvements in public services such as health care and education • worker's income benefits local businesses and may create more jobs and wealth – an example of the multiplier effect • new technology and training-skills and expertise.	• a leakage of profits as they are taken out of the host country and they may not be required to pay national or local taxes • exploitation of the labour force, with long hours of work • low pay for workers • highly skilled jobs may go to outsiders and may not be passed on to the local population • branch plants may close as other locations become more cost efficient and the resources that are being exploited may be dependent on world prices/demand and not local demand • may have poor work conditions where workers' health and safety may be compromised • loss of rural land/farmland for the building of new factories • through a lack of local environmental laws and regulations, air, land and water pollution may take place.

Table 13.1 Advantages and disadvantages of TNCs locating in LICs and MICs

TERM

multiplier effect: occurs when an initial injection of money into the local, regional or national economy causes a bigger final increase in local, regional or national wealth

Case study

A TNC and its global links – the Toyota Motor Company

The Toyota Motor Company (Japan) became the world's largest automobile manufacturer in 2012. In 2016, Toyota was narrowly overtaken by Volkswagen (Germany), with sales of 10.31 million automobiles against Toyota's 10.18 million. Toyota's sales span over 170 countries and the company has 53 manufacturing companies in 28 countries on six continents. In 2016, Toyota had over 344 100 employees worldwide and it was the 14th largest company in the world by revenue. Several of Toyota's manufacturing locations have taken advantage of government policies aimed at attracting the investment and jobs that a Toyota facility will bring to a country or region. Toyota does not simply have high-end functions in HICs and low-end production functions in LICs. It has Research and Development centres in the USA, Germany, France, UK, Spain, Belgium, Thailand, China and Australia, in addition to its Japanese facilities.

FACT

Toyota has a major impact on the economies of countries where it invests. Its US operations alone **directly** employ over 28 500 people, worth nearly $2.3 billion to the US economy. However, Toyota **indirectly** contributed to the support of more than 365 000 other jobs and provided over $12 billion to the total US economy in 2016.

Toyota has developed two new production systems which have helped to increase its profits and efficiency. Firstly, a lean manufacturing or Just-In-Time (JIT) system. It has now been adopted by many other TNCs. It allowed the company to reduce the number of parts it has to store on a factory site and produce only the exact number of parts it needs, based on actual customer vehicle orders and it minimises waste production. Secondly, the ***Jidoka*** (roughly translated as 'automation with a human touch') system. When a problem occurs on the production line, the equipment stops immediately, preventing defective products from being produced. It also allows workers to identify the cause of the problem and prevent its recurrence.

TERM

lean manufacturing/Just-In-Time (JIT): a method of industrial production aimed primarily at reducing flow times within production system as well as response times from suppliers and to customers

Self-test questions 13.2

1. Describe and explain how the proportions employed in primary, secondary and tertiary industries differ in LICs and HICs.
2. With the use of a case study, describe the characteristics of a TNC (transnational corporation) and the impact it has on a country in which it operates.

Sample question and answer

Explain why a large percentage of the population in many LICs is employed in the primary sector. [3]

Ensure you read the question carefully and refer to LICs only and why people tend to be found in primary sector industries and not in other sectors.

In many LICs, a large percentage of the people work in farming because many people are subsistence farmers **[1]** who have to grow the food that their families eat. They cannot afford machines to do the work on their farms so much of the work is done by people, often the family **[1]**. Also, many of the people do not have the chance to have an education/go to school and so they cannot get jobs that need them to read and write **[1]**. The governments have relatively small budgets which means that there is not much money to invest and establish new secondary and tertiary industries **[1]**.
(A maximum of 3 marks will be awarded.)

Exam-style questions

1 Give one example of the type of employment that could be found in:

a primary sector employment

b secondary sector employment

c tertiary sector employment. [3]

2 Explain why HICs have a large proportion of their workforce in the tertiary sector. [4]

3 What benefits and problems does the setting up of a TNC (transnational corporation) bring for both the HICs and the LICs involved? [6]

Chapter 14

Food production

Learning summary

By the end of this chapter, you should be able to:

- describe and explain the main features of an agricultural system: inputs, processes and outputs
- demonstrate knowledge of farming types: commercial and subsistence; arable, pastoral and mixed; intensive and extensive
- demonstrate knowledge of the influence of natural and human inputs on agricultural land use, including scale of production, methods of organisation and the products of agricultural systems
- recognise the causes and effects of food shortages and describe possible solutions to this problem, including natural problems, economic and political factors, food aid and measures to increase output
- demonstrate knowledge of case studies of a farm or agricultural system and a country or region suffering from food shortages.

14.1 An agricultural system

As in industry, farming can be seen as a system which has inputs, processes and outputs.

TERMS

system: a set of interrelated parts, or components, that are working together towards some kind of process

aspect: the direction that a slope faces

Inputs in farming

These can be either **natural** (physical) or **human** (economic and social).

Natural or physical inputs

- **Climate:** this includes temperature, precipitation and their impact on the length of the growing season. Plant growth starts at a temperature of 6 °C. As temperatures increase so does the rate of plant growth. Areas that have temperatures above 6 °C for most of the year have a long growing season. A crop such as wheat needs a growing season of 90 days. Precipitation in the form of rain needs to be high enough to provide plants with enough water to grow. The amount needed will vary according to the temperatures as more will be lost by evaporation, and therefore not available to the plants, if the temperatures are high.

FACT

Having water during the growth period of a crop is very important and less important when a crop, like a cereal, is ripening.

- **Relief (or topography):** this includes the height and shape of the land. How high the land is will affect the temperature and the amount and type of precipitation. Temperature drops at a rate of about 1 °C for every 100 m in height. This means, therefore, that the growing season shortens with height. Precipitation also increases with height, as does the amount of snow – again this contributes to the growing season becoming shorter with height. The shape of the land includes whether slopes are steep or gentle. A steep slope will be difficult to farm but more importantly will often have thin soils due to increased runoff and erosion. A gentler slope will have deeper soils and less erosion. It will also be easier and safer to use machinery on gently sloping land. The aspect of the land is important. In the Northern hemisphere, the south-facing slope will have warmer soils which will mean that seeds will germinate and grow faster. The growing season can be six weeks longer on a south-facing slope compared to a colder, north-facing slope at the same location. This is only true in the Northern Hemisphere.

- **Soils:** some soils are deeper and more fertile than others which will affect plant growth. A thin and infertile soil will not be very productive.
- **Drainage:** land needs to be well drained to allow most plants to grow and not find their roots waterlogged. Flat land is easy to plough but may become waterlogged and flooded. Gentle slopes allow water to drain away.

Human or economic and social inputs

- **Investment of money/capital:** the cost of land, buildings and machinery can be high on many farms. There is also the cost of seed and animals, artificial chemical fertilisers, pesticides and herbicides.
- **Labour:** some farms are labour intensive – they may need large numbers of people to carry out the jobs on the farm which may be impossible for machinery to do.
- **New machinery and technology** can help the farmer improve income and profits.
- **Markets:** the farmer must ensure that there is always a market for the crop or product and also look for potential new markets for new crops and animals.
- **Transport** can be an expensive part of farming as products need to be got to markets.
- The **cost and maintenance** of the farm buildings and machinery.
- **Artificial inputs** can increase yields considerably and more than pay for their costs in increased profits. Irrigation is an example and the use of artificial chemical fertilisers, pesticides and herbicides are other examples.
- **Subsidies** for crops and animals are now essential for some types of farming to survive in many areas of the developed world as in the EU, USA and Japan.

FACT

Subsidies are used to maintain farming communities in rural areas where there is little or no alternative employment.

- **Quotas:** governments can put limits on the amount of crop or product, like milk, that can be produced.
- The use of genetic engineering in crops and animals to produce genetically modified (GM) foods. This increases yields, growth rates and weights, resistance to disease, etc. The use of GM crops in many areas of the world, such as the EU, is very controversial.

TERM

genetically modified (GM) foods: foods produced from organisms that have had changes introduced into their DNA using the methods of **genetic engineering**. Genetic engineering techniques allow the introduction of new characteristics, as well as greater control over certain crop and animal characteristics, than previous methods such as selective breeding.

Processes in farming

The processes include all the activities that take place on a farm:

- On an **arable** farm, they include: **ploughing** and preparing the land, **harrowing** the soil to prepare it for planting, **planting** of seeds, **controlling weeds** and **pests**, **harvesting** the crop.
- On a **pastoral** farm, they include: **grazing**, **calving** and **lambing**, **milking**, **slaughter** and **shearing**, producing **fodder crops** such as silage.

Outputs in farming

The outputs include:

- **crops:** cereals, vegetables, fruit and flowers
- **animals:** cattle, lambs, pigs, chickens and turkeys
- **animal products:** meat, milk, wool, skins, eggs.

Farming types

Farming/agriculture types can be classified in several ways:

- **Specialisation:** it can be either **arable** or **pastoral** or a combination of the two – called **mixed farming**. Arable farming involves the growing of crops, pastoral farming is the rearing of animals.
- **Economic status:** it can either be **commercial** or **subsistence**. Commercial farming is the growing of crops or rearing animals for sale. Subsistence farming is where the crops or animals are grown to feed the farmer's family.

- **Intensity** of the use of the land: it can either be **extensive** or **intensive**. Extensive farming is characterised by very large farms, often where the land is not very productive and so large areas of land are needed for it to be profitable. Intensive farming is where the farm is smaller but the land is very intensively used either in terms of investment in the farm or the numbers of people working on the land.

FACT

Intensive farming normally has a higher yield, more capital input and greater profit per hectare.

- **Land tenure** or **ownership:** it can either be **shifting** (which includes **nomadic**) or **sedentary**. Shifting cultivation is where farmers move from one area to another. Sedentary is where the farming is located in a permanent location.

The combined influence of natural and human inputs on agricultural land use

Different types of farming tend to have different types and degrees of natural and human inputs. The different combinations of inputs influence the scale of production, the methods used in organising the farm and its activities and the outputs/products from the farm.

Commercial farming aims to sell the produce that comes from the farm, whereas, subsistence farming is where the produce is consumed by the family or community that grows it. Any surplus produce can be sold at local markets to make some money or it can be traded for other products. Subsistence farming is most common in LICs. It is normally small scale and may involve shifting cultivation, where small areas of land, often forest, is cleared and used for crop farming, or by nomadic peoples who keep animals like sheep, goats, cattle and camels and move over large areas grazing their herds.

The large-scale system of commercial farming

There are several types of large-scale commercial farming. **Plantation farming** is one of the most common and involves the growing of one crop (called **monoculture**), often over very large areas. Examples include sugar cane, bananas, rubber, tea, coffee and pineapples. Other types of large-scale commercial farming include the growing of cereals, such as wheat, maize and barley, and the rearing of livestock animals such as cattle and sheep.

Case studies

Examples of commercial farming

Sheep farming in Australia

Sheep are raised either as lambs for meat or as older sheep for wool. Commercial sheep farming in Australia is found on very large farms in marginal areas. Therefore, they are often found in areas of low rainfall (Figure 14.1), high temperatures and poor quality grazing.

TERM

marginal areas: areas where other animals and crops would not be as successful or as profitable due to physical and human factors

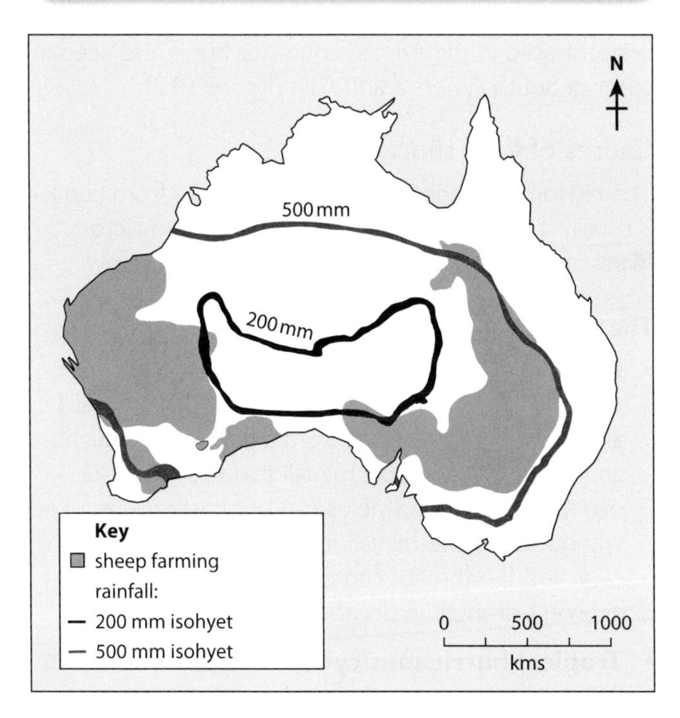

Figure 14.1 The location of sheep farming areas in Australia in relation to the 200 and 500 millimetre mean annual rainfall isohyets

Sheep farming in Australia is characterised as having **very low inputs** of **capital** per hectare – much of the land that is used is of relatively small value as it cannot often be used for arable farming, so it is cheap to buy. Farms may need up to 25 hectares of grazing land per animal as grazing land is so poor.

In terms of **labour**, it takes very few people to both look after large numbers of sheep, as they can be left out in the fields all year round, or gather the sheep together for shearing.

The **output** of this industry is that it produces about 620 000 tonnes of meat and 575 000 tonnes of wool in a typical year. Of this, about 68% of the wool and 39% of the meat are exported.

However, the industry faces several major challenges, most of them environmental:

- **periodic droughts:** some lasting several years have become increasingly common
- **weed infestation:** a variety of non-native plants have found their way into Australia and thrived, covering large areas of grazing land with plants inedible by sheep
- **destruction of natural habitats and soil erosion:** this is due to the grazing of sheep and in the worst cases sheep have been overgrazed which has destroyed the natural protective vegetation cover for the soil which is then exposed and easily eroded by water and wind
- **shortage of sheep shearers:** for example a very tough, hard, manual job which has lost workers to other, easier jobs, in the expanding Australian mining industry.

Rubber plantation farming

Plantations need a **high capital input** and **labour** input to firstly clear forest and then to drain and irrigate the land. The other inputs needed on a planation crop are fertilisers, pesticides and herbicides. The **processes** on a rubber plantation require the application of these inputs and the harvesting of the rubber from individual trees by hand.

The main threats to the rubber plantations in countries such as Malaysia have come from low prices, as 77% of the world's rubber is now synthesised from oil. Oil palms which can be grown in the same environment, but have lower production costs, need less labour, have higher yields and get higher prices than rubber.

Examples of small-scale subsistence farming

Shifting cultivation in the Amazon River basin

The most widely practised form of subsistence agriculture is **shifting cultivation**. It has been practised for several thousand years by many traditional societies in the tropical rainforests of the Amazon, the Congo River basin in Africa and in Indonesia and Malaysia. It involves the clearing of a small (about one hectare) area of rainforest by cutting it down, letting it dry and then burning the vegetation to both clear the land and provide nutrients for the soil. It is sometimes called slash and burn agriculture.

FACT

This form of agriculture is disappearing as the rainforests are cleared to be used for plantation agriculture and cattle ranching, mining, logging, roads and new settlements.

The fertility of the rainforest soils depends on the rapid and constant recycling of the large amounts of leaf litter that falls from the trees. For a short period of time, the soil remains fertile from the nutrients left by the forest. However, the soil quickly becomes less fertile due to the heavy tropical rains which erode the soil and remove the nutrients. When this happens, a new area of forest is then cleared and the small clearings are left to recover and be re-colonised by the forest. In some communities, this is done on an 18–25-year cycle.

Intensive subsistence rice farming

Rice has a high nutritional value and is the main food crop for much of the population of South and South East Asia. 'Wet' rice is grown on the fertile soils of the floodplains and 'dry' rice is grown on terraced fields cut into the valley sides of the rivers. Fields are flooded with water during the annual monsoon rains.

Rice growing is very **labour intensive** and it takes an average of 2000 hours to produce 1 hectare of rice. Most tasks, due to a lack of capital to buy machinery, are either done by hand or with the help of domesticated water buffalo. They include building the embankments around the fields to keep in the water, constructing and constantly maintaining the network of irrigation canals which bring the water to the fields, ploughing the fields to mix the rich soils with the water, planting the rice in small nursery fields, transplanting the rice into the main padi-fields, weeding, harvesting and threshing the rice crop (removing the rice from the stalk), drying and storing the rice, and planting other crops in the drier parts of the year.

Self-test questions 14.1

1. Name an example of large-scale commercial farming.
2. Describe the inputs, processes and outputs of one example of large-scale commercial farming.

14.2 Food shortages and possible solutions

From a total world population of over 7.5 billion people, 800 million in the LICs do not get enough food to eat and suffer from hunger. The problem is mainly concentrated in the African countries but is also seen in parts of South America and Asia (Figure 14.2).

Causes of food shortages

The reasons why these people are suffering from hunger is usually a combination of natural and human factors. Many of the natural factors are extreme climate events.

The natural factors

- **Drought and unreliable rainfall:** much of East Africa, including the north of Kenya, Sudan, Somalia and Ethiopia has suffered from a period of drought and very low, unreliable rainfall that has lasted for over ten years. Crop yields can be drastically reduced without adequate rainfall. In 2016, after two failed rains and the threat from El Niño, Ethiopia suffered its worst drought in decades.
- **Tropical hurricanes/cyclones/typhoons:** when these storms hit, with their high winds, torrential rainfall and storm surges, they can devastate farm land and crops.
- **Floods:** though often associated with tropical cyclones/typhoons/hurricanes, are usually the result of heavy rainfall, often associated with monsoons or El Niño events.
- **Pests and diseases:** there are many pests and diseases which can prey on crops, for example, locusts, and diseases, such as mildew.

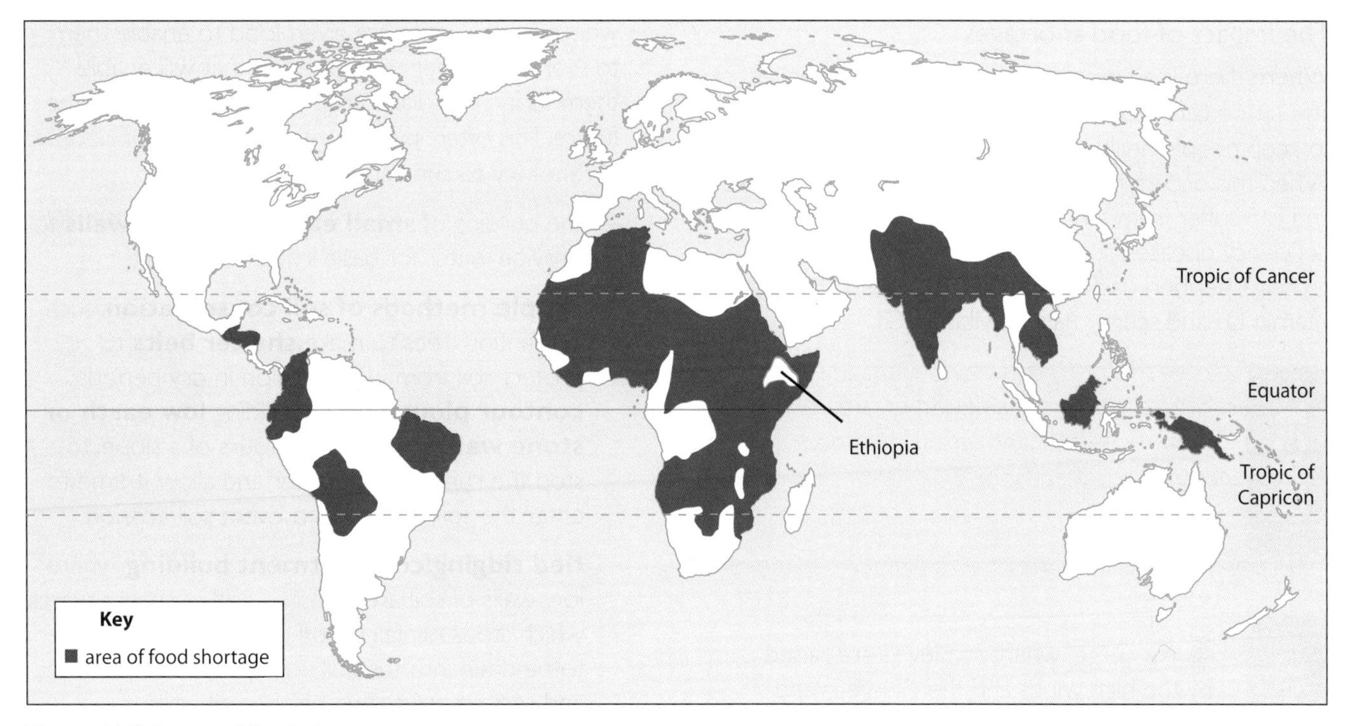

Figure 14.2 Areas of food shortage

Without the expensive pesticides and sprays to deal with them, poor rural communities may suffer severe crop losses and food shortages.

The human factors

- **Soil erosion and loss of soil fertility:** land that has been cleared of its natural protective cover of vegetation is much more easily eroded by rainfall. Soil fertility will be lost as land is eroded and soil minerals washed out of the soil. Also, overgrazing by grazing herds of domestic animals can remove the protective vegetation cover. Without applying either natural fertiliser, dung or expensive artificial chemical fertilisers, or allowing the soil to rest and recover (leaving the land fallow), many subsistence farmers have experienced a loss of fertility and lower crop yields.
- **Rural poverty:** in many LICs means that there is a lack of money to invest in irrigation, or buy expensive fertilisers. Many farmers are still using traditional farming practices such as ploughing up and down slopes, which increases soil erosion.
- **War:** Darfur in the south of Sudan, has suffered from many years of civil war. During these years of conflict, up to 2 million people have had to leave their homes and farms and it has often been impossible for farmers to grow crops and rear animals. By 2016, more than 70 000 people had died from a mixture of starvation and diseases often caused or made worse by malnutrition.
- **Increasing population:** as a result of rising population, more land is having to be cleared of forest, less fertile marginal land is being brought in to use and farms are being divided into much smaller units among farmers' children as they grow up and have their own families to feed.
- **Inability to invest capital and improve infrastructure:** the lack of capital available in LICs, particularly in rural areas, means that schemes to improve agricultural production, improve food storage, improve roads and transportation for distributing farm products are very hard to initiate and develop.
- **Volatile global food prices:** between 2001 and 2016, the price of many globally important crops fluctuated dramatically (Table 14.1). This has caused some serious problems.

Crop	2001 (price in $ per tonne)	2012 (price in $ per tonne)	2016 (price in $ per tonne)
Wheat	120	270	164
Rice	180	550	370
Maize	90	260	160
Barley	90	210	120
Sorghum	90	260	175

Table 14.1 Changing food prices

The impact of food shortages

Where there is a complete collapse of food supplies, starvation can occur. Where there is not enough food to keep people healthy, it can lead to malnutrition. When this occurs, people are less able to resist disease and can suffer from a number of protein and vitamin deficiency diseases, such as kwashiorkor (lack of protein), beriberi (lack of vitamin B1), rickets (lack of vitamin D) and scurvy (lack of vitamin C).

FACT

People who are malnourished are unable to work as hard and an already poor situation is made worse.

TERMS

storm surge: where sea levels are raised by the high winds and then surge inland flooding low lying areas like southern Bangladesh

starvation: the result of a severe or total lack of nutrients needed for the maintenance of life

malnutrition: this develops when the body does not get the right amount of the vitamins, minerals and other nutrients it needs to maintain healthy tissues and organ function

Solutions to food shortages

The governments of many HICs and the EU, often acting through agencies of the **United Nations**, such as the **FAO** (Food and Agriculture Organisation) and the **WFP** (World Food Programme) and many **NGO**s (Non-Government Organisations), such as **Action Aid** and **Oxfam**, give assistance and food aid to areas suffering from food shortages to improve both the standard of living and quality of life of the people affected. Food aid can be given on both a short-term and a long-term basis:

- **Short-term food aid** is delivered directly to those people affected during a crisis caused by natural events or war.
- **Long-term food aid** is often given to the government of an LIC to sell in local markets to add to that produced by local farmers. It may also be given as part of a '**Food for Work**' programme, where communities are given food to enable them to work on longer term projects that will enable them to increase local crop and animal yields in the future. This often involves introducing appropriate technology to rural areas. These include:
  - the building of **small earth dams** and **wells** to provide water for basic irrigation projects
  - **simple methods of soil conservation**, such as planting trees to make **shelter belts** to protect soil from wind erosion in dry periods, **contour ploughing** or building **low earth or stone walls** along the contours of a slope, to stop the runoff of rainwater and allow it time to enter the soil, helping to prevent soil erosion
  - **tied ridging/compartment building** where low walls of soil are built in a grid of small squares which stops rainfall runoff and again allows water to be drain into the soil – crops such as potatoes and cassava are grown on the soil walls
  - **strip or inter-cropping** which has alternate strips of crops being grown, at different stages of growth, across a slope to limit rainfall runoff as there is always a strip of crops to trap water and soil moving down the slope
  - **tier or layer cropping**, where several types and sizes of crops are grown in one field to provide year round protection from rainfall and increase food and crop yields
  - **improved food storage** which allows food to be kept fresh and edible for longer periods of time and protected from being eaten by rats and insects and affected by diseases.

Food Aid is not without its problems. It is very difficult and expensive to transport food from HICs to LICs. HIC food for sale in local markets may be cheaper than food grown by local farmers, making them less well off. This can be solved by providing money, not food, which can be used to buy local food and so increase local farmers' incomes.

Globally, the Green Revolution has had an impact on food shortages. The Green Revolution gained momentum with the development of High Yielding Varieties (HYVs) of five of the world's major cereal crops – rice, wheat, maize, sorghum and millet. These new hybrid varieties of cereals (apart from rice) were:

- resistant to drought
- higher yielding, often by two to four times, than traditional species of these crops

- had a shorter growing season, allowing more crops to be grown in a year in some areas.

TERMS

food aid: providing food and related assistance to tackle hunger, either in emergency situations or to help with long-term hunger alleviation and achieve **food security** (where people do not have to live in hunger or in fear of starvation)

appropriate technology: technology which is suited to the level of wealth, knowledge and skills of local people and is developed to meet their specific needs

Green Revolution: research and development of technology transfer initiatives occurring between the 1930s and the late 1960s, which increased agricultural production worldwide; the initiatives resulted in the adoption of new technologies

High Yielding Varieties (HYVs): crops with a higher yield per area, an improved response to fertilisers and a high reliance on irrigation and fertilisers; crops mature more quickly in contrast to conventional varieties of crops

mechanisation: the process of changing from working largely, or exclusively, by hand or with animals to doing that work with machinery

The Green Revolution had several **positive results**:

- farm incomes increased, increasing the standard of living of many people in rural areas which, in turn, meant that families had money to pay for the education of their children; giving them access to qualifications that could allow them to get a better job in future
- increased yields mean that crops can now be exported
- stopped food shortages in some areas and improved people's diets
- increased employment on farms and in food processing industries
- paid for machinery, fertilisers, pesticides, herbicides and irrigation.

However, the Green Revolution also had its **problems and critics**, such as:

- the HYV crops need bigger, more expensive inputs of fertiliser, pesticides and herbicides
- the mechanisation of jobs on the farms has increased unemployment, increasing poverty for some and forcing some people to migrate to cities in search of jobs
- much of the farm land is now being used for growing one HYV crop. This has had two impacts on local diets, making them more deficient. Firstly, other local food crops are not being grown, removing important minerals and vitamins from people's diets. Secondly, HYV crops are often lower in minerals and vitamins than the local varieties they have replaced, which means that they do not provide people with the same level of nutrition compared to their traditional crop varieties.

The use of irrigation and the problem of salination

Irrigation is an artificial application of water to the soil usually for assisting in growing crops. In crop production, it is mainly used in dry areas and in periods of rainfall shortfalls. There are several types of irrigation:

- **Surface irrigation** is where water moves over and across the land by simple gravity flow in order to wet it and to infiltrate into the soil. Surface irrigation can be subdivided into furrow, border strip or basin irrigation. It is often called **flood irrigation** when the irrigation results in flooding or near flooding of the cultivated land. Where water levels from the irrigation source permit, the levels are controlled by dikes, usually plugged by soil. This is often seen in terraced rice fields (rice paddies), where the method is used to flood or control the level of water in each distinct field.
- **Drip irrigation**, also known as trickle irrigation, is where water is delivered at or near the root zone of plants, drop by drop. This method can be the most water-efficient method of irrigation, if managed properly, since evaporation and runoff are minimised. Drip irrigation can also deliver fertiliser to the crops, a process known as **fertigation**.
- **Subsurface drip irrigation (SDI)** uses permanently or temporarily buried dripper line or a drip tape located at or below the plant roots. It is becoming popular for row crop irrigation, especially in areas where water supplies are limited or recycled water is used for irrigation.

The problem of salination

Overuse and poor irrigation practices have led to increased salt content in the soil, reducing the productivity of the land. Irrigation salinity is caused by water soaking through the soil level adding to the ground water below. This causes the water table to rise, bringing dissolved salts to the surface. As the irrigated area dries, the salt remains.

In New South Wales, Australia, irrigation salinity is solved through a salt interception scheme that pumps saline groundwater into evaporation basins, protecting approximately 50 000 hectares of farmland in the area from high water tables and salinity.

Other problems caused by irrigation

- Increased competition for water, from individual farmers, communities and even countries.
- Over extraction of water can lead to the dropping and depletion of underground aquifers.

A country suffering from food shortage – Madagascar

Madagascar, in southern Africa, is currently facing a **locust plague** that could affect the livelihoods and food and nutrition security of 13 million people. A plague started in 2012 and continued into 2016. It has had a dramatic impact on agricultural production. Rice production declined by 21%, resulting in a national rice deficit of 240 000 tonnes in 2013/14. Maize production in 2013 was also below the national requirement, and an estimated 28 000 tonnes of maize had to be imported to help bridge the deficit.

Madagascar is also prone to frequent natural disasters that have a significant effect on the livelihoods of the most vulnerable rural populations. Since 2009, the combined effect of **droughts** and **cyclones**, as well as **political instability**, has reduced food production and worsened the living conditions of thousands of households, mainly in the south-western part of the country where more than 80% of the population live below the poverty line.

The FAO and the WFP are joining together to assist vulnerable communities in Madagascar through:

- the provision of food assistance to the most vulnerable people with special emphasis on the needs of children and pregnant and nursing women
- the construction or rehabilitation in the south of the island of community infrastructure such as irrigation canals, dams and water catchment systems; this will be done through food-for-work and cash-for-work programmes
- the provision of adequate support to farmers to increase the production of staple crops such as rice and maize
- the promotion of short-cycle crop varieties, such as peppers, beans and pineapples, which are particularly adapted to the climatic conditions
- the introduction of more efficient water management.

Self-test questions 14.2

1. What is meant by the term Green Revolution?
2. Explain how both natural and human factors may cause food shortages.

Sample question and answer

1 a **What is the difference between subsistence and commercial farming? [2]**

b **Identify an input, a process and an output on a commercial farm. [3]**

c **For a named area or country which you have studied which suffers from famine, explain why there is a shortage of food. [5]**

a Subsistence farming is where crops are grown and animals are reared for the farmer's family only **[1]**. Commercial farming is where the crops and animals and their products are produced for sale for money **[1]**.

b An input could be fertiliser **[1]**, a process could be applying the fertiliser **[1]** and an output could be a crop such as wheat **[1]**.

c In Somalia, in 2011, many subsistence farmers suffered from famine and food shortages as a result of three years of drought **[1]**, which greatly reduced the amount of crops they could grow **[1]** and their animals died from a lack of food and water **[1]**. After these three years, they ran out of seeds **[1]** and they did not have the money to buy any more seeds **[1]** and they could not find any water for their animals **[1]**. Also there was a war going on in the country **[1]**, which meant it was difficult for them to receive any aid or help **[1]** so many had to leave their farms.
(A maximum of 5 marks will be awarded.)

TIP

In questions where extended answers are needed, as in part (c) here, you will need to **develop** your ideas. A question like this provides good differentiation with most students being able to provide simple statements relating to drought or poverty, but you must develop these ideas in relation to your chosen example by including place-specific information to gain the top Level 3 marks.

Exam-style questions

1 What is meant by the terms nomadic and sedentary farming? [2]

2 a Describe how irrigation and drainage may improve crop production on a farm. [3]

b What problems can be caused by irrigation? [3]

3 Explain the ways in which farmers in LICs might increase the amount of food they produce from their land. [4]

Industry

Learning summary

By the end of this chapter, you should be able to:

- demonstrate an understanding of an industrial system: inputs, processes and outputs (products and waste)
- demonstrate knowledge of industry types: manufacturing, processing, assembly and high technology industry
- describe and explain the factors influencing the distribution and location of factories and industrial zones
- demonstrate knowledge of a case study of an industrial zone or factory.

15.1 Industry as a system

As in farming, industry can be seen as a system where there are:

- **Inputs:** these can include raw materials, energy, transport, labour, money (capital investment) and government policies.
- **Processes:** these can include the processing raw materials, assembling component parts, packaging and administration.
- **Outputs:** these can include the finished products, the profits and waste.

Types of industry

Manufacturing industries are those industries which involve the manufacturing/conversion of raw materials, such as iron ore, into a finished product, such as iron and steel. The manufacturing industry accounts for a significant share of the industrial sector in many HICs and MICs. Their final products can either serve as a finished good for sale to customers or as an intermediate good that can be used in the production process. For example, providing steel to the car manufacturing and construction industries. They are often major wealth-producing sectors of an economy.

Processing industries use raw materials, which are treated or prepared in a series of stages, often using chemical processes. Processing industries include oil refining, petrochemicals, water and sewage treatment, food processing and pharmaceuticals. In processing industries, there are normally a number of ingredients, often bulk materials rather than individual units, which are changed/processed in some way to produce a finished product, for example, petrol and diesel fuel for use in vehicle engines.

Assembly industries put together and combine component parts in what is sometimes called a progressive assembly, until the final product is assembled. By continuously moving the parts to the assembly work and moving the semi-finished assembly from work station to work station, a finished product can be assembled faster and with less labour than by having workers carry parts to a stationary piece for assembly. The worker is sometimes replaced by programmable robots, as in the welding and painting parts of a vehicle assembly line. Assembly lines are common methods of assembling complex items such as automobiles, household appliances and electronic goods.

High technology or hi-tech industries

Hi-tech industries have several characteristics which make them different from many of the older traditional industries:

- They have been set up in the last 25 years.
- They have processing techniques which normally involve the use of microelectronics.
- They normally produce high value products such as computers, microchips, electronic equipment or medical products.
- They use the most advanced manufacturing techniques.
- They employ a highly skilled workforce and carry out a lot of research and development.

Many hi-tech industries are divided into two sections:

1 Research and product development
2 The manufacturing of the product, which normally involves the assembly of small, easily transported component parts.

Hi-tech industries are normally put into the footloose group of industries because they:

- do not require bulky raw materials
- use small, light component parts
- have products that tend to be small and light
- use electricity as their power source
- need a small, skilled labour force
- are non-polluting and so can locate in, or near, residential areas.

Do they favour any particular location?

Hi-tech industries need to attract a highly skilled workforce, so they often locate in areas with a pleasant living and working environment. There are several areas around the world which have become particularly popular. For example, 'Silicon Valley' near San Francisco, California, USA; 'Silicon Island' in Penang, Malaysia; Electronic City in Bangalore, India; and 'Silicon Strip' in England, UK.

FACT

Silicon Valley has many companies making and using silicon chips, computers, semi-conductor devices and computer-controlled machinery. The area is now the home to many of the world's largest hi-tech corporations, such as Apple, eBay, Google and Facebook, and employs about a quarter of a million people in information technology.

Business and science parks

Within these areas, hi-tech industries often group together in newly developed business and science parks. These are usually found on the edges of towns in a spacious, semi-rural situation, surrounded by countryside – greenfield sites. They have modern buildings, landscaped gardens and some have woodland, lakes and ponds – often over 70% of the park area can be used for this. The science parks are associated with universities which provide the research facilities and many of the highly skilled workers.

TERMS

footloose: an industry which is not dependent on a particular location, unlike many other types of industry which need to be beside their raw materials, ports, fast and easy transport, etc

greenfield site: an area of land, usually agricultural or amenity land, which is being considered for urban or industrial development

There are several other advantages for a hi-tech business setting up in one of these parks:

- They can exchange ideas with neighbouring businesses.
- They can share the cost of maintenance and support services like waste and rubbish removal, electricity lines, water and gas links.
- They provide a large pool of skilled workers.

Hi-tech industries in HICs and MICs

There are a number of reasons why many large companies have located hi-tech industries in new HICs and MICs, such as South Korea and Malaysia:

- The large and expanding market for goods in many MICs
- There is a large and educated workforce available
- The labour supply is relatively cheap compared with Europe and North America
- The low cost of land for building on
- The increasing ease in the global transport of components parts and products by both air and sea
- Government help and assistance in the setting up of new industries such as subsidies, low taxes and development grants.

Many of the reasons for locating hi-tech industries are to do with **globalisation**. For more on globalisation, see Chapter 13 *Development*, Section 13.5.

Self-test questions 15.1

1 Using examples, explain what is meant by inputs, processes and outputs in an industrial system.

2 With reference to one type of manufacturing or processing industry that you have studied, describe the main features of this industry.

15.2 Factors affecting the location of industry

In deciding where to locate a factory, or industrial zone, a number of factors need to be taken into account. These factors can be put into two groups – **physical** factors and **human and economic** factors. All these factors combine to produce the total costs of production.

TERMS

costs of production: the costs involved in the process of manufacturing need to be below the price charged to customers so that a factory or business can make a **profit** and remain in business

break of bulk location: a place where goods are transferred from one mode of transport to another

Physical factors

There are a number of physical factors that affect the location of industry:

- **Easy access to raw materials:** raw materials are often very bulky, heavy and expensive to transport. An industry that uses large amounts of bulky raw materials will find it much easier to locate near the source of the raw materials or at a location where they can be cheaply transported to. For example, a deep water port – a break of bulk location – where goods are transferred from ship to truck/lorry or railway. Bulky cargo is unloaded and then often processed at that location, so saving on very significant transport costs if the materials were to be transported inland by road or rail.

FACT

Many iron and steel works tend to locate on a deep water coastal location so that they can more cheaply bring together the bulky raw materials, iron ore, coal and limestone, that they require.

- **Easy access to cheap sources of power:** many industries need large amounts of power and therefore a location beside a cheap source of power, such as fast-flowing water for mills or a coalfield, is both useful and often much cheaper. In the past, one of the major sources of power was coal and therefore a location on or near a major coalfield was a perfect location. For example, South Wales in the UK and the Ruhr in Germany.

FACT

Hydro Electric Power (HEP) is a cheap source of electricity and so is often used by power hungry industries – such as an aluminium smelter.

- **A site that is cheap to buy or rent and is easy to build and expand on:** a large, flat site is easier to build on and for a factory to expand on in the future. A large site may be expensive to buy so cheaper land is also an advantage. As a result, a large, flat floodplain or a coastal plain are popular sites for factories. The site does need to be well drained though, so choosing either a river floodplain or a coastal plain needs careful thought.
- **A geographical situation that allows easy transport routes to be set up:** a good, natural routeway such as where valleys meet (called a confluence), rivers meet or a port, provide the ways in which raw materials, component parts or finished goods can be transported easily to and from factories. Valleys can carry roads, railways and canals so that where they meet in a confluence or where they link to the coast gives that location an advantage over others.

FACT

Large industrial complexes can often be found near important ports, such as Rotterdam in the Netherlands.

Human and economic factors

There are also a number of human and economic factors that affect the location of industry:

- **Availability of labour:** different industries need different types of workers. Some industries need large numbers of relatively unskilled workers – some types of farming, for example, market gardening and sorting and packing vegetables. These are called **labour intensive industries**. Other industries need relatively few workers but they must be highly skilled – such as the IT industry – often called **mechanised industries**.
- **Availability of capital** (finance/money) to invest in the factory. New factories are often highly expensive to build and bring online. A new oil refinery, for example, can cost between $1–4 billion.

> **FACT**
> Volkswagen Phaeton car factory in Dresden, Germany, cost 186 million Euros (roughly $208 million) in 2002.

- **A market where the products of the factory can be sold:** the size and the location of a market have now become more important than raw materials. Large HIC cities like New York, Paris and London provide very large markets for many products so locating near these cities or with an easy, fast form of transport, like a motorway, is a prime location.
- **Availability of cheap, fast and efficient transport:** transport costs can make up a larger proportion of production costs and therefore finding the cheapest forms of transport for moving raw materials and finished goods is very important. Bulky raw materials like crude oil, iron ore and wheat can be most cheaply moved by bulk carriers. Container ships can carry other goods relatively cheaply and efficiently as they are easy to transfer to road and rail.

> **FACT**
> It now costs $1500 to transport a container from South East Asia to Western Europe. This means that transport costs can be as little as 1% of the final cost of a product that is transported in this way, compared to 18% 20 years ago.

- **Government policies affecting the location of industry:** many national governments and the EU have a wide range of policies which they can use to encourage industries to move to particular locations. A common policy is to decentralise their own government departments. The process allows wealth to be distributed more fairly across a country and contribute to the regeneration of city and town centres. In the UK, for example, the relocation of the government's Office for National Statistics out of London to Newport, in South Wales, took over 600 jobs from London to Newport. In addition, governments can also provide incentives such as lower company taxes, subsidised wages, lower rents and improved infrastructures like improving roads and railways.
- **Economies of scale:** a business with many small factories may not be very profitable compared to those which have just one large factory location and so many businesses have closed their smaller plants and built larger ones to put all the industrial processes in one site.
- **Changes in technology:** the use of robotic machines run with computer software to do repetitive jobs has transformed many factories, for example car assembly plants. The internet, video conferencing and fax machines have released workers from their normal workplace. This means that many workers in tertiary industries can work from home. It has also led to the practise of outsourcing. IT software companies, banks and insurance companies have been able to close offices in HICs where labour costs are high and open up in India where labour costs are much lower.

> **TERMS**
> decentralisation: the process of redistributing or dispersing functions, powers, jobs and people away from a central location or authority
>
> outsourcing: a practice used by different companies to reduce their production costs by transferring portions of their work to outside suppliers rather than completing it internally

TIP

Many of the reasons for relocating industries are to do with **globalisation**. For more on globalisation, see Chapter 13 *Development*, Section 13.5.

- **Living and working environment:** many businesses now look at attracting workers by offering them better living environments. This often involves moving out of large urban areas to suburban and rural environments. This offers workers a number of opportunities to improve their quality of life, including a cleaner, less polluted, quieter, environment, lower rates of crime, cheaper housing costs, less time spent commuting to work and better schools.

The changing location and nature of industry

There are many reasons why businesses may fail and industries close, or move location within a country, or to another part of the world:

- The resources or raw materials that primary industries depend on may become **exhausted** or too expensive to get out of the ground. For example, the exhaustion of a coalfield, or an iron ore deposit.
- The replacement of manual workers with **mechanisation**. Car assembly plants have replaced most of the manual jobs, like paint spraying and welding, with robots.
- A **fall in demand** for a product. Products as diverse as car models, trainers, food products, mobile phones, TV sets and clothing all change rapidly and customers stop buying older models. This may result in the closure of a factory.
- **Increases in production costs**, in wages, transport costs and the cost of raw materials can all make a factory uncompetitive. The minimum wage law in the EU has seen many industries move out to MIC and LIC locations where labour costs are much lower – going from $10 an hour in the EU to $2 a day in Vietnam, China and India.
- **Foreign competition:** countries with lower costs of production can undercut the prices of their competitors and cause the collapse of some industry. For example, the textile industry in the UK could not compete with the lower costs of factories in India and Bangladesh. The shipbuilding industry in Europe found it very difficult to compete with lower cost producers in Japan and South Korea.
- **Lack of money for new investment:** industries facing competition may find it hard to find the money to pay for new factories and machinery to make them more competitive. The UK car industry found this a real problem in the 1970s and 80s. They often had to close and sell the brand name to foreign competitors – such as selling Jaguar, Land Rover and Range Rover to Ford in the USA (who then sold it to the Indian auto company Tata Motors in 2008), Rolls Royce and the Mini to the German BMW group, Bentley to Volkswagen.

TERM

agglomeration economies: the benefits that companies and businesses obtain by locating near each other. As more companies in related fields of business cluster together, their costs of production may decline significantly as they attract more suppliers as well as customers.

Self-test questions 15.2

1. Using examples, describe and explain the factors influencing the distribution and location of manufacturing/processing industries.
2. With the use of a case study you have studied, describe the changing location of an industry through time.

Case studies

An industrial zone – Silicon Strip, England, UK

'Silicon Strip' is located to the west of London in the UK, along the M4 motorway. The reasons for locating beside the M4 include the following:

- Being close to the M4, industries will have a fast motorway link between the cities of London, Bristol and Cardiff.
- The busiest international airport in the world at Heathrow is just to the west of London and beside the M4.
- It is the location of several government research establishments – many have been sold by the government and are now in the private sector.
- A large, skilled labour force nearby, so that they benefit from agglomeration economies by being so close together – they can swap both workers and ideas easily – and the large population of this part of England is within easy commuting distance.
- Close to universities which provide potential new, highly qualified graduate workers, research facilities and expertise.
- An attractive working and living environment – less air, noise and visual pollution.

An industrial zone – the iron and steel industry, South Wales, UK

The iron and steel industry in South Wales in the UK has undergone enormous change through time, reflecting the changing global factors influencing the location of this particular industry.

The early location of the industry

In the early 19th century, the industry was located where it could find its three essential raw materials – iron ore, coal and limestone. In several places in the UK, these were found in almost the same location – South Wales was such an area. By 1850, there were 35 iron works dotted around the South Wales coalfield. Another factor in favour of their location was the fact that the valleys in which they were located led down to the coast where iron products could be transported and then exported easily.

The 20th century

A hundred years later, South Wales only had two iron and steel works. There were several reasons which caused this enormous change:

- the raw materials became exhausted – the iron ore deposits and the easily accessible coal deposits were used up and so these now had to be imported
- the works inland suddenly found themselves at a real disadvantage as the iron ore and coal had to be unloaded at the coast and put on trains to be transported to the works
- this added enormously to the production costs and made the inland works non-competitive as their products became much more expensive than those at coastal works.

The 21st century

Two huge, fully integrated sites were built on the South Wales coast. These were on large, flat sites to accommodate the very large buildings, over 1 km long and to give plenty of room for future expansion. The main factor governing the future of both plants was the world price for steel and the demand for steel. In 2001, the demand for steel fell and as a result one plant was closed. In 2002, demand for steel increased as the Chinese economy underwent enormous expansion, which made the Port Talbot works in South Wales very profitable.

Then global recession in 2008–2009 saw a slowdown in the global economy and some HICs actually went into recession, which dropped the demand for steel. In the case of Port Talbot, its TNC owner, Tata, based in India, claimed it was losing $1.4 million a day and put it up for sale or possible closure in 2016, with the loss of 4000 jobs at the iron and steel works and the potential loss of another 10000 more jobs in its supply chain and in the local area. Tata claimed it could not compete with the Chinese producers as its production costs were higher due to higher energy costs and business rates.

Sample question and answer

1 a **What is meant by high technology industry?** **[1]**

b **What factors attract high technology industries to an area?** **[4]**

a A high technology industry is a science-based industry that uses the most advanced techniques in its production methods. This may often mean the manufacturing of silicon processing chips and other types of micro-electronics. **[1]**

b The factors that may attract high technology industries to an area may include the fact that an area might have a very skilled and highly qualified labour force **[1]**. An area like this may have a university nearby producing highly qualified students or there may be a lot of high tech industries there already **[1]**. It should have very good transport links by both roads and rail and an international airport **[1]**. It should also be a nice place to work – possibly out in the countryside with good views **[1]**. The land on which the factory is going to be built should be cheap to buy, flat so that it is easy to build on and have plenty of room to expand **[1]**. (A maximum of 4 marks will be awarded.)

TIP

Use the mark allocations and answer spaces provided on the paper as a guide to the length of answer required and the number of points to be made. Some students write overlong answers to questions worth few marks at the expense of including detail in those requiring extended writing.

Exam-style questions

1 What is a meant by the term footloose industry? [3]

2 How may the growth of high technology industries in MICs, such as Thailand and Malaysia, benefit both the people and the economy of the country? [5]

3 Using an example that you have studied, describe how a national government can influence the location of a new factory. [4]

Tourism

Learning summary

By the end of this chapter, you should be able to:

- describe and explain the growth of tourism in relation to the main attractions of the physical and human landscape
- evaluate the benefits and disadvantages of tourism to receiving areas
- demonstrate an understanding that careful management of tourism is required in order for it to be sustainable
- demonstrate knowledge of a case study of an area where tourism is important.

16.1 Growth of tourism

Tourism can be defined in many ways. A simple definition is that it is 'travel away from home for recreation and pleasure that involves at least one overnight stay'. The World Tourism Organisation defines **tourists** as people who 'travel to and stay in places outside their usual environment for not more than one consecutive year for leisure, business and other purposes not related to the exercise of an activity remunerated from within the place visited'.

These definitions place 'tourism' apart from the term **recreation**, which can be defined as 'the use of a person's leisure time for relaxation and enjoyment that does not involve travelling away from their home'.

Tourism has become the most popular global leisure activity:

- International tourist arrivals grew by 4.4% in 2015 to reach a total of 1184 million in 2015.
- Around 50 million more tourists (overnight visitors) travelled to international destinations around the world in 2015 compared to 2014.
- The year 2015 marked the 6th consecutive year of above-average growth, with international arrivals increasing by 4% or more every year since the post-crisis year of 2010.
- International tourist income was worth over $1.4 trillion, over $4 billion a day, to global export earnings. In the same year, there were between 5–6 billion domestic tourists.
- During the period 2000–2016, tourism has been growing at a rate of between 3 and 5% per year.
- Tourism's contribution to the global gross domestic product (GDP) is estimated at about 10% and its contribution to employment is estimated to be 9% of the overall number of jobs worldwide (direct and indirect).
- An ever-increasing number of destinations worldwide are opening up to, and investing in tourism, turning it into a key driver of socio-economic progress through the creation of jobs and businesses, export revenues, and infrastructure development in the countries and areas in which it is found.
- In the last 60 years, tourism has experienced continued expansion and diversification, to become one of the largest and fastest-growing economic sectors in the world. During this period, tourism has shown virtually uninterrupted growth (Table 16.1).

	1950	1980	1995	2014	2015
International tourist arrivals (millions)	245	278	527	1133	1184
International tourist receipts ($ billion)	2	104	415	1245	1400

Table 16.1 Growth in tourism, 1950–2015

FACT

According to United Nations World Tourism Organisation long-term forecast *Tourism Towards 2030*, international tourist arrivals worldwide are expected to increase by 3.3% a year between 2010 and 2030 to reach 1.8 billion by 2030.

Tourism is the world's biggest service industry. Over the last decade it has also been the world's fastest growing industry. Tourism is vital for the economy and employment in many countries, such as the UAE, Egypt, Greece and Thailand, and many island nations, such as the Bahamas, Fiji, the Maldives and the Seychelles. The service industries for tourism include transportation services, such as airlines, cruise ships and taxis; hospitality services such as accommodation, including hotels and resorts; and entertainment venues, such as amusement parks, casinos, shopping malls, various music venues and the theatre.

FACT

Over 80 of the world's 196 countries now earn more than $1 billion annually from international tourists.

A tourist development can act as a catalyst for further economic growth.

Reasons for the growth of tourism

Tourism has been in existence for over 400 years when rich Europeans, Japanese and North Americans visited spas to benefit from the mineral waters that they contained. The early 1900s saw social, economic and cultural changes which led to the emergence of a larger, middle-class population. These people could afford to take one or more days off work and take advantage of an improved transport network, particularly the railways.

Since 1945, several social and economic changes have taken place to encourage tourism, including:

- a **rise in incomes** which gave people, after they had paid for their basic needs, spare 'disposable' money which they could spend on leisure activities and tourism
- **increased leisure time** caused by a shorter working week, flexitime, paid annual holidays, earlier retirement with a pension
- **increased mobility** as a result of private car ownership, improved roads, a decrease in the cost of air travel combined with greater numbers of airports, the expansion of budget airlines, like Air Asia, Ryanair, and Easyjet and the increased numbers of flights to a wider range of destinations
- **increased media coverage** by television, magazines and the internet of different holiday destinations and types of holiday
- governments have used **major sporting events** such as the winter and summer Olympic Games, World Championship Athletics, Football, Rugby and Cricket World Cups to advertise tourist opportunities
- **increased international migration** encouraging more people to visit relatives and friends abroad.

16.2 The benefits and disadvantages of tourism to receiving areas

Advantages

There are many economic, social and cultural advantages of tourism.

Economic

- A growth in income which will have an impact at both a national and a local level as it will provide extra finance for new developments in infrastructure, education, health care, etc. Both the tourist industries and the employed workers will usually pay taxes to their governments which increases government income. This helps pay for major infrastructure developments such as in health care, education, water and energy supply and roads. All of which may be used by the local populations.
- An increase in foreign currency which helps pay for goods and services imported from abroad.
- Increased employment opportunities in the many jobs created directly and indirectly by tourism.
- It can encourage other developments to take place in an area – cumulative causation.
- By increasing employment opportunities, it can help reduce migration, especially from rural areas. This employment can be in small cafés, hotels, souvenir shops, tour guides, local taxis, etc. Many of these jobs will be in the **informal sector** (see Chapter 13) which helps the people of LICs in particular.

Social and cultural

- an increased understanding of different peoples, cultures and customs
- increased cultural links with other countries
- increased foreign language skills for both visitors and hosts
- increased social and recreational facilities for local people
- the preservation of traditional heritage sites and customs.

Disadvantages

There are many economic, social and cultural disadvantages of tourism:

Economic

- seasonal unemployment – if people come for summer sun or winter skiing the rest of the year may mean few or no tourists and little or no employment.
- leakage of tourist income – airlines, hotels and tourist activities in LICs are often foreign owned which can mean that 60–75% of tourist income may either never come to, or may leave, a LIC.
- many tourists may spend most of their money in, have most of their meals in, and do trips organised by their hotels so, although the companies and individuals that own the hotels benefit, they have little impact on the wider local economy
- many of the jobs provided by tourism in an LIC are low paid and low skilled. Many of the higher skilled and better paid jobs will be taken by foreigners
- some locations may become over dependent on tourism. Should a natural or human disaster occur, they may have little alternative income
- water shortages caused by tourist complexes, hotels and golf courses using large amounts may lead to local farms and villages not having enough
- traffic congestion and pollution from litter, increased sewage, etc. especially at honeypot sites where there are large numbers of tourists in one location
- damage to the physical landscape – ranging from ski areas increasing soil erosion to damage to coral reefs
- increasing competition from other new tourist destinations.

Social and cultural

- the demonstration effect – local people may copy the actions of some tourists and in doing so, their traditional values may be abandoned
- an increase in prostitution and the development of 'sex tourism'
- young people may truant from school to work in the informal tourist industry and earn money as unofficial guides or selling souvenirs
- people leave family farms to work in the tourist industry and this makes it more difficult to run the farms without their help
- people may be moved from their houses and land to make way for tourist developments; local landowners may sell large areas of land and coastline to non-local or foreign buyers who may then deny access to local people
- house and land prices may rise as non-locals buy them and put them out of the reach of local people.

TERMS

honeypot site: a location attracting a large number of tourists who, due to their numbers, place pressure on the environment and local people

demonstration effect: local people may copy the actions of some tourists in different dress, diets, habits and, possibly, alcohol and drug abuse

Self-test questions 16.1

1. Suggest reasons why there has been an increase in international tourism.
2. How might the growth of tourism be likely to improve the lives of people who live in LICs?

16.3 Sustainable tourism

Most areas of the world which are developing tourism are working towards ensuring that the industry will be there in the long term; in other words, that it remains sustainable.

Ecotourism is a sustainable form of tourism which allows people to visit natural environments and traditional cultures while enabling local people to share in the economic and social benefits of tourism. At the same time, measures are taken to protect the natural environment, the local way of life and the traditional culture. Ecotourism is encouraged in the National Parks in Kenya and the protection of the environment is managed in several ways:

- restricting tourist numbers to both the parks and to certain areas of the parks
- a limited number of tourist firms are licensed to use the Parks and their activities are regulated in several ways. For example, minibuses are not allowed within 25 metres of animals. Unfortunately, the wild animals can be disturbed and prevented from hunting, mating or resting, or be separated from their young – all essential if they are to survive in a hostile environment
- limiting or preventing the destruction of natural vegetation and habitat that is cleared for tourist development
- ensure that any building developments are low level and made out of local materials and in local styles
- use local labour in as many activities as possible and provide training for local people
- educating tourists with regard to environmental and conservation issues in the parks
- restricting access to sensitive areas of the Parks and at certain times of the year
- employing local people to check and to clear up any tourist rubbish regularly
- ban tourists from any hunting activities.

TERMS

ecotourism: tourism that minimises harmful impacts on the environment while at the same time using tourism to help local communities

cultural tourism: visiting and learning about different cultures

sustainable carrying capacity: the maximum population size that the environment can sustain indefinitely, given the food, habitat, water and other necessities available in the environment

FACT

There are about 900 000 Maasai in Kenya and northern Tanzania and Amboseli National Park contains a large number of this total.

Self-test question 16.2

1 For a named area which you have studied, describe and explain the impact of tourism on its environment.

Case study

Tourism in an LIC – Kenya, East Africa

Tourism in Kenya has developed by using:

- its beautiful natural landscapes and ecosystems, including palm-fringed tropical beaches, many of which have coral reefs close to the shore, which it has tried to carefully conserve
- its different cultures, from the coastal Swahili people to inland tribal peoples like the Maasai.

Tourism is Kenya's second largest source of foreign exchange revenue after agriculture. The main tourist attractions are the 19 National Parks and Game Preserves where tourists take photo safaris. Amboseli National Park is the largest park at 390 km^2 and is at the centre of an 8000 km^2 ecosystem that spreads across the Kenya–Tanzania border. A major attraction for tourists interested in cultural tourism is the Maasai people, who live in this and other Kenyan Parks. The Maasai are an indigenous (native) African group of semi-nomadic people located in Kenya and its neighbour Tanzania. Due to their distinctive customs and dress and their location near the many game parks of East Africa, they are among the most well-known African ethnic groups internationally.

Government policies, such as the preservation of the National Parks and reserves, often mean keeping the Maasai out of their traditional grazing areas. Along with an increasing population of Maasai, this has made the traditional Maasai way of life increasingly difficult to maintain. To overcome these problems, many projects and new forms of employment in tourism for the Maasai people have been introduced. This includes employment as security guards, waiters and tourist guides, and providing help to establish small businesses, such as small shops selling Maasai made bracelets, clothing, etc.

The impact of tourism on the Kenyan environment

Tourism can have a major impact on the natural environment. The Kenyan National Park environments are fragile and sensitive – both the natural landscapes and the animals that live in them. As a result, they can be changed and damaged by the thousands of tourists who visit them every year.

When visiting the Park as part of a Safari (meaning journey), most tourists will take one of the many tourist buses to get into the Park and close to the animals. Minibuses are meant to keep to well-defined trails but sometimes drivers may go off the tracks to get closer to animals so that tourists can get better views and photos. This can increase the driver's tips and increase their incomes. During the wet season, the tracks can also get very muddy so drivers drive outside them and widen them, some end up 50–60 metres wide as a result.

Tourism has brought great advantages to Kenya but it has also brought disadvantages. On the **positive** side, it provides jobs in areas where employment would be very limited and the income from tourism helps raise the standard of living of the local people and improve their quality of life.

On the **negative** side, The National Parks and Game Reserves are often part of the traditional grazing areas of the nomadic tribes. These tribes move their animals over very large areas so that they do not overgraze any area. However, the Park boundaries stop the tribes using quite large areas of land, which means that they have smaller areas to put their animals on and the land can be overgrazed as a result. This means that the tribal people lose income and see their traditional environment become degraded. Many tribal people therefore have to live in more permanent settlements earning money from selling products they make or from putting on dance performances for tourists. Recently, the Kenyan government has worked more closely with the tribal peoples to give them a share of the tourist income.

The Kenyan Coast

The beautiful, natural coastal environment of Kenya is also a very sensitive and fragile environment. Tourists are attracted by the beautiful beaches, fringed by coconut palms, blue skies, warm climate and clear, warm blue seas. Thousands of tourists visit the coral reefs every year and this can have a major negative impact on the local area and population:

- One of the main impacts has been the building of hotels and tourist facilities on the coast. Some of these developments are very large and result in pollution of the air, water and also noise, as well as visual pollution.

- Leakage of money from tourism – many of the tourist companies are foreign owned and so the profits disappear abroad – only 10–20% of the money may stay in Kenya.
- New social problems may occur such as prostitution, crime, drugs and alcohol abuse.
- Local children may skip school to work in the tourist areas.
- Tourist developments also produce a lot of waste from rubbish to waste water and sewage. Unfortunately this has not always been properly dealt with in the past and much of it was put into the sea where it polluted the beaches and reefs – destroying the very things the tourists had come to see!
- The building of hotels may mean that local people lose their homes and jobs – many of the hotels are built on the coast where the local people have their houses, settlements and where they work. It also affects the fishing industry.

Many of these problems can be solved by careful management, educating and guiding tourists, as well as the operators who take people onto the reefs. Numbers of boats and tourists can be controlled and rules made up to protect the reef – such as not allowing corals and shells to be taken from the reefs either by tourists or by local people to sell in shops.

There are many advantages and some disadvantages of tourism on the coast (Table 16.2).

However, many of these problems can be overcome by thoughtful, sensitive management and careful control of tourist numbers so that a sustainable carrying capacity is observed. Coral reef pollution can be managed with guidelines, rules and regulation. Numbers of visitors and how they conduct themselves can be regulated to ensure that the reefs provide a long-term income for local people and businesses. The major parks, such as Amboseli and the marine parks near Mombasa have all introduced laws, rules, regulations and advice to increase their sustainability.

Advantages	Disadvantages
• Most of the people who live on the coast are in small fishing villages – the people depend on the coral reefs for fishing as a source of food and to sell the fish as an income. The fishermen can greatly increase the income by using their boats to take people on trips to see and dive on the reefs. They also have a whole new market for what they catch in the new hotels and restaurants. • The hotels, restaurants and other facilities also provide hundreds of new jobs, from building to maintenance, making furniture, cleaning and cooking, jobs in a wide range of water activities, in restaurants and cafés, etc., all of which add enormously to the local people's incomes and may provide career opportunities. • The opportunities are wide-ranging as tourists have a wide range of tastes and demands. • Tourists can be quite demanding in wanting the best facilities, like satellite TV, and so there is a lot of competition to provide these and, again, offers more opportunities for more jobs. • Local crafts people also have a much bigger market for their goods as well as local farmers who can provide the food for the hotels and restaurants. • Shops now provide a wide range of goods, which greatly increase local incomes. • Transport also provides many jobs with coach and bus companies needing drivers, engineers and cleaners.	• The local people who live on the coast are mainly Swahili and they follow the Islamic religion. This can lead to problems when tourists are unaware of the local culture, customs and values and can offend local people by their dress and activities. Young local people can be influenced by the behaviour of tourists and copy them, such as in their clothes – in the worst cases, by drinking alcohol and taking drugs – the demonstration effect. • At the worst level, contact with the tourists has led to prostitution and the introduction of diseases like HIV/Aids. • The coral reefs are really sensitive to any pollution – they are easily killed by any pollution, even sun tan cream from people swimming, or by being touched or walked on.

Table 16.2 Advantages and disadvantages of tourism on the coast

Sample question and answer

1 a What is meant by the term ecotourism? [2]

b State three benefits which tourism brings to an LIC. [3]

c Suggest reasons why some people are worried about the continued growth of the tourist industry in some areas of the world. [4]

TIP

Try to avoid giving general statements in your answer, such as 'bad habits' or 'locals would lose their culture' as these will not gain credit. Another general statement that is often used is 'improves standard of living' but the student does not go on to say how. Better answers would include ideas such as 'increased economic growth, enables spending on health care or education, the development of infrastructure, more business for local shops', all of which can be further expanded.

a Ecotourism is a type of tourism which lets people visit natural environments without damaging the environment **[1]**. There will be laws to protect the place from being damaged **[1]**. It also helps local people to get jobs looking after and being guides for the tourists so that they benefit from the tourism **[1]**.
(A maximum of 2 marks will be awarded.)

b Tourism can bring many benefits to an LIC. They include tourists bringing in valuable foreign exchange **[1]**. This allows the people in the LIC to spend money on building schools and hospitals **[1]**. The money from tourism can also be used to build roads and water and electricity systems **[1]**. Local farmers will have a bigger market for their crops as hotels will need food for the tourists **[1]**. Local people will be able to sell craft items as souvenirs to the tourists **[1]**.
(A maximum of 3 marks will be awarded.)

c People may be worried about the growth of tourism because it may have a bad effect on the natural landscape and animals in many ways. For example, tourism can cause the pollution of the sea and rivers **[1]** through sewage coming out of hotels and tourist areas **[1]**. They leave a lot of litter and rubbish behind which may get eaten by animals and make them ill **[1]**. There may be a lot of traffic congestion on local roads **[1]** and the cars can cause a lot of air pollution **[1]**. Tourists may be badly behaved and noisy **[1]** and use a lot of water for their showers and baths **[1]**.
(A maximum of 4 marks will be awarded.)

Exam-style questions

1 Tourism is important in many countries. How can it be developed so that it is sustainable? [5]

2 Describe the possible disadvantages of tourism for people who live in LICs. [5]

3 For a named area which you have studied, explain why the tourist industry has developed there and the advantages it has brought to the area. You should refer to the area's physical and human attractions. [7]

Chapter 17

Energy

Learning summary

By the end of this chapter, you should be able to:

- describe the importance of non-renewable fossil fuels, renewable energy supplies, nuclear power and fuelwood; globally and in different countries at different levels of development
- evaluate the benefits and disadvantages of nuclear power and renewable energy sources
- demonstrate knowledge of a case study of the energy supply in a country or area.

17.1 Energy resources

Global energy consumption

In 2015, non-renewable fossil fuels supplied 86% of the world's energy. Of the three main fossil fuels, **oil** provided 32% of world energy, **coal** 30% and **natural gas** 24%. Of the remaining 14%, **nuclear energy** provided 4% and **hydro-electric power (HEP)** provided 7%; all the other renewable energy supplies combined provided less than 3%.

TERMS

non-renewable fossil fuel: a fuel that is either finite or **non-sustainable**. This is because their use will eventually lead to them running out. Non-renewable fossil fuels include coal, oil, natural gas and peat.

renewable energy supply: a resource that can be used continually without running out – it is a **sustainable** resource. Wind, water, geothermal, wave, tidal, biogas, **biofuels** (like ethanol) and solar energy are examples.

biofuels: fuels produced from living organisms or from their by-products, such as food waste

FACT

Some would argue that uranium, the raw material that fuels nuclear power, is a non-renewable resource and so this could be added to the fossil fuels.

Figure 17.1 shows the relative global importance in the consumption of different types of energy (converted into millions of tonnes of oil equivalent) from 1989–2014. The graph shows that the non-renewable fossil fuels (coal, oil and gas) continue to dominate energy supplies at a global level.

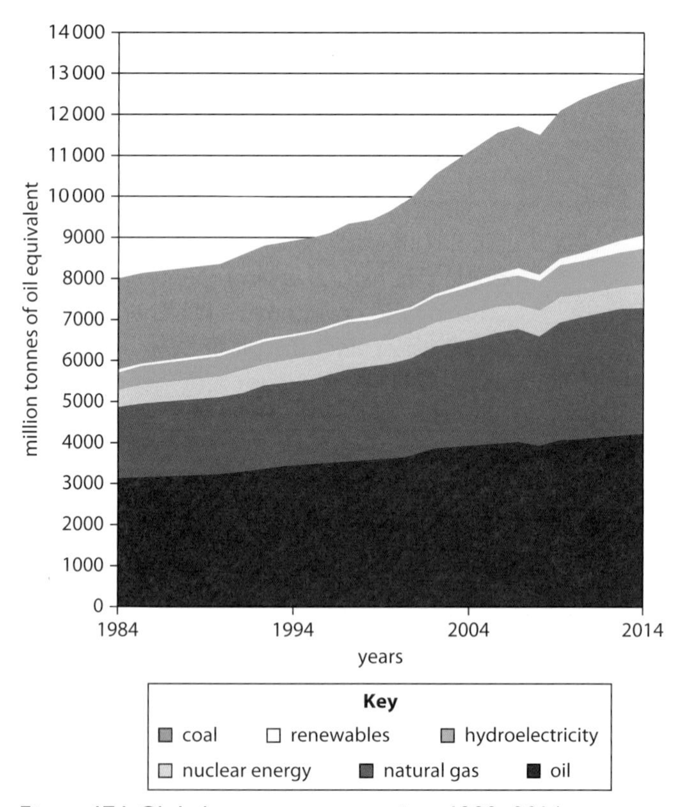

Figure 17.1 Global energy consumption, 1989–2014

In the period through to 2035, it is currently projected that the non-renewable fossil fuels will remain the dominant source of energy powering the global economy. They will provide around 60% of the growth in energy and will still account for almost 80% of total energy supply in 2035 (down from 86% in 2015). **Natural gas** is the fastest growing fossil fuel and its share is gradually increasing. **Oil** will continue to grow steadily, although the trend in the decline in its share will continue. In contrast, the growth of **coal** is projected to slow sharply, so that by 2035 the share of coal may be at an all-time low, with gas replacing it as the second-largest fuel source.

Among non-fossil fuels, those **renewables**, including biofuels but excluding HEP, are projected to grow rapidly, so that their share will rise from around 3% today to 9% by 2035.

TIP

Questions on energy frequently use graphs to compare and monitor over time the changing use of the different energy types. When writing your answer, double check that you have correctly identified values on the axes such as years and percentages, to ensure that your answer is accurate.

The uneven spread of energy use

The use of these energy resources is not evenly spread across the countries of the world – there is a very uneven distribution. Currently, the richest 25% of the world's population in the HICs use over 75% of the world's available energy resources.

The huge contrast between LIC, MIC and HIC use of energy can be seen in Table 17.1.

Country	Oil consumption (tonnes of oil equivalent per person per year)
Niger	152
Bangladesh	216
Congo	292
India	630
Indonesia	850
Brazil	1 450
China	2 350
UK	2 750
USA	6 920
Qatar	19 200

(Source: World Bank, 2013)

Table 17.1 Energy consumption

Figure 17.2 shows that the pattern of consumption (converted to percentage) is not equal across the world. Note the high percentages of oil and gas consumption in the Middle East.

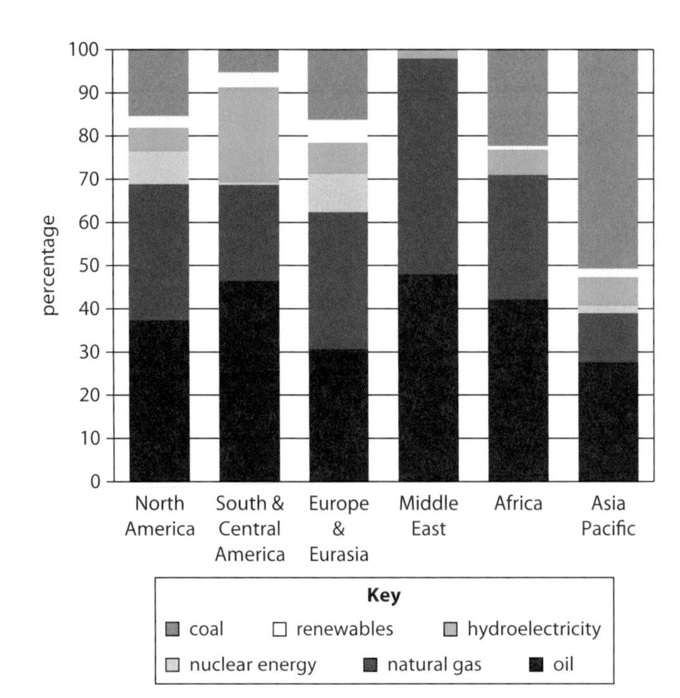

Figure 17.2 Regional energy consumption, 2015

17.2 Non-renewable fossil fuels

Coal

Most coal is used for producing electricity in thermal power stations. It has advantages and disadvantages as a source of energy (Table 17.2):

TIP

For more information on the Enhanced Greenhouse Effect, see *Chapter 19 Environmental risks of economic development*, Section 19.3 The enhanced greenhouse effect and global warming.

TERMS

greenhouse gas: a gas in the atmosphere that is capable of absorbing infrared radiation, thereby trapping and holding heat in the atmosphere

acid rain: rain made up of water droplets that have absorbed certain gases in the atmosphere and have become slightly acidic; usually due to atmospheric pollution and, most notably, the excessive amounts of certain sulphur and nitrogen gases released by burning fossil fuels, cars and industrial processes

Advantages	Disadvantages
• World reserves of coal will last at least 118 years at the rate they are currently being consumed. • Improved technology has improved output per miner making it cheaper and it has also improved the efficiency and the cleanliness of emissions of coal fired power stations • Apart from generating electricity, it has other uses and can be used for heating and making coking coal (used in the iron and steel industry).	• The cheapest and most accessible sources of coal have been used up, and the cost of production has risen as a result in many areas. • It causes air pollution through its production of **carbon dioxide** (a greenhouse gas) and **sulphur dioxide** which produces acid rain. • The greenhouse gases it produces contributes towards the **enhanced greenhouse effect (EGE)**, leading to an increase in global warming, the melting of ice caps and a rise in sea level, which will cause areas of coastal lowland to be flooded. • **Open cast mining** harms the natural environment and deep coal mining is dangerous for miners. • Coal is heavy and bulky to transport so most thermal power stations have to be on or beside coal fields or beside a deep water port, as a **break of bulk** location (see Chapter 15), which can import the coal in bulk ore carriers.

Table 17.2 Advantages and disadvantages of coal as a fuel source

Advantages	Disadvantages
• More efficient to burn, transport and distribute by pipeline and tanker. • Less harmful to the environment than coal – gas is even cheaper and cleaner than oil. • Can be used for generating electricity – gas is a very popular fuel for thermal power stations. • Oil provides the raw material for the petrochemical industry.	• In 2016, **world reserves** of oil may only last 51 years and natural gas 81 years at the rate they were being consumed in 2014. • '**Peak oil**' has been reached – the world now consumes more oil than it finds in new oil fields. • Danger of **pollution through oil spills** – e.g. the Gulf War in Kuwait in 1990 when several hundred oil wells were set alight causing massive air pollution, while oil spillages at sea kills aquatic life and may have a massive impact on fishing industries as in the Gulf of Mexico oil spill in 2010. • When burnt, gas and oil give off **nitrogen oxide** and **sulphur dioxide** respectively, which contribute to **acid rain**. • **Prices can fluctuate widely** – oil went from \$150 to \$30 a barrel between 2008 and 2016. In 2016, it stabilised at around \$40–50 per barrel. • Oil and gas pipelines are **targets for terrorism**. • **Political decisions** can cause supply problems, e.g. the turning off of gas supplies to Europe by Russia in 2006, and to the Ukraine in 2009, 2014 and 2015.

Table 17.3 Advantages and disadvantages of oil and natural gas as fuel sources

Oil and natural gas

These are the main sources of energy for many HICs (Table 17.3). In early 2017, the USA was the world's largest producer and consumer of oil (9.0 million barrels, approximately 1 259 000 tonnes, per day in 2017) and natural gas (3.7 million barrels, approximately 518 000 tonnes, per day in 2017).

17.3 Nuclear energy

No other source of energy has caused more controversy (Table 17.4). Nuclear energy is mostly used in countries that do not have their own large supplies of fossil fuels, such as France, Japan, South Korea and Belgium. In 2016, nuclear energy accounted for 11.5% of global electricity production. Currently, there are 441 nuclear reactors in operation, with 67 new reactors under construction. The five countries with the world's highest nuclear electricity generation capacity in terms of percentage in 2016 were: France (77%), Slovakia (57%), Hungary (54%), Ukraine (49%) and Belgium (48%).

Self-test questions 17.1

1. Using examples, describe the differences between non-renewable and renewable sources of energy.
2. Why is nuclear energy a controversial source of energy?

17.4 Renewable energy

There has been a rising awareness globally that the development of renewable energy resources is critical, not only for addressing the problems of **climate change**, but also for creating new economic opportunities and for providing energy access to the billions of people still living without the modern energy services available to the HICs. In recent years, the advances in renewable energy technologies, the global increases in renewable energy capacity and the rapid reduction in the cost of renewable forms of energy

Advantages	Disadvantages
• It is not a bulky fuel – 50 tonnes per year for a power station compared to 540 tonnes of coal per hour for a large coal fired thermal power station. • Nuclear waste is very small in quantity and can be stored underground. • It does not produce greenhouse gases, does not produce carbon emissions or contribute to acid rain. • It stops countries relying on imported oil, coal and gas. • There are relatively large reserves of uranium. In 2015, Kazakhstan, Canada and Australia were the largest suppliers of uranium ore. • Nuclear power stations have relatively low running costs.	• It can be very dangerous if there is a **nuclear accident** and radioactive materials are released in to the environment, as happened in 1986 when a reactor at **Chernobyl**, in the Ukraine, exploded. The explosions and the resulting fire sent a plume of highly radioactive fallout into the atmosphere and over a large area. Four hundred times more fallout was released than had been by the atomic bombing of Hiroshima. The 2005 report attributed 56 direct deaths (47 workers and nine children with thyroid cancer) and estimated that there may be 4000 extra cancer deaths among the approximately 600 000 most highly exposed people. In 2011, when the **Fukishima** reactor exploded in Japan, over 200 000 people had to be evacuated from the immediate area and the long-term effects are at present unknown. • **Nuclear waste** can remain dangerous for several thousand years and so there are problems in storing it. • The **cost of shutting down (decommissioning)** nuclear reactors is very high and there is constant debate as to who will pay for this – national governments or electricity companies.

Table 17.4 Advantages and disadvantages of nuclear energy sources

Advantages	Disadvantages
• Clean and non-polluting • Renewable and sustainable.	• Both tides and waves do not operate all the time and so they need back-up energy systems • High cost of tidal dams and high production costs, at the present time.

Table 17.5 Advantages and disadvantages of wave and tidal power

have been due to changes in government policies, which have encouraged **increased investment** in the different forms of renewable energy and this has reduced costs through **economies of scale**.

The most rapid growth and the largest increase in renewable capacity in the last few years has been in **wind**, **solar photovoltaic (PV)** and **HEP**. Renewable energy, including HEP, is likely to account for over a third of the growth in power generation, causing their share of global power to increase to 16% by 2035.

In LICs, renewable energy systems offer a real opportunity to provide modern technology and increase access to energy. The falling costs of solar PV have made it the most economical source of power for off-grid electrification and its costs are likely to fall by around 40% over the next 20 years. Over 90% of the population of many LICs (over 2 billion people) do not have access to electricity, which most people in HICs take for granted. A similar number of people depend on fuels such as wood and charcoal which they have to cut and gather or use the dung of their animals to cook their daily meals. A growing population means that it is becoming increasingly difficult for many people to find sufficient and sustainable supplies of energy.

The development of sustainable and renewable energy resources would both greatly help these people and provide an alternative to the finite non-renewable fossil fuels that so much of the world depends on for its energy.

Hydro-electric power (HEP)

HEP generates the highest proportion of renewable energy and 6·5% of the world's total energy. In some countries, however, it forms a very high proportion of their total energy use – in Norway 96% of electricity, Paraguay 93% and Brazil 86%. The major producers of HEP are China, Canada, Brazil, the USA and Russia.

TIP

For a case study on HEP (**Three Gorges Dam**, China) and the advantages and disadvantages of HEP, see Table 17.11.

Wave and tidal power

Waves and tides have high energy levels but it has proved difficult to design a wave power generator that can withstand large storm waves and saltwater (Table 17.5). Tidal barrages, such as the Sihwa Lake Tidal Power Station in South Korea and the Rance Estuary in northern France, have the potential to produce large amounts of electricity. The Sihwa Lake station is the world's largest, with a total power output capacity of 254 MW, surpassing the 240 MW Rance Tidal Power Station.

Geothermal energy

Geothermal energy is used commercially in over 70 countries (Table 17.6). In volcanic areas, heat comes close to the Earth's surface from the magma beneath the surface. Rainwater infiltrating the ground becomes heated and may rise to the surface as steam or as hot springs. It can then be used to heat buildings or in the production of electricity. Alternatively, water can be pumped underground, where it is heated, and then brought back to the surface and used in the same way.

FACT

In Iceland, geothermal energy heats 95% of the buildings in the capital city of Reykjavik.

The USA leads the world in geothermal electricity production with over 3000 MW of installed capacity from 77 power plants. The Philippines and Iceland are the only countries to generate a significant percentage of their electricity from geothermal sources. In both countries, 27–30% of their power comes from geothermal plants. However, in 2016, geothermal power supplied less than 1% of the world's energy, but this is increasing. The largest share of new geothermal power capacity came on line in Kenya, producing over 13% of Kenya's electricity in 2016 by using the geothermal heat of the African Rift Valley. It has highlighted the potential for geothermal energy in East Africa.

Advantages	Disadvantages
• Renewable, clean and non-polluting, as well as cheap • Geothermal plants use very small areas of land – existing geothermal plants use 3.5 square kilometres per megawatt (MW) produced, compared to 32 square kilometres for coal facilities and 12 square kilometres for wind farms • It can be used at a variety of scales; a large geothermal plant can power an entire city while smaller power plants can supply more remote sites, such as rural villages.	• High cost of construction and maintenance • The steam contains sulphuric gases • Although geothermal sites are capable of providing heat for many decades, locations may eventually cool down.

Table 17.6 Advantages and disadvantages of geothermal energy

Wind power

Wind power is the conversion of wind energy into a useful form, such as electricity, using wind turbines. A group of wind turbines is a **wind farm** (Table 17.7). The land between wind turbines may be used for agricultural or other purposes. A large wind farm may consist of several hundred individual wind turbines. The largest wind farm in the world is the Gansu Wind Farm Project, located in an arid desert area in western China. It has several thousand turbines, distributed over an extended area, and is expected to grow to 20000 megawatts by 2020, at an estimated cost of $17.5 billion.

Much of the new wind farm capacity in Europe is being located offshore, such as the London Array, a 175 turbine, 630MW offshore wind farm located 20km off the Kent coast in the Thames Estuary in the UK. This wind farm is intended to reduce annual carbon dioxide emissions by about 900000 tonnes, equal to the emissions of 300000 passenger cars.

FACT

Global wind power capacity expanded from just over 6000MW in 1996 to almost 456000MW in 2016.

Advantages	Disadvantages
• Clean, renewable, sustainable and pollution free • It is cheap • Provides an income for the landowner on whose land the wind turbines are located • Can be used in off-grid locations in remote areas, or island locations.	• Wind does not blow all the time and so needs to have a back-up power source • Large turbines can cause visual pollution (as a result, many are now being put offshore) • Expensive to build and many are needed to replace conventional power stations • Can be noisy and can also kills birds if they fly into the rotors • Can disrupt radio and TV signals.

Table 17.7 Advantages and disadvantages of wind power

Advantages	Disadvantages
• Clean, renewable, sustainable and non-polluting • Many LICs are in the tropics and have a lot of sunshine and therefore a lot of potential for using solar power • Can be used in remote locations where it would be expensive to build a network of electricity lines.	• Expensive to produce the stations and the photovoltaic cells • The weather – it needs long hours of sunshine, which is not always possible in many parts of the Earth.

Table 17.8 Advantages and disadvantages of solar energy

Solar energy

Solar power can generate electricity either by providing heated water for a thermal generator or by using photovoltaic cells which convert sunlight to electricity. It is commonly used to provide hot water and thermal energy for cooking (Table 17.8).

Biomass

Biomass refers to plant matter that is grown to generate electricity. Biomass may also include biodegradable wastes that can be burnt as fuel. As it is cheap and easy to construct and maintain, it has a great deal of potential for use in rural areas of LICs. (Table 17.9). The *Gober* gas model is used in both India and Pakistan and generates biogas from cow dung. Owing to its simplicity in construction and its use of cheap raw materials in the villages, it is seen as one of the most environmentally sound energy sources for the rural needs of LICs.

Advantages	Disadvantages
• It is cheap • Renewable and sustainable – the animals are constantly producing dung • Replaces fuelwood (see the section on 'fuelwood') and so helps prevent deforestation.	• It produces methane, which is a greenhouse gas • The dung that is used for some bio-digesters sometimes cannot be used for fertiliser afterwards.

Table 17.9 Advantages and disadvantages of biomass

Biofuels

Many countries, such as those in the European Union, are highly dependent on oil and its derivatives – petrol, diesel and kerosene – for their transport fuels and many have to import either the crude oil or its derivatives. To reduce their dependency on fossil fuels imports, and to cut greenhouse gas emissions, many countries have adopted measures to encourage the production and use of sustainable biofuels (Table 17.10). When used as a vehicle fuel, the biofuels are sometimes referred to as **biodiesel**.

Globally, biofuels are most commonly used to power vehicles, heat homes and for cooking. **Agro-fuels** are biofuels produced from crops. There are two common ways of producing liquid and gas agro-fuels.

1 Grow crops high in sugar content (such as sugar cane, sugar beet and sweet sorghum) or starch content (such as corn/maize) to produce ethyl alcohol (ethanol). Ethanol accounts for nearly three-quarters of global biofuels production, and is dominant in North America and South and Central America. In Brazil, the fuel sold at filling stations for vehicles is 22–26% ethanol, but newer vehicles can run on ethanol alone.

FACT

In 2015, 788 million litres of biofuel was used in the UK – 3% of total road and non-road mobile machinery fuel. The main source for the biodiesel was used cooking oil and for the bioethanol, sugar beet imported from France.

Biodiesel is most important in Europe and Eurasia.

Advantages	Disadvantages
• Renewable and sustainable • Jatropha can be grown on infertile, marginal land that crops cannot be grown on and it is drought resistant.	• The effect of a drop in oil prices – as oil prices drop, biofuel may become a more expensive option • The '**food versus fuel**' debate – many farmers are now growing biofuel crops instead of food crops as they are being paid more money for the biofuel crops. The EU has put a cap of 7% on the contribution of biofuels produced from 'food' crops, and is putting a greater emphasis on the production of advanced biofuels from waste biomaterials • Increased carbon emissions from burning biofuels • The deforestation of land to plant biofuel crops, especially oil palms as in Indonesia, and the soil erosion that may result as well as the impact on soil being washed in to rivers and lakes. There is also the loss of rainforest habitat in Indonesian and Malaysian rainforests being cleared for the growing of oil palms • Human rights issues as people are moved off their land and farms by big landowners.

Table 17.10 Advantages and disadvantages of biofuels

2 The second is to grow plants that contain high amounts of vegetable oil, such as oil palm, soybean, algae, jatropha or pongamia pinnata. These oils can be burnt directly in a diesel engine, or they can be chemically processed to produce fuels such as biodiesel.

Fuelwood

Fuelwood is sometimes regarded as a biofuel, but more often it is seen as a separate fuel source. In LICs, about 2.5 billion people rely on fuelwood for cooking and heating. It is sometimes burnt directly or used to produce charcoal, which is then burnt. In Africa, trees are often called the '**staff of life**' as they provide many communities with many of their basic needs – fuel, shelter, food and shade. However, demand is now outstripping supply in many areas of LICs where it is the main fuel source. In Mali in West Africa, with a population of 12 million, 50 million tonnes of wood are cut from forests every year, which is greater than the rate of regrowth and replanting. People, especially women, are having to travel further away from their homes to cut and collect the fuelwood. Much of it is usually burnt inside or close to houses. As these are not properly vented, it is a major air pollutant for many families and currently accounts for 1.5 million deaths from respiratory illnesses in LICs every year.

Deforestation also leads to increased soil erosion and a decrease in water quality as soil gets into rivers streams and lakes. In turn, this can increase flood events in both size and number as there is a lack of interception of water on the valley sides so that surface runoff reaches the rivers more quickly and in larger amounts. Plus, river channels become filled with sediment so that they cannot contain as much floodwater and so flood more easily.

Case study

Energy supply in a country or area – Three Gorges Dam, China

The **Three Gorges Dam** spans the Yangtze River in China and is now the largest power station in the world in terms of installed capacity (Table 17.11). Inside the dam, there are 32 huge water turbines, each with a capacity of 700 megawatts (a typical nuclear reactor in a nuclear power station generates about 650 MW). Since it was fully completed in 2012, the Three Gorges Dam power station has the capacity to produce 22 500 MW of electricity when the turbines are running at full capacity.

Advantages	Disadvantages
Note: many of these also apply to other HEP sources. • The HEP is renewable, clean and non-polluting. • It is cheap (after the initial cost of the dam). • The dams also help with flood control. • They provide water for the local population and for farming (irrigation) and industry. • They can also be stocked with fish and support a local fishery. • They can be used for recreation and attract tourists. • The new source of electricity may attract manufacturing industry and new jobs will be created.	**Note:** many of these also apply to other HEP sources. • Dams are expensive to build. • The lakes created cover large areas of natural habitats and farmland. • They destroy wildlife habitats. • People may have to move (1.3 million people in the Three Gorges Dam area in China) and whole towns and communities may disappear along with historical and archaeological remains. • They may trap sediment carried by the river and gradually fill up; some Californian dams last only 25 years as a result. The Aswan Dam in Egypt prevents fertile alluvium flooding over the floodplain of the Nile and this affects farming below the dam. • The dams may collapse – although this is rare, a dam collapsed in Indonesia in March 2009 killing 55 people. • A long period of drought may mean that they do not have enough water to power their turbines – Venezuela suffered power shortages due to a severe drought in 2009, 2010 and 2016. • The reservoirs behind the dams create large areas of still water, ideal for mosquitoes to breed – malaria may then appear. They are also ideal for the bilharzia snail to breed and so bring a new problem of disease into an area. • The visual impact of dams and their reservoirs may change the look of a natural landscape.

Table 17.11 Advantages and disadvantages of hydro-electric power

Self-test questions 17.2

1 With the use of a case study, explain why hydro-electric power (HEP) may have several disadvantages.

2 What are the possible disadvantages of using fuelwood as a source of energy?

Sample question and answer

1 a What is meant by the term 'renewable energy'? [1]

b Give two advantages that using a renewable form of energy to generate electricity has instead of using a non-renewable form of energy. [2]

TIP

In part (a), **do not** compare renewable with non-renewable – it is a 1-mark question and simply requires a definition. In part (b), **do not** describe the disadvantages of non-renewable forms of energy.

a The term 'renewable energy' means a type of energy which can be used over and over again and will not run out. For example, the wind. **[1]**

b Two advantages that using a renewable form of energy to generate electricity has is that it will usually be environmentally friendly and not pollute the environment, compared to burning non-renewable fossil fuels **[1]**. It will also never run out, unlike a non-renewable type of energy like coal **[1]**.
(A maximum of 3 marks will be awarded).

Exam-style questions

1 a What is meant by 'fossil fuel'? Give an example. [2]

b Describe the advantages of using hydro-electric power (HEP) rather than other sources of energy. [4]

2 Explain why there are advantages and problems in using nuclear power. [5]

3 For a country or area that you have studied, describe the ways in which forms of renewable energy are being developed. [6]

Water

Learning summary

By the end of this chapter, you should be able to:

- describe methods of water supply and the proportions of water used for agriculture, domestic and industrial purposes in countries at different levels of economic development
- explain why there are water shortages in some areas and demonstrate that careful management is required to ensure future supplies
- explain the impact of lack of access to clean water on local people and the potential for economic development
- demonstrate knowledge of a case study of water supply in a country or area.

18.1 Supplying and using water

Of the water on the Earth, 97% is salt water and only 3% is fresh water. Of this freshwater, slightly over two-thirds of this is frozen in the form of ice sheets/caps (such as in Antarctica and Greenland) and glaciers (69%). The remaining unfrozen freshwater is found mainly as groundwater (30%), with only a very small amount present above ground, 0.3%.

The average annual precipitation on Earth's land surface is about 814 mm. This figure gives the impression that there is an abundant water supply but most of this is not available for use, as 56% is lost by evaporation and transpiration by forests and 5% by rain-fed agriculture. The remaining 39% is the available annual renewable freshwater for human uses and the environment. In 2014, this was equal to about 5800 m³ per person per year or 16 000 litres per person per day.

Large differences in precipitation input exist between continents, regions, countries and within countries:

- On a continental scale, South and Central America with the Caribbean region is the wettest continent, averaging more than 1600 mm/year. North America averages 637 mm/year, Africa 677 mm/year, while Europe has 545 mm/year.
- On a country scale, the country with the lowest precipitation worldwide, is Egypt, with only 51 mm/year. However, thanks to the Nile River bringing a large amount of freshwater into Egypt each year from its neighbouring countries to the south, the renewable freshwater resources per person, 1900 litres per day, was much higher than in Libya (300 litres per day).

FACT

The countries with the highest precipitation are usually islands, such as São Tomé and Principe in Africa with 3200 mm/year.

Depending on a person's diet and lifestyle, about 2000–5000 litres of water is used to produce one person's daily food and their daily drinking water and sanitation requirements. So, theoretically, there is more than enough water available worldwide, even taking into consideration the water that is needed to produce clothes and other consumable and non-consumable goods. However, freshwater is very unevenly distributed and a large amount of it is not easily accessible. Increasingly, this is leading to water shortages as population and the demand for water increases.

Methods of water supply

Surface freshwater is unfortunately limited and unequally distributed in the world. Almost 50% of the world's lakes are located in one country, Canada. In addition, pollution from various activities into rivers and lakes leads to surface water that is not suitable for drinking. Therefore, treatment systems (either large scale or at the household level) must be put in place.

Structures such as **dams** may be used to store water. In addition to providing a water supply, dams can be used for power generation (HEP), irrigation, flood prevention, water diversion, navigation, etc. If properly designed and constructed, dams can help provide a sustainable water supply. Large-scale dam projects may present various challenges to sustainability, including their negative environmental impacts on wildlife habitats, fish migration, water flow and quality, and socio-economic impacts resulting from resettled local communities. A sustainability impact assessment should therefore be performed to determine the environmental, economic and social consequences of the construction.

Groundwater accounts for greater than 50% of global freshwater and is a major source of water. The water emerging from some deep groundwater sources may have fallen as rain on the land surface many decades, hundreds, thousands or in some cases millions of years ago. Soil and rock layers naturally filter the groundwater as it percolates downwards. Groundwater can be accessed by digging **wells**, or drilling **boreholes**, down to the water table and either pulling up the water using a container from the bottom of a well or pumping it up.

The Great Artesian Basin in Australia is the largest and deepest artesian basin in the world, stretching over 1 700 000 square kilometres. The Great Artesian Basin provides the only reliable source of freshwater through much of inland Australia and underlies 23% of the continent. The basin is 3000 metres deep in places and is estimated to contain 64 900 cubic kilometres of groundwater.

Figure 18.1 shows the main features and characteristics of an artesian basin. A basic requirement is the presence of a layer/bed/strata of porous or permeable rock absorbing precipitation, with a layer/bed/strata of non-porous/impermeable rock below and above it, which prevents water percolating out of the porous/permeable rock. The water that percolates into the porous/permeable rock will eventually fill all the open/pore spaces in the rock and it will become fully saturated with water, forming an **aquifer** (See Chapter 9).

TERMS

renewable freshwater: surface water and groundwater resources

water table: the level underground at which the rocks or soil are completely saturated with water. It forms the upper limit of groundwater, and it can go up and down based on a number of factors, including the amount of precipitation and rates of evapotranspiration.

artesian basin: this is found where the groundwater is confined or trapped in an area underground

If a well or borehole is dug/drilled into this saturated rock, water will enter the well or borehole and it can be pumped/drawn up to the surface. A borehole drilled at **A** on Figure 18.1 will find that the water comes to the surface without being pumped, as it is experiencing hydrostatic pressure from the weight of water pushing down on the water below it in the aquifer. This will continue to occur until the top of the aquifer, the water table (shown by the dotted line), falls below the height

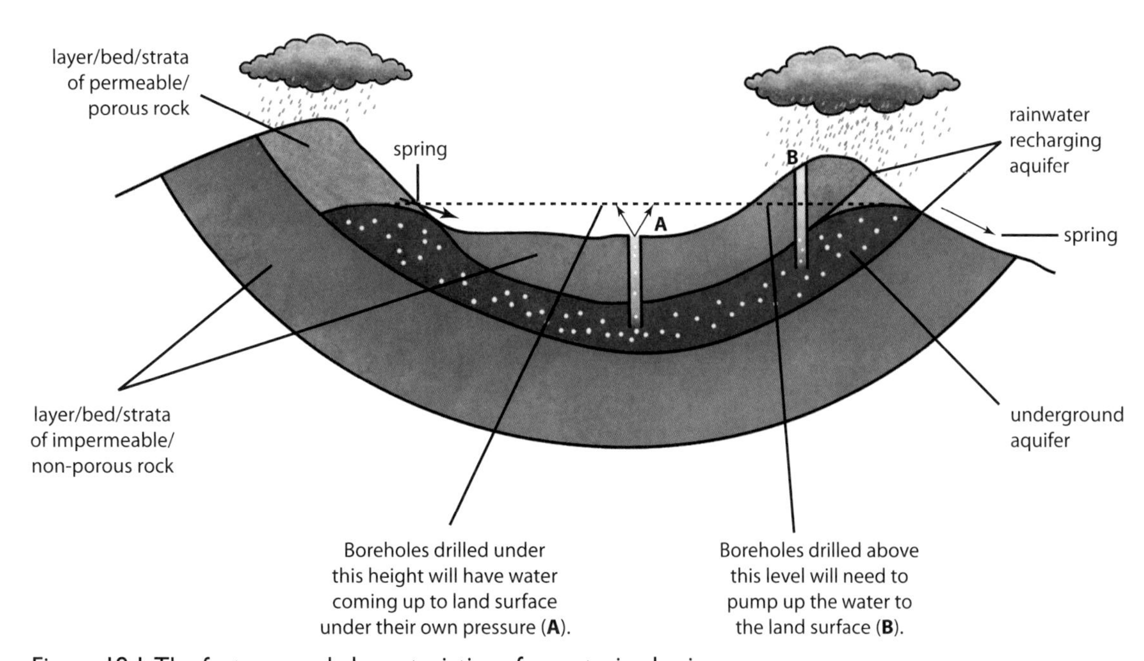

layer/bed/strata of permeable/ porous rock
spring
A
B
rainwater recharging aquifer
spring
layer/bed/strata of impermeable/ non-porous rock
underground aquifer
Boreholes drilled under this height will have water coming up to land surface under their own pressure (A).
Boreholes drilled above this level will need to pump up the water to the land surface (B).

Figure 18.1 The features and characteristics of an artesian basin

of the land surface into which the borehole has been drilled. When this happens, the water will have to be pumped to the surface.

In contrast, as the land surface at borehole at **B** in Figure 18.1 lies above the water table (the dotted line), the water will have to be pumped to the surface. If so much water is pumped out of the aquifer that the water table falls below the bottom of the borehole, the borehole will dry up. This is a frequent occurrence in many parts of the world where aquifers are over-used, or where the aquifers have not been recharged with precipitation as a result of long periods without rainfall or in areas experiencing droughts. Where the level of saturated rock (the water table) reaches the surface, the water will appear on the surface as a spring.

Three main factors determine the source and amount of water flowing through an underground aquifer/groundwater system:

- precipitation
- the location of rivers and other surface water bodies
- the evapotranspiration rate.

A long-term management plan is necessary when extracting groundwater since the effects of groundwater removal can sometimes take several years before becoming apparent. It is important to integrate the management of an aquifer/groundwater supply within adequate land and urban management.

Rainwater harvesting – collecting water from precipitation – is one of the most sustainable sources of water supply since it is easier to control and manage compared to other surface and groundwater sources, and it directly provides water of drinking quality. Although, theoretically, given the Earth's surface and precipitation, rainwater harvesting can meet global water demand, it can most practically be used as a supplement to other sustainable water supply systems given a level of uncertainty (especially with climate change) on its supply, and competing land-use.

Reclaimed water, or water recycled from human use, can also be a sustainable source of water supply. It is an important solution to reduce stress on primary water resources such as surface and groundwater. Reclaimed water must be treated to provide the appropriate quality for a given application (such as irrigation, industry use, human use, etc.). It is often most efficient to separate what is called greywater from blackwater, thereby using the two different water sources for different uses.

TERMS

recharged (aquifer): this can take place either naturally, as rainwater percolates from the surface into rocks and soil underground, or artificially by humans by flooding an area and allowing the water to infiltrate into the ground and percolate down to the underground aquifer, or by injecting water underground via boreholes or wells.

spring: a starting point for a stream or river

greywater: water from domestic activities such as washing

blackwater: water that contains human waste

Desalination has the potential to provide an adequate water quantity to those areas of the world that are freshwater poor, including small island states. However, the energy demands of reverse osmosis, a widely-used procedure used to remove salt from water, are a challenge to the adaptation of this technology as a sustainable one.

FACT

The cost of desalination averages around $0.81 to supply one cubic metre of water compared to an average of $0.16 to supply one cubic metre from other sources of water, such as from freshwater dams and reservoirs.

Bottled water is a 21st-century phenomenon whereby mostly private companies provide drinking water in a bottle for a cost. In some areas, however, bottled water is the only reliable source of safe drinking water for a population. In these same locations, often in LICs, the cost is prohibitively expensive for the local population to use in a sustainable manner. When the costs of producing bottled water are taken into account, it falls short in many situations of being a sustainable water supply. Economic costs, the pollution associated with its manufacturing (plastic, energy, etc.) and transportation, as well as extra water use, makes bottled water an unsustainable water supply system for many regions.

FACT

It takes 3–4 litres of water to make less than 1 litre of bottled water.

Water use

The amount of water used in different countries varies enormously. HICs use more water than LICs, and what the water is used for depends on the country. Table 18.1 shows the percentage of total water use for domestic, industrial and agricultural purposes in several countries.

In general, LICs and MICs will have most of their water used in agriculture and relatively little in industry or domestic use. In Bangladesh, Kenya and Egypt, agriculture is a large part of the economy so a large percentage of their water is used for that purpose. HICs tend to have a higher percentage for industrial use. There are exceptions, however. The USA, Australia and Japan are HICs, but still have a high amount of water used for agriculture because it is still an important activity. Iceland shows the impact of a harsh climate.

On a global scale, **agriculture** uses the largest amount of freshwater. It represents roughly 70% of all water withdrawal worldwide, with various regional differences. The productivity of irrigated land is approximately three times greater than that of rain-fed land. Irrigation is therefore an important factor in maintaining sustainable agricultural systems. In addition, global food production is expected to increase by 60% from 2000 to 2030, creating a 14% increase in water demand for irrigation.

Agriculture is also responsible for some of the surface and groundwater degradation because of the surface runoff of water from fields (from agro-chemicals and soil erosion). Agriculture has a dual role in providing a sustainable water supply. Firstly, by using water efficiently for irrigation and, secondly, by protecting surface and groundwater sources of water supply. Techniques for maintaining a sustainable water supply in agriculture include:

- using organic farming practices which limit the chemical substances that could contaminate water
- the efficient delivery of water to crops
- using micro-irrigation systems, zero tillage (planting seeds without ploughing/tilling the soil – often done by using seed pumps which push the seed into the ground), rainwater harvesting and drip irrigation methods, which allow water to drip slowly to plant roots by using pipes, valves, tubes and emitters.

You may be asked to select figures from a table like Table 18.1. Ensure you do this accurately and select figures that help confirm points you are making in an answer.

Self-test questions 18.1

1 Name and describe two methods of water supply.

2 Explain what is meant by the term 'sustainable sources of water supply'.

Country	Domestic	Industrial	Agriculture
UK	22	75	3
USA	13	46	41
Australia	15	10	75
Iceland	34	66	0
Kuwait	44	2	54
Japan	20	18	62
Brazil	28	17	55
Malaysia	17	21	62
Egypt	8	6	86
Kenya	17	4	79
Bangladesh	10	2	88

Table 18.1 The percentage share of total water usage per country

18.2 Water shortages and management of water supplies

There are several reasons, both natural and human, why some areas have a shortage of water whereas others can have a surplus:

- the amount of precipitation received
- the amount of evaporation/evapotranspiration taking place
- temperatures – the higher the temperature, the greater the amount of evaporation/evapotranspiration taking place
- the type of land use, for example natural forests compared to farming – farming may consume large amounts of water for irrigation
- the level of economic development – richer HICs may consume larger amounts of water
- population density – more densely populated areas may use more water
- the presence or absence of water-bearing rocks or aquifers may mean that more or less water is available to people
- how close people live to rivers will affect how easy it is to obtain water
- political decisions – where a river flows through several countries, they may not agree on how much water each country may take (e.g. the Nile in north east Africa).

The **Nile Basin Initiative** is a regional intergovernmental partnership that seeks to develop the River Nile in a cooperative manner, share substantial socio-economic benefits and promote regional peace and security. It includes all the countries through which the Nile flows – Burundi, Democratic Republic of the Congo, Egypt, Ethiopia, Kenya, Rwanda, South Sudan, Sudan, Tanzania and Uganda. It provides a platform for multi-country dialogue and information sharing, as well as joint planning and management of water and related resources in the Nile Basin. These countries have a total population of 437 million, of which 238 million (54%) live within the Nile Basin.

As with many of the world's rivers, the Nile Basin is facing ever-growing challenges of possible future water shortages and the need for careful water management and pressures:

- climate change which is expected to increase the likelihood of extreme events (prolonged droughts and floods) which will adversely affect the food, water and energy security of the basin countries
- high population growth rates (in seven of the 11 countries, population will double in the coming 20–25 years)
- the demands of faster economic growth across the basin
- the upper area of the Basin faces huge soil loss due to land degradation
- the wetlands in the midsection are increasingly threatened by commercial agriculture and land conversion
- the extreme end of the Nile Delta in Egypt is threatened by seawater intrusion and soil salinisation
- in all the regions of the basin, there is on-going significant loss of biodiversity.

Water availability

Large differences in water availability exist between continents, regions, countries and within countries:

- At a continental scale, it would seem that all continents have sufficient water resources to cover the daily requirements of their populations. South and Central America has the largest volume of freshwater resources per person, 86 600 litres per day, and in North America, 55 500 litres per day.
- At a country scale, China has 5500 litres available per person per day, but there are very large differences between the dry north and the humid south of the country. India has only 4200 litres per person per day, but there are also large differences between the dry northwest and the wetter east of the country. In the Northern Africa and the Arabian Peninsula regions, renewable freshwater resources are only 750 and 230 litres per person per day respectively. Iceland has the largest volume of renewable freshwater resources available per person, 1.4 million litres per day, while Kuwait has the smallest volume per person, 16 litres per day.

Globally, **water consumption** is not equal. The populations of HIC countries use much more water than those in LICs. The average person in North America uses over 1600 cubic metres of water a year compared to less than 200 cubic metres for a person in Sub Saharan Africa. On a daily basis, in the HICs actual water use can be up to 800 litres per person per day, against only 10 litres per person per day in the LICs

In addition to these spatial differences, there are the **temporal/seasonal differences**. In India, for example, most of the renewable water resources are generated during the three-month monsoon period during which 70–95% of its annual precipitation occurs.

Many countries have the benefit of receiving freshwater from their neighbouring countries, almost two-thirds of all countries of the world have rivers flowing into them from neighbouring countries. Seven countries share the Amazon River basin in Southern America, eight countries share the Mekong River basin in Southern and Eastern Asia, 11 countries share the Nile River basin in Africa and 19 countries share the Danube River basin in Europe.

To maximise use of the water in rivers, there are thousands of **dams**, globally. Canada and Russia have the largest dam capacities, over 800 km^3 each. However, when scaled to country population, Suriname with over 42 500 m^3/person largely exceeds Canada's (25 000 m^3/person) and Russia's (5700 m^3/person) capacity. In addition to access to surface water, about 2 billion people worldwide depend on **groundwater**, which includes about 300 transboundary **aquifer** systems. **Desalination** is a growing source of freshwater in many countries. However, 44 countries are landlocked, depriving them from the potentially important sources of desalinated water in the future.

FACT

About half of the world's desalinated water is produced in the seven countries of the Arabian Peninsula and about 10% of the total is produced in the USA.

Water demand

The demand for freshwater has tripled over the last 50 years, as the world's population grows, and demand is increasing by 64 billion cubic metres a year. Not only does the demand for domestic water grow, so does the demand from industry as countries develop and from agriculture as the area of irrigated farm land increases.

In areas near large conurbations or in semi-arid areas, the demand for water may far exceed the availability of water. This can lead to **water shortages** which can have a severe impact on a local population and restrict the development of industry, agriculture and urbanisation.

Where river catchments flow through several countries, it can lead to conflict and argument over the use of a finite supply of water, as in the Nile and Jordan Rivers where countries upstream of Egypt (such as Kenya, Sudan, Ethiopia and Tanzania) and Jordan (Lebanon, Syria and Israel) would like to further develop their agriculture, industry and urban areas to improve standards of living and quality of life.

Managing water supply

In the UK, water supply is a major issue. Areas that receive high amounts of rainfall, in the north and the west, tend to be sparsely populated while one-third of the UK population of over 65 million live in South East England, which is also the driest area in the UK.

There are many ways to manage the water supply:

- making sure broken water pipes are mended
- using reservoirs and dams in one area to transfer water into large urban areas
- making sure that the water supply is of good quality by reducing fertiliser use on farms and industrial pollution.

Managing water demand

In **agriculture**, drip-feed irrigation systems can be used rather than flood or sprinkler systems, this can save up to 70% of the water. About 70% of all water use worldwide is for agriculture irrigation, but 60% of this water is currently wasted by inefficient systems of irrigation.

Industry can also look to recycle waste water. For example, when using water for cooling in steel-making, the water can be recycled again and again in the process. Total industrial water use in the world is about 22%, with HICs using 59% and LICs using just 8%.

Domestically, the amount used varies between HICs and LICs. In **HICs**, such as the USA, every person uses approximately 300–380 litres of water a day and in a typical home, approximately 50% of water is used in the bathroom. The toilet makes up 26%, while the shower and sink use 23%. Outside of the home, 35% of water use is for lawn or garden care. In the UK, every person uses approximately 150 litres of water a day, a figure that has been growing every year by 1% since 1930. Of this, 63% is used in the bathroom and only 7% outside the home. In **MICs and LICs**, the figure is considerably less, often due to the lack, and the cost, of piped water to the home. In Kenya, domestic use is less than 50 litres per person per day and in Cambodia, Ethiopia, Haiti and Uganda, it is less than 20 litres per day.

Domestically, water can be conserved in a number of ways:

- **having a shower not a bath:** a bath typically uses around 80 litres, while a short shower can use as little as 25 litres – bath water can be recycled and used to flush the toilets.
- **turning off the water while brushing your teeth:** a running tap wastes over 6 litres per minute – if the entire adult population of England and Wales remembered to do this, 180 mega litres a day could be saved – enough to supply nearly 500 000 homes.
- **installing more efficient appliances** such as washing machines and dishwashers – the most water-efficient dishwashers use 7 litres of water, the least efficient 21 litres – a dishwasher can be much more efficient then hand washing dishes.
- **collecting rainwater** to use on the garden rather than using tap water.

In LICs, the demand for water can be met using various methods:

- **wells**, dug by hand – however, the supply can be unreliable and sometimes the well itself can be a source of disease, so they need to be covered or protected to stop animals falling into them
- **gravity-fed schemes** where there is a spring on a hillside, or from a river or lake further up a valley – the water can be piped, or small channels can be constructed from the spring/river/lake down to the villages/fields
- **boreholes**, which require more equipment to dig, but they can be created quickly and usually safely – they may require a hand pump to bring the water to the surface.

In addition to locating new supplies of water, strategies can be used to **reduce the need for water**. These include:

- collecting rainwater landing on the roofs of buildings
- recycling waste water to use on crops
- improving irrigation techniques
- growing crops that are drought-resistant or less dependent on a constant water supply.

18.3 Access to clean water

In 2000, the member countries of the UN signed the Millennium Declaration, which later gave rise to the **Millennium Development Goals (MDGs).** Goal 7 included a target that challenged the global community to halve, by 2015, the proportion of people without sustainable access to safe drinking water and basic sanitation. Around 2.6 billion people gained access to an improved drinking water source between 1990 and 2015, and 96% of the global urban population was using improved drinking water sources, compared with 84% of the rural population. However, eight out of ten people still without improved drinking water sources live in rural areas.

The LICs did not meet the MDG target. Although 42% of their population gained access to improved drinking water sources by 2015, 663 million people still lacked improved drinking water sources. Nearly half of all people using unimproved drinking water sources live in sub-Saharan Africa, while one-fifth live in Southern Asia (as of 2015). The 44 LICs have faced the greatest challenges in meeting the MDG target, and many have been affected by political conflict but have nevertheless made progress. Between 1990 and 2015, the proportion of people in these LICs using improved drinking water sources increased from 51% to 69%, but the use of piped water in homes only increased from 7 to 12%.

FACT

Seven out of ten of the 159 million people relying on water taken directly from rivers, lakes and other surface waters live in sub-Saharan Africa, eight times more than any other region.

In many LICs, up to 4 out of 10 schools and health care facilities lack basic water, sanitation and hygiene facilities. In Africa, 42% of health facilities do not have access to an improved water source within 500 metres.

In 2015, the World Health Organisation (WHO) estimated that 663 million people still lacked improved drinking water sources and that 5.9 million children under five die annually from untreated water-related causes. For example, diarrhoeal disease is the third leading cause of death among children under five. It is estimated that more than 340 000 children under five die annually from diarrhoeal diseases due to poor sanitation, poor hygiene or unsafe drinking water – that is almost 1000 per day.

Inequalities in access to clean water – Namibia, south west Africa

Figures collected at a national/country scale can hide the inequalities in access to drinking water and sanitation between rural and urban areas. Namibia, in south west Africa, provides one such example.

Namibia is an MIC, but it has the highest income inequality in the world. Even though the country is relatively rich and has a good economic growth compared to most other African countries, almost half of the population lives below the international poverty line of $1.25 per day. The climate is hot and dry with erratic rainfall. Namibia shares several large rivers, but they are far away from the population centres and the cost of tapping them for drinking water supply is prohibitive.

Clean water is not easy to obtain in many poor urban areas, such as the township of Mondesa (population of about 7000, mostly immigrants from rural areas) near Swakopmund, where water is provided from standpipes with pre-paid metres. Each standpipe serves 250 people and is accessed using pre-paid tokens that the people buy to fill a water container, weighing 20–30 kg, and this has to be carried back to the home each day. Users can spend up to 10% of their income on this water. As a result of the high cost of water, many households in the informal settlements cannot afford a flush toilet and many inhabitants resort to '**flying toilets**'. These are plastic bags which people use to relieve themselves after which they discard them onto the streets, alleys, ditches or even rooftops.

Self-test questions 18.2

1. Explain the reasons, both natural and human, why some areas may have a shortage of water and others can have a surplus.
2. With the use of a case study, describe how water usage can be managed in a HIC.

TIP

When using case studies to answer high-level questions, make sure that the points you select from the case study are relevant to the answer required.

Case study

Water supply – the California State Water Project, California, USA

HICs have the money and resources to transfer large quantities of water long distances which has enabled them to develop areas with water shortages, for example, in the south west of the USA. Southern California has developed enormously and it could not have done this without transferring water from the north of the state to the south with the development of the **California State Water Project**. The Project provides additional water to approximately 25 million people and about 303 500 hectares of irrigated farmland.

The Project is a **water storage and delivery system** of reservoirs, aqueducts, power stations and pumping plants (Figure 18.2). It has involved the construction of 21 dams and almost 1300 km of canals, pipelines and tunnels.

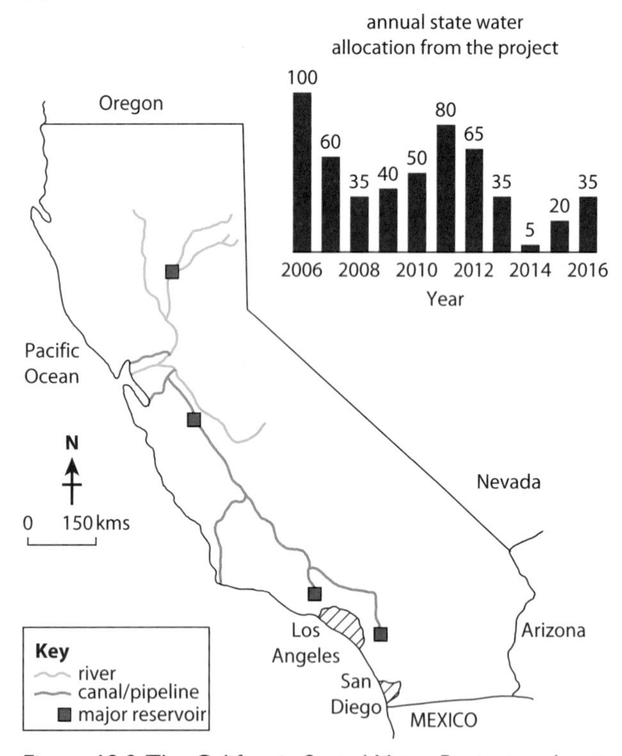

Figure 18.2 The California State Water Project and water allocation, 2006–2016

The Project's main purpose is to supplement the already available water supply in California which was not capable of meeting the needs of the state. It does this by storing water and distributing it to 29 urban areas and farm land in California. About 70% of the water provided by the Project is used for urban areas and industry in Southern California and the San Francisco Bay Area, and 30% is used for irrigation in the Central Valley.

The northern third of California has 70% of the state's water but 80% of the demand for water is from the southern two-thirds of the state. Water demand in California is mainly from agriculture, which uses 80% of the state's available water, but the expansion of the large urban areas of Los Angeles and San Diego, has further increased demand so that the state has looked further away for its water – to the Colorado River.

The last time the Project delivered 100 percent of the water it had allocated to its customers, as a result of pumping restrictions to protect threatened and endangered fish species, was in 2006 (see graph on Fig 18.2 for the percentage it has managed to deliver from 2006–2016).

The Colorado River runs through seven states. Its source is in the state of Colorado, in the Rocky Mountains, and from California it crosses into Mexico. In 1922, the water in the river was divided between the states, as part of the **Colorado River Compact**. Since that time, water demands have increased enormously, through an increase in both population and area of irrigated farm land.

Although the Project has been committed to deliver 20.35 million litres annually, the annual flow of the Colorado River has only averaged 17.25 trillion litres since 1930! Added to this, **evaporation** from the many reservoirs created by dams built along its course has meant that there is a further annual loss of 2.45 trillion litres. Periodic **droughts**, including a moderately dry year in 2012 and a record drought in 2013 in the south west of the USA have also meant that the Colorado River has fallen well below its average flow in several years. This has meant that the Project has been unable to meet its projected delivery of water in recent years.

The Project has raised several **environmental concerns** caused by its removal of water during the dry season, which has affected the Sacramento–San Joaquin River Delta, a sensitive estuary region. Fish migration has been badly affected as a result of low water flows due to water abstraction.

The differences in the cost of the water supplied to the Project's various users has been a frequent source of controversy. Although the overall average cost of Project water is $119 per 1000 m^3, agricultural users pay far less than their urban counterparts. Agricultural users pay as little as $36 per 1 000 m^3 for their water, which is mostly used for irrigation, while the urban areas of southern California pay $241 per 1000 m^3, six to seven times as much.

The increase in demand has also been due to the setting up of another water supply project in south west USA, the **Central Arizona Project (CAP)**, costing $4 billion. This involved diverting water away to central Arizona, including two of the fastest growing cities in the USA: Phoenix and Tucson. Each year, 1.85 trillion litres of water is now delivered to farms, cities and Native American Indian reservations over a distance of 570 km. In the past, Arizona never took up its full allocation from the Colorado and it was used by the other states, especially California. Now that Arizona is taking more water, California has to find a way to make up for the shortfall between what it is now allowed – 5427 million cubic metres, and what it was taking in the past – 6416 million cubic metres.

As with many other areas of the world, California and south west USA now has to carefully look at its future options and develop careful **resource management strategies**. This may produce a sustainable future. In this particular example of water resources, there are several strategies that can be put in place for California and elsewhere:

- **reducing the leakage of water** from pipes and aqueducts as well as the loss of water by evaporation from aqueducts – this would stop up to 25% of losses
- **recycling water** – that is used in industry (greywater) and from sewage (blackwater) – this does not have to be treated to the same standard as drinking water, and can be used to irrigate gardens and golf courses and flush toilets
- **reducing water subsidies** – at the present time, farmers in south west USA only pay 10% of the actual cost of the water they use for irrigation – the federal government subsidises the rest of the cost – if subsidies were reduced, farmers may then look to use irrigation water more efficiently – such as using **drip irrigation systems** which are 100 times more efficient than flood irrigation
- **growing less water-dependent crops** rather than rice and the fodder crop, alfalfa
- several cities in California are looking at using **desalination plants** to produce water.

Sample questions and answers

1 **What are the problems in supplying water in both HICs and LICs? [4]**

2 **In what ways can water supply be managed? [4]**

Ensure that you read the question carefully. Here, you must refer to problems that apply to **both** HICs and LICs.

1 The problems can include the quality of available water **[1]**, the distribution of available water supplies **[1]**, the seasonal changes in supply **[1]** and the existence of leaking or broken pipes when transporting water **[1]**.

2 Ways include the use of reservoirs and dams to store water for use **[1]**, the use of pipelines/canals to transfer water from one area to another **[1]**, maintaining a good quality water supply by reducing pollution of rivers and lakes, reducing fertiliser use on farms, etc. **[1]** and mending leaking/broken water pipes **[1]**.

(A maximum of 8 marks will be awarded).

Exam-style questions

1 Using one or more examples, describe how water supply is being improved in LICs. [6]

2 Explain the positive and negative effects of a named water development project. [5]

3 Why does water usage differ greatly from country to country? [4]

Environmental risks of economic development

Learning summary

By the end of this chapter, you should be able to:

- describe how economic activities may pose threats to the natural environment, locally and globally, including soil erosion, desertification, enhanced global warming and pollution
- demonstrate the need for sustainable development and management
- understand the importance of resource conservation
- demonstrate knowledge of a case study of an area where economic development is taking place causing the environment to be at risk.

19.1 Economic activities and threats to the natural environment

There has often been an unfortunate link between many economic activities and threats to the natural environment. The link between these threats and level of economic development, has been shown by the **environmental Kuznets curve** (Figure 19.1). The hypothesis behind this curve puts forward the idea that in the early stages of economic development, environmental degradation will increase until a certain level of income is reached (known as the turning point) and then environmental improvement will occur. This relationship between per capita income and environmental degradation is often shown as an inverted U-shaped curve.

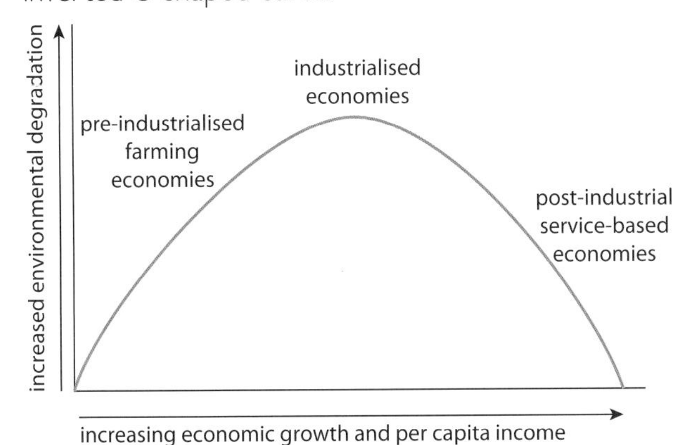

Figure 19.1 The environmental Kuznets curve

Possible explanations for this curve were said to be the result of the progress of economic development, from a clean, agriculture-based economy to a polluting industrial economy and, finally, to a clean tertiary / service-based economy. In addition, it was suggested that people with a higher income have a higher preference for improved environmental quality.

Soil erosion

The degradation of the land through soil erosion is the result of the interaction of several natural physical processes and human activities. As with all the threats, its impact is both increasing and accelerating and in a world of 7.5 billion people, it is having a negative effect on food production.

> **FACT**
> Approximately 20 million square kilometres of the world's land surface is in a state of degradation. Water and wind erosion account for more than 80% of this degradation.

Soil erosion is a global problem and it largely related to clearing the land and making it available for agricultural use. The major causes of soil erosion include the following:

- **Deforestation** (clearing forests, woodlands and shrub land) to obtain timber, fuelwood and other products is currently far in excess of the rate of natural re-growth. This has become an increasing problem in semi-arid environments, where fuelwood shortages are often severe. In rainforest areas, deforestation has been carried out for several reasons, including logging for valuable timber, plantation agriculture, cattle ranching and construction of new settlements. The forest clearances have badly affected the habitat for thousands of species of plants and animals and destroyed the delicate food webs and food chains that exist in the natural environment. The natural home for many indigenous peoples have also been lost.

TIP

For more on the impacts of deforestation, see Chapter 12 *Climate and natural vegetation*, Section 12.3 *Deforestation in the tropical rainforest*.

- **Overcultivation**, where low-income farmers attempt to increase the yield from their land. Farmers are not allowing the land to rest and recover (leaving the land **fallow**) for a period of time (1–18 years) and are instead planting often the same crop in the same land every year and the soil is losing fertility.
- **Rising populations** are forcing farmers into farming more marginal areas on desert fringes, where they would not normally choose to farm. This can work in years when there is abundant rainfall but when the rains fail, crops fail and the soil quickly degrades.
- **Overgrazing** (when too many animals are kept on the same land for too long and the animals eat too much of the grass/fodder plants so that they cannot re-grow and die) is resulting in a decrease in the vegetation cover, and this is a major cause of wind and water erosion.

FACT

Agricultural mismanagement, often in the form of overgrazing, has affected more than 12 million square kilometres worldwide. This means that 20% of the world's pastures and grazing lands have been damaged. The situation is most severe in the continents of Africa and Asia.

- The trampling of the ground by animals leading to **soil compaction**, destroying the structure of the soil and leaving it open to wind and water erosion. As the world's population has increased, so has the worldwide animal population. Herds of cows, goats and sheep concentrate in certain areas, stripping the vegetation back and exposing the soil to erosion. Great pressure is put on cultivated areas around the boreholes and wells where the animals drink.

Several **processes** contribute to soil erosion:

- **water erosion** when water flows over the surface of the ground and removes small particles of soil. It accounts for nearly 60% of global soil degradation
- **wind erosion**, when dried soil particles are exposed to high winds and are picked up/entrained
- **over** abstraction **of groundwater**, which may lead to soils drying out and being easily removed by heavy rain and wind, or resulting in their being a lack of water in the soil for crops to germinate and grow
- **salination** is an increasingly common problem in semi-arid areas where high rates of evaporation draws groundwater up to the surface of the soil.
  - Groundwater often contains a lot of dissolved salts and these salts are left on or near the surface of the soil when the groundwater is evaporated.
  - These salts build up and make the soil too toxic for most of the common plants being cultivated by the world's farmers.
  - Salination has been a major problem in parts of Australia following the removal of the natural vegetation for farming.
- **climate change** will probably increase the problem of soil degradation as an increase in the number and severity of floods will cause more water erosion – also, an increase in the number and severity of droughts will cause more wind erosion.

TERMS

groundwater abstraction: the process of taking water from a ground source, either temporarily or permanently

desertification: the degradation of land in arid, semi-arid and dry, sub-humid areas

Desertification

Desertification is caused primarily by human activities and climatic variations. It should be noted that the term desertification does not refer to the expansion of existing deserts. It occurs because semi-arid and arid ecosystems, which cover over one-third of the world's land area, are extremely vulnerable to over-exploitation and inappropriate land use.

19.2 Sustainable development and management

Managing soil erosion

Using **appropriate technology** can help to reduce or prevent soil erosion. These methods involve building physical barriers such as embankments and wind breaks, or maintaining a protective vegetation cover over the soil. The methods include planting trees to make **shelter belts** to protect the soil from wind erosion in dry periods and **contour ploughing** to prevent or slow water movement over the surface of the soil. On steep slopes and in areas with heavy rainfall, such as the monsoon in South East Asia, contour ploughing is insufficient and **terracing** is used where the slope of the land is broken up into a series of flat steps, with small walls at their edge. The use of terracing also allows otherwise unsuitable steep valley sides to be cultivated. **Bunding** is the name given to the building of low stone walls, often just 30–40 cm high, along the contours of a slope to stop the surface runoff of rainwater, giving it more time to infiltrate the soil. This both helps prevent soil erosion and increases the amount of water in the soil, making it available for crops when it would otherwise be lost and flow away from the farm land.

Compartment building or **tied ridging**, where low walls of soil are built in a grid of small squares, stops rainfall runoff and allows water to be drained into the soil. **Strip** or **inter cropping**, where alternate strips of crops at different stages of growth are grown across a slope, help to prevent rainfall runoff. **Tier** or **layer cropping**, where several types and sizes of crops are grown in one field, providing protection from rainfall and increase food and crop yields from a unit area of land.

TIP

The use of **appropriate technology** is also discussed in Chapter 14 *Food production*, Section 14.2 *Food shortages and possible solutions*.

Soil erosion can also be prevented by educating farmers in sustainable cropping techniques. This involves:

- maintaining a protective cover crop for as long as possible
- keeping in place the stubble and roots of the last crop after harvesting, to protect and stabilise the soil
- planting a protective crop of grass or alfalfa (used for grazing animals and making hay and silage) – their roots bind together small soil particles, preventing the wind and rain eroding them away
- increasing the organic content of the soil by adding organic manure which, apart from adding nutrients to the soil, allows the soil to hold more water
- taking care over the use of heavy machinery, or keeping cattle, on wet soils.

Managing desertification

Burkina Faso, Africa

Burkina Faso is located in the Western Sahel of Africa and lies across a transitional zone of arid desert and semi-arid grassland. Many farmers used quite intensive farming methods, unsuited to and not sustainable in these drier areas and this led to increasing desertification during the 1990s, resulting in increased hardship for rural communities and families. Many people and families had no option but to leave the land and were forced to migrate further south.

In an attempt to remedy some of the problems caused by desertification, aid agencies such as the **Eden Foundation (Oxfam)** and **USAID** intervened both to supply the equipment needed for sustainable farming and to educate local people and communities on sustainable farming techniques. Farmers were taught how to use **drip irrigation techniques**, using far less water than flood irrigation. This system drips water slowly to where each seed is planted. It reduced the amount of water lost to evaporation by the traditional flood irrigation method, where much water was lost as it was spread in large quantities across the whole field.

To raise crop production, the technique of **microdosing** was introduced. This involves the application of small, affordable quantities of fertiliser with the use of a bottle cap, taken from an old water, cola or lemonade container, either during planting or 3 to 4 weeks after germination. This technique maximises the use of fertiliser and greatly improves crop productivity. Microdosing has increased sorghum and millet yields by up to 120%, and farm and family incomes by as much as 50% in more than 200 000 households in Africa.

FACT

Microdosing has triggered the reintroduction of fertiliser use not only in Burkina Faso, but in Zimbabwe, Mozambique, South Africa, Niger and Mali.

TIP

For more on overcoming desertification, see Chapter 12 *Climate and natural vegetation*, Section 12.4 *Characteristics of hot desert ecosystems.*

Pollution and management strategies

Globally, many urban areas have, and are, experiencing urban degradation. The environmental problems include air pollution, noise and visual pollution, inadequate waste management and the pollution of land, rivers, lakes and the coast.

TERMS

urban degradation: damage to the urban physical environment; it often has a detrimental effect on the health and well-being of people living in these areas

noise contours: lines joining places of constant noise level measured in Leq; 57 dBA Leq is classed as significant community noise annoyance

The most common **air pollution** sources are thermal power stations burning fossil fuels, vehicle emissions and various industrial sources. Air pollution is also affected by weather conditions where calm air conditions can allow pollution levels to build up, as air pollutants are not dispersed and moved away by the wind. In some urban areas, as in **Los Angeles** and **Mexico City**, air pollution can be trapped at ground level for long periods of time.

Every year in **Mexico City**, about 4 million tonnes of pollutants are released into the atmosphere from the urban area. In an effort to counter this problem and to reduce pollution, car tax discs in Mexico City have different colours, indicating the day of the week each car is not allowed to drive within the city. This policy is intended to force car drivers onto public transport for that day.

Land pollution includes the dumping of household, industrial and commercial wastes including toxic/hazardous waste in urban areas. Apart from contaminating the land surface, toxic chemicals can also leak from waste dumps and pollute rivers and sources of groundwater.

The source of most outdoor **noise pollution**, globally, is through the use of machines and transportation systems, motor vehicles, aircraft, and trains. Poor urban planning may give rise to noise pollution, for example placing industrial and residential buildings adjacent to each other, which can result in noise pollution in the residential areas.

Noise pollution can be remedied/mitigated in a number of ways:

- **Road noise** can be reduced by the use of noise barriers, limitation of vehicle speeds, alteration of roadway surface texture, limitation of heavy vehicles, use of traffic controls that smooth vehicle flow to reduce braking and acceleration, and tyre design. The costs of building in remedies/mitigation can be modest, provided the solutions form part of the planning stage of a road/transport/urban/industrial project.
- **Aircraft noise** can be reduced by using quieter jet engines. Altering flight paths and the time of day that the runway is used has benefitted residents near airports. In 2015, there were approximately 474 000 aircraft movements at **Heathrow Airport**, London, UK (the world's busiest international airport), handling 75 million passengers. An analysis of the 2015 summer traffic data for Heathrow Airport revealed that there were 1354 average daily movements of aircraft landing or taking off. The area affected by noise pollution around the airport is estimated to be 102.5 km² with a population of 258 300 (Figure 19.2). Noise exposure is shown in the form of noise contours.

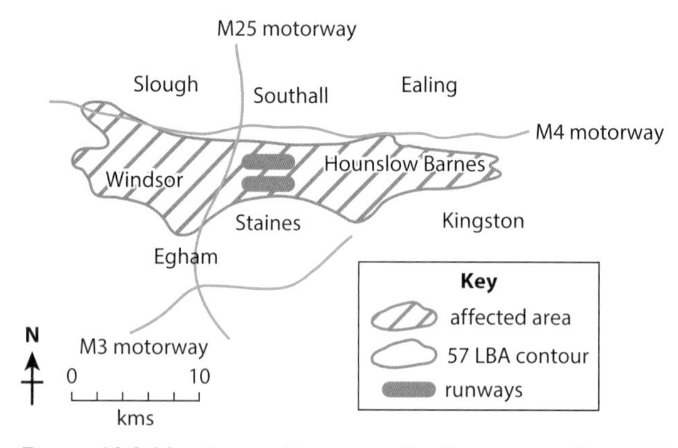

Figure 19.2 Heathrow Airport and urban areas affected by noise pollution

- **Industrial noise** remedies involve the redesign of industrial equipment and physical barriers in the workplace. In recent years, **Buy Quiet** programmes and initiatives have been set up to combat occupational noise exposures. These programmes promote the purchase of quieter tools and equipment and encourage manufacturers to design quieter equipment.

- **Visual pollution** refers to the impacts of pollution that impair a person's ability to enjoy a view. Visual pollution disturbs the visual areas of people by creating negative changes in the natural environment. The overcrowding of an area can cause visual pollution. It can be caused not only by (giant) billboards, business signs, street signs, telephone and utility poles, and electricity wires but also by the piles of rubbish sometimes seen on beaches, in rivers, along roadsides, or overflowing rubbish/garbage containers, plastic bags stuck in fences or trees, cigarette butts outside restaurants. It can also include open cast mines, dog poo, rubbish dumps, mobile phone towers, etc.

Wind and solar farms highlight how subjective the issue of visual pollution can be. On one side, there are enthusiastic supporters who see the farms as aesthetically/emotionally appealing in relation to their sense of beauty, and appreciate their ability to capture a clean, renewable energy. Opponents, however, say both sets of structures are anything but soft on the eye.

In the 1980s, it was calculated that the people of the USA were subjected to around 2000 visual advertisements a day. This figure had increased to about 5000–6000 by 2015. To counter this visual pollution, bans on billboards now exist. For example, in the states of Vermont, Maine, Hawaii and Alaska in the USA, and some 1500 urban areas around the world. In 2006, the city of **São Paolo**, in Brazil, introduced a '**clean city law**' banning the use of all outdoor advertisements, including on billboards, buses and trains, and in front of stores. This resulted in 15 000 billboards in the world's seventh largest city being taken down.

The city of **Pune** (population 5 million), in India, pursued a similar policy in 2011, banning signs and advertisements that inconvenience traffic or pedestrians. In the process, it also shifted the focus back towards its people and their relationship to their city by issuing rules for the size and structure of the advertisements to avoid defacement of the city. No advertisements are allowed in public places like gardens, hilltops, hill slopes and rivers as well as historic and religious places.

Many cities and other urban areas have former industrial sites that have left a legacy of pollution. In many HICs, resources exist to remove the pollutants and regenerate these former polluted sites to be re-used for a variety of uses. The **O2 site in Greenwich**, in east London, UK, was identified as a potential brownfield site (a site with pre-existing urban industrial uses). It was cleared and used as the site for the **Millennium Dome** structure in the late 1990s, now re-named and used as the **O2 Arena**. It is now a large entertainment district, including an indoor arena, music club, cinema, exhibition space, piazzas, bars and restaurants. The land was previously derelict and contaminated by toxic sludge from East Greenwich Gas Works, which operated from 1889–1985, as well as from tar distillation works and a benzene plant. The remedial actions needed for this development to take place cost $33 million. The UK government also spent $19 million cleaning up a part of the 2012 Olympic site in east London that was 'grossly contaminated' with toxic waste left behind under a chemical storage facility, which was bulldozed to make way for the main stadium.

Water pollution is a major problem in many urban areas. In India, about 80% of urban waste ends up in the country's rivers and poorly planned urban growth, poor management and a lack of government accountability all contribute to a major pollution problem. In **Delhi**, 75–80% of the pollution of the city's major river, the Yamuna River, is the result of raw sewage. This is combined with industrial runoff and rubbish thrown into the river, totalling over 3 billion litres of waste per day, a quantity well beyond the river's capacity to assimilate it.

Remedying this situation, as in many LICs and MICs, will take huge amounts of investment and resources that are currently not available.

Self-test questions 19.1

1. What are the major causes of soil erosion?
2. With the use of a case study, describe two methods used to remedy some of the problems caused by desertification.

19.3 The enhanced greenhouse effect and global warming

The **greenhouse effect** is the naturally occurring process by which the Earth's atmosphere is warmed as certain gases within the atmosphere absorb some of the radiation being emitted by the Earth. The most common greenhouse gases are carbon dioxide, methane, nitrous oxide, water vapour and chlorofluorocarbons (CFCs). These greenhouse gases allow the incoming short-wave solar radiation from the Sun to pass through them, but they are then very effective in trapping the outgoing long-wave terrestrial radiation being emitted from the Earth.

Without these greenhouse gases, life on Earth as we know it could not exist. The greenhouse gases combine to raise the average temperature of the Earth by 33°C. Before the rapid growth of the human population and its various activities in the last 200 years, the atmosphere was fairly balanced with the carbon dioxide being produced by animals and humans equalling the amount being taken up in plants as part of the process of photosynthesis. However, the term enhanced greenhouse effect (EGE) is used to describe the build-up of certain greenhouse gases in the last 200 years by human actions and activities, which have contributed to global warming (Figure 19.3). The rise in air temperatures has had an impact on ocean temperatures (Figure 19.4) and land surface temperatures (Figure 19.5).

TERMS

Enhanced Greenhouse Effect (EGE): the impact on the climate from the additional heat retained due to the increased amounts of carbon dioxide and other greenhouse gases that humans have released into the Earth's atmosphere since the Industrial Revolution

global warming: a gradual increase in the average temperature of the Earth's atmosphere and its oceans

coral bleaching: the corals are killed and only their bleached skeletons remain

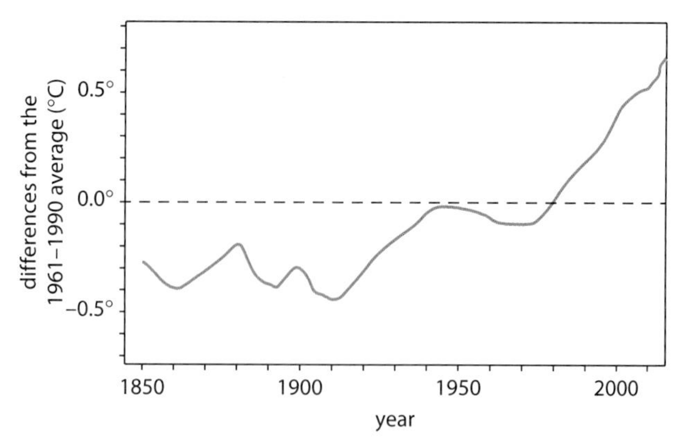

Figure 19.3 Changes in average global temperatures, 1850–2015

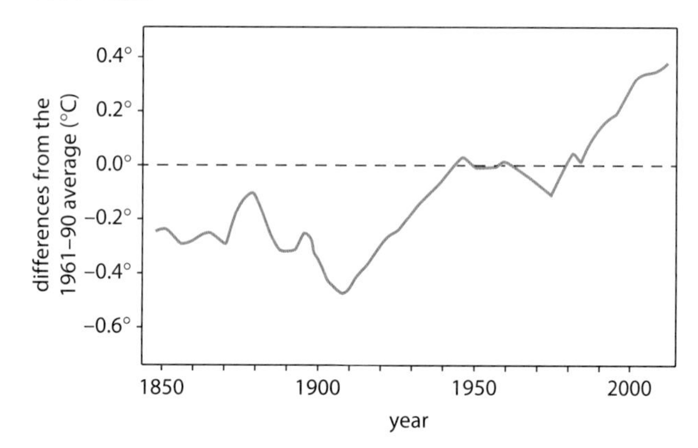

Figure 19.4 Global average sea surface temperatures, 1856–2012

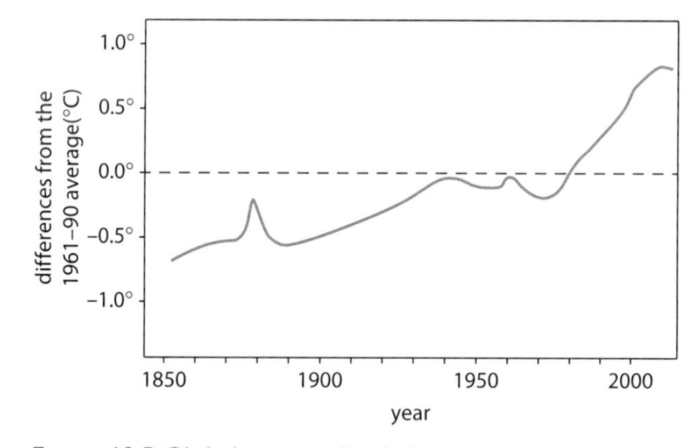

Figure 19.5 Global average land air temperatures, 1856–2012

One of the most significant changes in the last 60 years has been the rise in **carbon dioxide** from 315 ppm to 407 ppm in 2016 (Figure 19.6), resulting in increasing amounts of heat being retained in the Earth's atmosphere and leading to a global rise in temperatures.

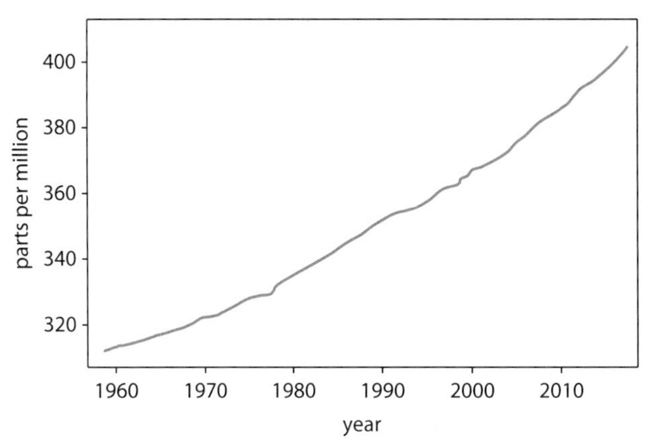

Figure 19.6 Atmospheric carbon dioxide concentration at the Mauna Loa Observatory in Hawaii, 1856–2012

The rise in carbon dioxide has been caused by a number of factors:

- **Clearing of forests/deforestation:** this is often done by burning the forests, which increases the amount of carbon dioxide in the atmosphere and it also has an additional impact as the trees convert carbon dioxide to oxygen.
- **Industrialisation:** since the mid-nineteenth century, industrialisation has put large amounts of carbon dioxide into the atmosphere from the burning of fossil fuels.
- **Emissions:** from internal combustion engines and jet engines which contain carbon dioxide.
- **Melting of permafrost:** in Arctic areas which releases carbon dioxide from the organic matter that was previously frozen and is now melting as a result of global warming. This can be accelerated by forest fires, as in Fort McMurray in Alberta, Canada, in 2016 where the fire covered an area of over 1610 km^2.

The effects of fires like this are to melt the frozen organic matter that lies under the forest in its permanently frozen ground, called permafrost. Such fires burn away the protective forest blanket and release all the stockpiled carbon into the atmosphere. Global warming has led to longer fire seasons and the northern forests in Canada and Alaska have been burning in the last few decades at a rate unprecedented in the previous 10 000 years. In 2015, over 34 000 km^2 of forest were destroyed by burning in this region.

Methane is at least as great a problem as it traps 29 times more heat than carbon dioxide. Global methane levels had risen to 1850 parts per billion (ppb) by 2016 (Figure 19.7), the highest value in at least 800 000 years. The concentration of methane in Earth's atmosphere has increased by about 150% since 1750.

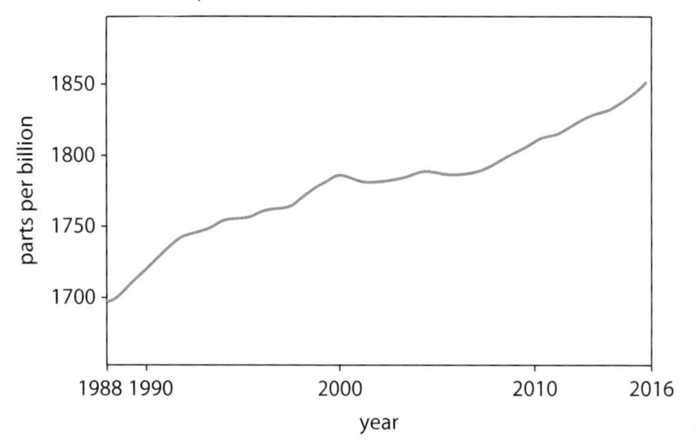

Figure 19.7 Atmospheric methane concentration at the Mauna Loa Observatory in Hawaii, 1988–2016

Methane may be produced from several sources:

- **cattle**, which emit 100 million tonnes per year by burping and flatulence
- **wet rice fields** and **natural wetlands:** both of which release methane by decomposing organic matter.

Nitrous oxide is naturally emitted by the bacteria in soils and in the oceans. However, its production has been increased by human activities (Figure 19.8), and the amount in the atmosphere has risen by more than 15% since 1750. Agriculture is the main source of human-produced nitrous oxide, brought about by the increased cultivation of the soil, the use of nitrogen fertilisers and the increased production of animal waste, all of which stimulate naturally occurring bacteria to produce more nitrous oxide. When compared to carbon dioxide, it has 298 times the ability per molecule of gas to trap heat in the atmosphere.

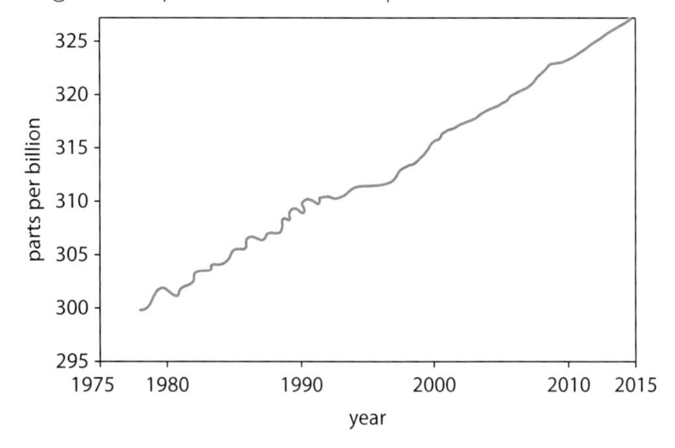

Figure 19.8 Atmospheric nitrous oxide concentration, 1977–2015

The impacts of the EGE and global warming

Average global temperature has risen by around 0.8°C since the early 20th century. This may not sound like much, but some regions will experience a much more extreme response than the global average. More importantly, even a small but permanent increase in temperature can have a significant impact upon large-scale environmental features like ice sheets or the forest cover.

The amount of Arctic sea ice, snow cover and the volume of glacier ice globally have all decreased and surface ocean temperatures have increased (see Figure 19.4). The warmer ocean temperatures can kill large areas of coral reefs. The phenomenon of coral bleaching has increased over the years due to ocean warming (93% of the EGE heat is absorbed by the oceans). Corals simply have not been able to adapt to the higher temperatures of the ocean.

FACT

The 2015–16 global coral bleaching event, only the third of its kind in recorded history, affected approximately 38% of the world's coral reefs and killed over 12 000 km² of coral reef.

Although coral reefs represent less than 0.1% of the world's ocean floor, they help support approximately 25% of all ocean species of plants and animals. As a result, the livelihoods of 500 million people and income worth over $30 billion are at stake.

Other impacts of the EGE and global warming include:

- **rising sea levels** – threatening low-lying countries such as the Maldives, Bangladesh and the Netherlands
- **changing climatic patterns** – a poleward shift of climatic belts and the impact this will have on farming crops and animals
- **more extreme weather events** such as storms, hurricanes, floods and droughts.

Self-test questions 19.2

1. Describe three impacts of the enhanced greenhouse effect (EGE) and global warming.
2. Explain why, in many countries, industrial waste is released into rivers, lakes and seas.

Sample questions and answer

1 'Water and air may be polluted by economic development.' With the use of a case study or examples of a place which has suffered from pollution, describe the causes of the pollution and its effects on the natural environment. [6]

2 With the use of an example or case study, describe the impact on the natural environment from desertification. [5]

TIP

The answer should include a detailed description of the causes and effects of the pollution on the natural environment **only**.

1 Use a named example with accurate statements, including some place-specific references, such as smoke and fumes from power stations and industrial facilities [1]; nitrogen oxide from vehicles [1], sulphur dioxide from burning fossil fuels [1], and the production of polluted dust and particles [1]; affecting tree and plant growth [1], river and lake water quality [1], seawater quality [1].
(A maximum of 6 marks will be awarded.)

2 Using a named example or case study you can describe the impacts on: soils becoming exhausted/**lacking** of nutrients [1]; the effects of overgrazing by livestock on the natural flora/habitat [1]; the loss of vegetation leading to more rapid surface water runoff [1]; leading to increased soil erosion [1]; by water runoff and the wind [1]; less moisture retained in the soil [1]; greater potential for more and bigger flash floods [1], causing erosion and degradation of the natural environment [1].
(A maximum of 5 marks will be awarded.)

Examination-style questions

1 With reference to examples or case studies, explain the causes of

a deforestation

b increasing levels of carbon dioxide in the atmosphere. [6]

2 Explain the threats to the natural environment by certain forms of industrial activity. [4]

3 In what ways can certain agricultural practices damage the natural environment? [4]

4 What is meant by the term pollution? [2]

5 Describe a way in which pollution can be caused by the use of transport. [4]

Geographical skills

20.1 Geographical skills for examination

Geographical skills are important for geographers and you are likely to be tested on them in your examinations. You must answer **all** of the questions. The paper will be **skills-based** and you will be able to demonstrate your ability to handle various ways of depicting geographical information, such as topographical maps, other maps, diagrams and graphs, tables of data, written material and photographs. You will also be able to demonstrate your skills of analysis and interpretation and your application of graphical and other techniques. Any questions that require knowledge and understanding are likely to be based on topics from the three main syllabus themes dealt with in chapters 1–19.

TERMS

topographic maps: a type of map characterised by large-scale detail and quantitative representation of relief, usually using contour lines, but a variety of methods can be used

analysis (analyse): to study or examine something in detail, in order to discover more about it

interpretation (interpret): to give or provide the meaning of something

application (apply): the action of putting something into operation

Equipment

You should ensure that you take the following items with you:

- a pencil, eraser, ruler and a protractor (preferably 360°)
- a calculator
- It is also advisable that you should use a ruler, and/ or a straight edged piece of paper for measuring distance or for assisting with cross sections on the large-scale topographic map – an extra A4 sheet of plain paper should be given to you for this purpose.

20.2 Reading and interpreting maps

A question based on a topographical map can be taken from any area of the world.

Before answering the question, you should familiarise yourself with the **map**, the **scale** and the **key**. This will take no more than one minute, but it will make the understanding and answering of the question that much easier. All the maps will have a full key which you will need to refer to for answering some of the questions. Look at the key and look at how features have been grouped together in the key – the types of roads, types of land use and vegetation, for example. This brief 'once over' of the map will help familiarise you with the location.

The scale of the maps and measuring distances

The maps will be on a scale of either **1:25 000** or **1:50 000**. This means that **1 cm** on the maps will be either **250 cm** or **500 cm** on the ground, respectively. Note that map scales are always given in millimetres. This makes the calculation of distances relatively easy. For example, a distance of 4 cm on a 1:25 000 map will equal 1000 m (or 1 km) and on a 1:50 000 map will equal 2000 m (or 2 km).

The distances you are asked to measure will normally be straight, along roads, or with easy-to-measure distances, along fairly straight roads. You can either measure the straight-line distances with your ruler or the edge of a straight piece of paper. If the distance to be measured is **curved** (Figure 20.1), you should divide the curve into straight sections and rotate the paper after each straight section to follow the next straight section. In Figure 20.1, this is done in four stages. The question here was to measure the distance along the A30 road from the crossroads with the B6 (1) to the crossroads with the B12 (4). When each stage has been measured, the completed straight-edged piece of paper is laid along the linear scale line on the map extract and the distance read off in kilometres/metres. This method avoids complicated mathematical calculations which can arise when rulers are used.

Grid references

You need to be able to give and to read **4-figure** and **6-figure** grid references so that you can locate places on the map.

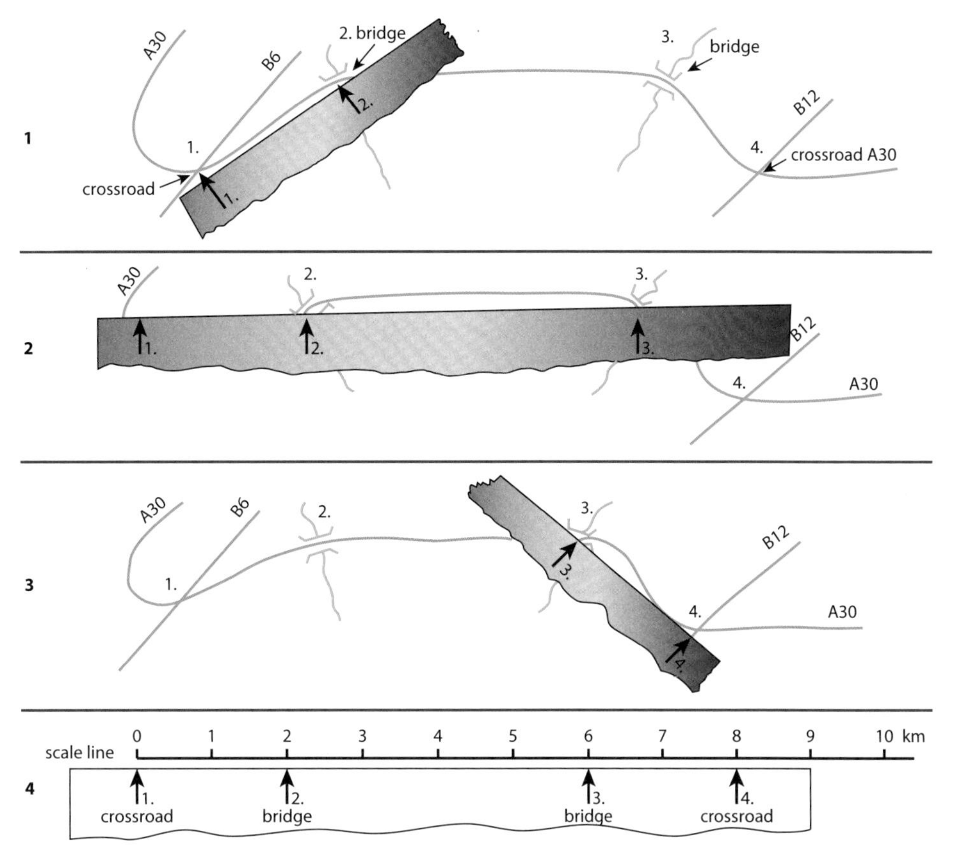

Figure 20.1 Measuring the distance along a curved feature, such as a road

TIP

There will **always** be questions that give you a grid reference and ask you to locate a feature, either a name or a conventional sign that you need to look up on the key provided with the map.

- A **4-figure grid reference** will locate a **grid square**. These are used to locate a fairly large feature like a town, a river valley, a coastline or a particular land use, such as a sugar plantation.
- A **6-figure grid reference** will locate an **exact point** and this will usually be to find a conventional sign or a building, such as a power station, and you will need to search the key to find out what it is.

TERMS

grid reference: the referral to a grid of squares which helps the map-reader to locate a place

eastings: the vertical lines on a map

northings: the horizontal lines on a map

To find a **4-figure grid reference** you use the first two figures/numbers to go along the **bottom** of the map, **from left to right**. You use the second two figures/numbers to go **up** the side of the map. To minimise the possibility of you making a basic mistake on this when under timed conditions, there are two practical techniques:

1. In the first method, move a finger from one hand along the bottom line until you find the correct vertical line (an easting) and **keep your finger to the right of this line**; then move a finger from your other hand **up** the **left** side of the map until you find the correct horizontal line (a northing) and **keep this finger above this line**. Then bring your fingers together to find the correct grid square.
2. In the **second method**, use a ruler and pencil to draw the vertical and horizontal lines onto your copy of the map, so the precise point can be located.

A **6-figure grid reference** uses the four figures from a 4-figure grid reference, for example 06 67, and then adds two more: 06**5** 67**5**. This allows you to find an exact point in a grid square (on Figure 20.2, this is the letter **A**), such as a building in the centre of a town. In finding the third and sixth figures, you have to imagine that the grid line has been split into 10 equal parts. This means that 06**5** will be halfway between the 06 and 07 lines. Again, use your fingers to keep a place on the bottom of the map and up the side, and then bring your fingers together to find the point. You might find it useful to also make a pencil mark on both lines to help with the accuracy.

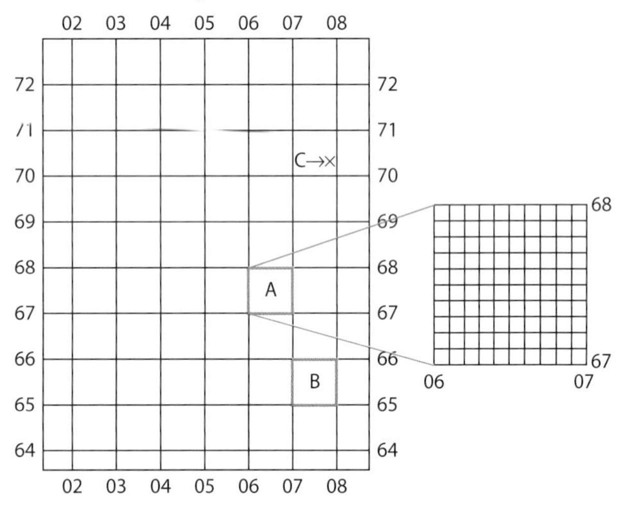

Figure 20.2 Grid squares and references

The following explanation and Figure 20.2 will help explain all this further.

3 In Figure 20.2, two grid squares are located – **A** (grid ref 06 67) and **B** (grid ref 07 65). To find **A**, go **along** the bottom of the map, until you find the 06 vertical line. The square we are looking for will be in the column between the 06 and 07 lines, so keep your finger here. Going **up** the side find the 67 line. The square we are looking for will be in the row between the 67 and 68 lines, so keeping your finger here, bring your two fingers together and they should meet in the **A** square. Try the same exercise with square **B**.

4 In Figure 20.2, **C** is used to locate a point marked with an asterisk (*), which has a 6-figure grid reference of 07**9** 70**2**. To find the 6-figure grid reference, we first go along the bottom line (or the top line if that is nearer) and find the 07 line and go along 9 tenths of the way between the 07 and 08 lines. Keep a finger on this location, or put a pencil mark on the line. Then go **up** the side to find the 70 line and go **up** 2 tenths of the way between the 70 and 71 lines. Again, put your finger, or a pencil mark, here. Then bring your fingers together and you will find the asterisk!

Compass directions and bearings

You need to be able to use and give **directions** on the map. **North** is always up, towards the top, on these maps, so **East** is always **to the right** and **West** to the left (and **South** towards the bottom) of the map.

The order can be remembered by using one of these made-up phrases – **N**ever **E**at **S**hredded **W**heat – or – **N**aughty **E**lephants **S**quirt **W**ater!

You will need to know the **16** points of the compass (Figure 20.3), but you do not need to know their degree equivalent.

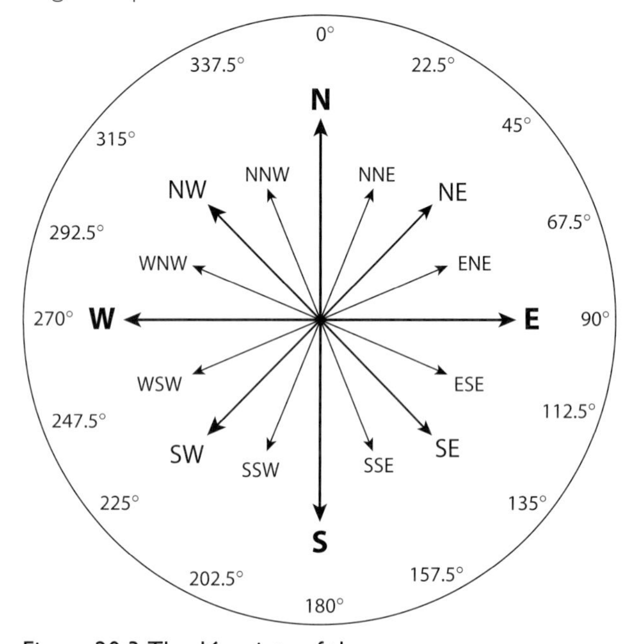

Figure 20.3 The 16 points of the compass

TERMS

compass: an instrument used for navigation and orientation that shows direction relative to the geographic directions or 'points'

grid bearing: a bearing measured from grid north

To be even more accurate in giving a direction, a grid bearing is used. This uses the 360 degrees of a circle, so that North is a **grid bearing** of 0°, East is 90°, South is 180° and West is 270°. You may be asked to find the bearing of a place from another place, or the direction a river flows, or the direction a road takes, for example, from a road junction in a town. Using Figure 20.2 as an example:

1 The **compass direction** of the letter **B** from **A** is **South East** (SE or SSE).

2 To find the **grid bearing** you will need your **protractor** and put it on the map so that the 0° is pointing vertically, north, up the map and the centre of the protractor is on the letter **A**. You can then calculate the degree bearing of a location by finding how many degrees around from 0° is the letter **B**. It is **150°**.

Finding the height on a map

Height is usually shown in three ways on the topographical maps: by contour lines, spot heights and trigonometrical stations.

TERMS

contour line: a line that joins places on the map with the same height. The **contour interval** is the height between contours. This can be either **5** or **10** metres.

spot height: an exact point, or spot, on the map with a height measurement written beside it

trigonometrical station: a small, blue/black triangle with an exact height measurement written beside it

Describing the shape and slope of the land

You should be able to describe the size and shape of a feature on the map in real terms using the scale of the map to work out its true size. You need to be able to draw inferences about the physical and human landscape by interpreting the map evidence such as patterns of relief, drainage, settlement, communication and land use.

A lack of contour lines on a map means that the land is **flat**. If the contour lines are very close together the land is **very steep**, if they are further apart the land is **gently sloping**. You will often be asked to describe how steep are the sides of valleys or hills and the contour lines allow you to describe this. Sometimes land is so steep, or almost vertical, that a special sign is used, for a cliff, on a map. This can be found in the key. Depending on the shape of the land, contours take on distinct patterns. A **hill** will have a series of circular shaped contours and a **valley** will have a V-shaped set of contours, usually with a river in the bottom of the valley.

You may be expected to identify basic landscape features such as river valleys and areas of uplands and to give brief descriptions of them using appropriate geographical terms such as a **high ridge**, or **plateau** (a high flat area), a **scarp slope** (another name for a steep slope), and a **floodplain** and simple descriptions such as **broad**, **flat**, **steep-sided**, **deeply cut** (by a river), **gently sloping**.

You should be able to recognise differences in the **density of drainage** (this simply means the length of streams, rivers or drainage channels you find in a grid square. If there are lots of rivers and streams in the square, it has a **high drainage density**; if it has few, it has a **low drainage density**. You should also be able to describe the shape and form of river channels as they are shown on large-scale maps – such as **meanders** and **oxbow lakes** and **deltas**, and the **physical** features of coastlines such as **cliffs**, **headlands and bays**, **beaches**, **spits**, **bars** and **tombolos**.

TERMS

cross section: a drawing that shows what the inside of something looks like after a cut has been made through it

Cross sections

You may be asked to include a drawn cross section, like the one below, of parts of the map for you to mark on and describe certain places on the map. They allow you to see areas of flat, steep and gentle slopes and locate rivers and towns, etc. You will use them to analyse and describe certain areas of the map.

The cross section will usually be run along one of the vertical (eastings) or horizontal (northings) grid lines drawn on the map between two places.

Figure 20.4 is part of an exam-style question – a basic outline of a cross section is provided and you have to add certain pieces of information after looking at the map.

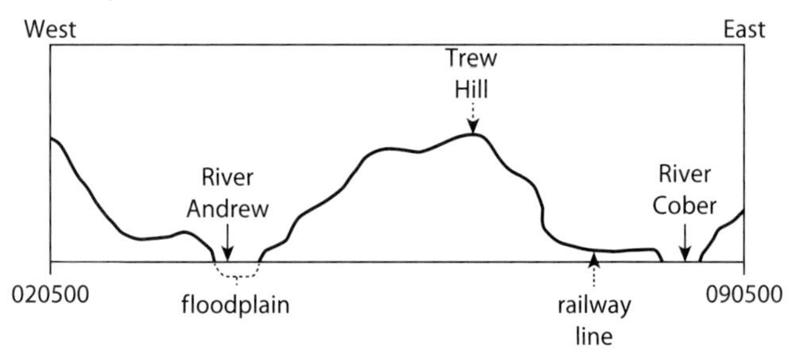

Figure 20.4

Patterns of settlement

You should be able to recognise and analyse the main patterns of settlement (**dispersed**, **nucleated**, **linear**) and be able to draw simple sketch maps illustrating these patterns. You should also be able to interpret and describe features of urban morphology on the maps and be able to describe the **functions** of and **services** provided by settlements. You should also be able to give reasons for the **site** and **growth** of individual settlements.

Transport/communication networks (the provision of roads, railways, ports and airports) should be recognised in terms of their type and density in relation to the **physical** geography (i.e. keeping to areas of flat land and valley floors, avoiding steep slopes and mountainous areas) and **human** geography (i.e. linking settlements together, taking traffic around a large town or city rather than through the middle – called a by-pass) features. Your explanations should be based **entirely on map evidence**, showing the interaction between humans and their physical environment. For example, differences in land use between upland and lowland areas as a result of climate differences due to height; differences in land use within a town; differences between dense settlement on river floodplains and sparse settlement on steep upland slopes.

Practice describing variations in land use as part of your preparation for the examination.

Sample question and answer

1 **Figure 20.4 shows a cross section along northing 50, from 020 500 to 090 500. On Figure 20.4, use labelled arrows to show the positions of:**

a the River Andrew's floodplain

b Trew Hill

c the railway line [3]

TERMS

urban morphology: the geographical shape of urban development

isoline map: (the prefix '*iso-*' means 'equal') a map with continuous lines joining points of the same value; examples include equal height (contour lines), temperature (isotherms), rainfall (isohyet) and barometric pressure (isobars)

choropleth map: a thematic map in which areas are shaded or patterned in proportion to the measurement of the statistical variable which is being displayed on the map, such as population density or per-capita income; each colour is associated with a different, quantitative value

20.3 Maps, diagrams, graphs, tables of data, written material

Questions on all the Geography papers may be set using some or all of the following resources.

Maps

Maps may include isoline maps and choropleth maps, as well as the topographic maps. A choropleth map provides an easy way to visualise how a measurement varies across a geographic area or it can show the level of variability of a phenomena, such as population density, within a region.

More useful points about the maps that may be used:

- The maps will contain many of the answers so spending time to understand, interpret and analyse them is important.

- The maps will, on purpose, be a variety of different scales.
- Many will be world/global maps, but others will show individual continents, countries, regions, cities, towns, rivers, coastlines, etc.
- The common theme for **all** the maps is that they have been carefully prepared and designed for the question – to only show the information you need to answer the question.
- Usually, you will be asked to **find**, **identify** and then describe, or compare, **analyse** and explain important features of the human and physical landscape.

TERMS

describe: write what something is like or where it is

compare: describe two situations and present **both** the similarities and differences

explain: write about why something occurs or happens and make plain, or intelligible, a concept or idea

- You may be asked to recognise patterns on the maps and work out relationships.

For example, **population distribution and density** – often this is on a world scale and areas of high and low population density may be picked out. You may then be asked to identify areas of:

1 high and low population density and give reasons for their differences – hot or cold, wet or dry, long or short growing seasons, fertile or infertile land, mountainous or flat, fertile river valleys like the Nile in Egypt or the Ganges in India, natural resources like coal, oil and gas fields or other minerals like iron ore
2 population migration/movements – how many people, whether it is international (between countries) or national (within a country), economic, forced or voluntary
3 road and rail transport networks
4 settlement sizes, shapes and patterns
5 relief (the shape and height of the land)
6 river – including floodplains, deltas (arcuate/fan or digital/birds foot), meanders, oxbow lakes
7 coastal – including headlands and bays (which you can then relate to resistant/hard and less resistant/soft rocks and differential erosion), spits, bars, tombolos, beaches, wave-cut platforms and cliffs
8 volcanoes and earthquake features, etc.

The maps may have diagrams added to them such as small **bar** or **line graphs**, or **population pyramids** to illustrate the age and sex structure of a country or contrasting countries.

Graphs and diagrams, tables of data and written material

You may be asked to extract specified geographical information and analyse information from graphs, diagrams, pictograms, tables of data and written material.

The types of graph used may include:

- population pyramids
- line graphs
- bar graphs
- triangular graphs
- pie graphs
- radial graphs
- divided bar graphs
- scatter graphs
- wind roses
- histograms
- kite diagrams
- flow diagrams.

In the exam-style questions that follow, you will see examples of these and how they may be used in questions. You may be asked to **describe variations** and **identify trends** in the information on the graphs.

Graphs may show, for example, temperature, rainfall, birth rate, death rate, energy use, rainfall distribution, river discharge, etc. Also you may be asked to **plot information on the graphs** when axes and scales are provided.

Population pyramids

You may be asked to describe the broad features of the **population structure** to show comparisons and contrasts between the male and female populations, the working (economically active) and non-working (non-economically active) population and the young (young dependant – 15 and under) and old age (old dependant – 65 and over) groups. A single pyramid of one country may be used or you may be asked to compare the pyramids of two countries – normally a typical LIC and a typical HIC.

There follow some examples of exam-style questions using different graphs and diagrams, tables of data and written material:

Exam-style questions

1 Describe the main features of the population pyramid of an LIC, shown in Figure 20.5, and suggest reasons for the features found on the pyramid. [6]

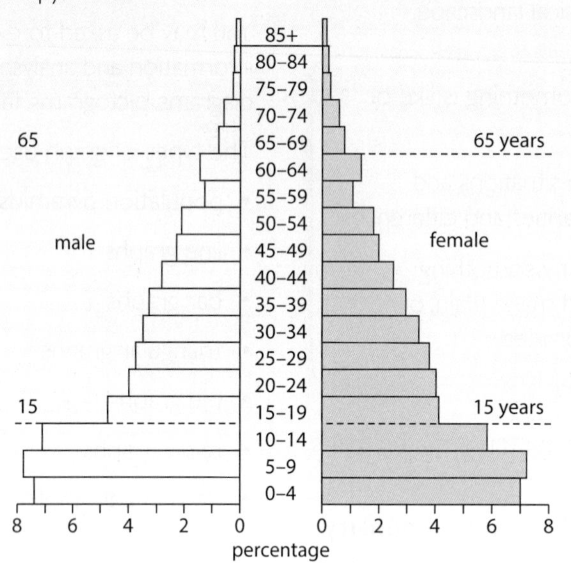

Figure 20.5

2 Study Figure 20.6, a population pyramid of a typical LIC, and Figure 20.7, a population pyramid of a typical HIC.

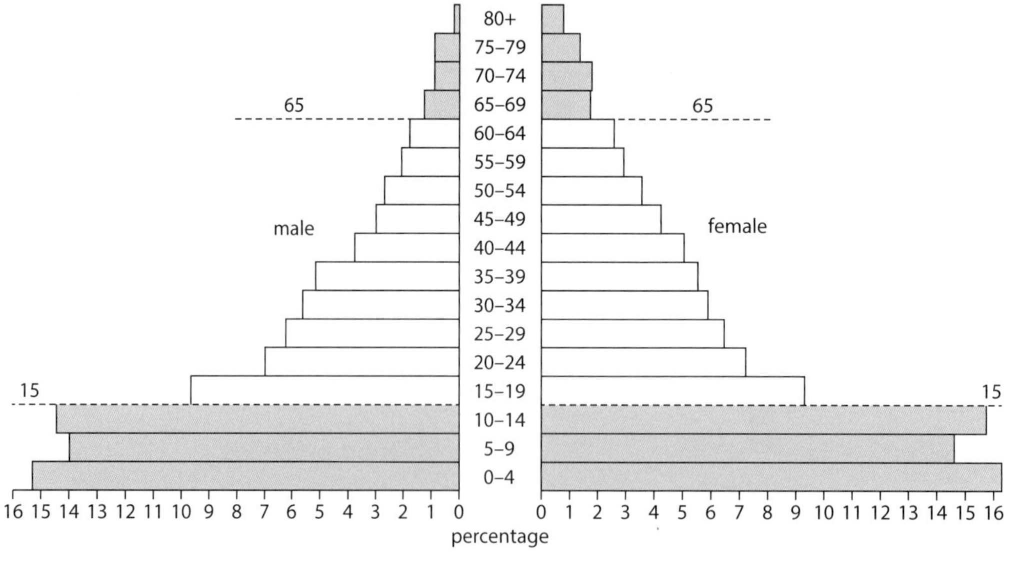

Figure 20.6

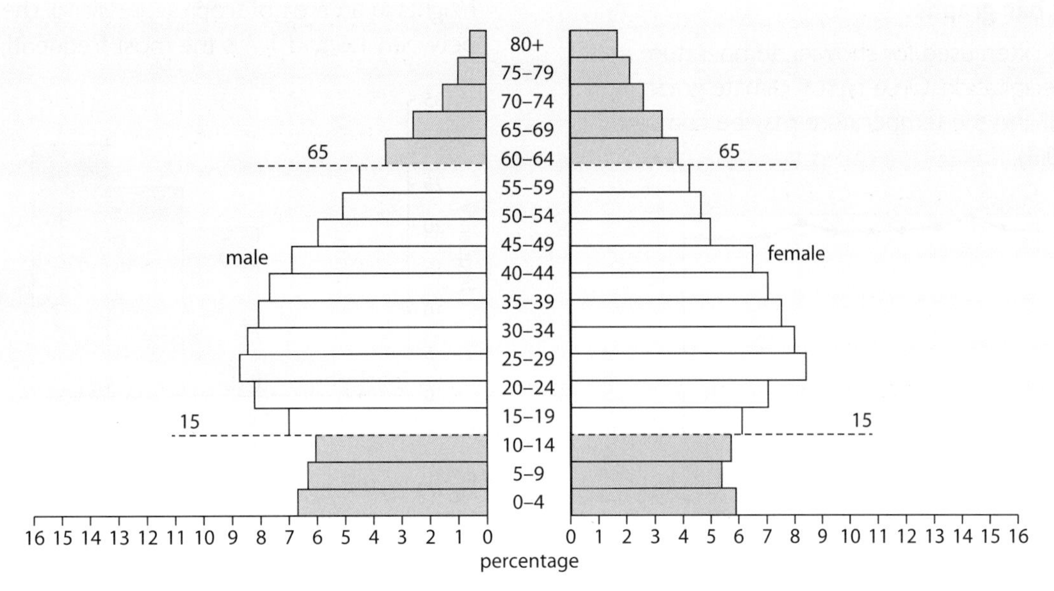

Figure 20.7

a What evidence is there in the population pyramids that:

  i People in HICs may expect a longer life expectancy than people in LICs?

  ii Countries which are LICs may have a higher birth rate than HICs? [2]

b Supporting your answer with figures from the two population pyramids, describe how the economically dependent population, people aged below 15 and above 65, of the LIC differs from that of the HIC? [3]

c Describe the ways in which the economically dependent population may be supported in a typical LIC and a typical HIC. [4]

Exam-style questions

1 Study Figure 20.8, a climate graph for a town in southern Ghana, in West Africa.

  a i What was the highest temperature reached during the year? [1]

  ii How much rain fell in the month of December? [1]

2 Describe the main features of the tropical climate shown in Figure 20.8. [4]

3 Explain why the climate may be uncomfortable for people from a colder temperate latitude travelling to this area. [2]

Line and bar graphs

These are often used for showing temperature and rainfall/precipitation. On a typical climate graph, both the rainfall and the temperature may be combined (Figure 20.8).

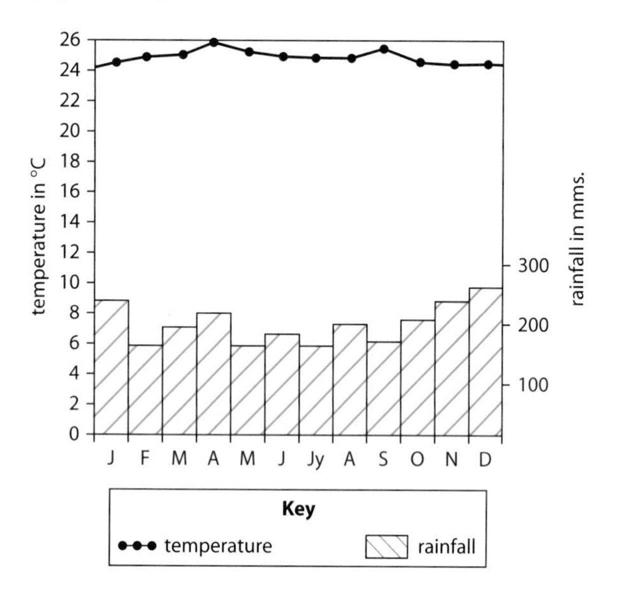

Figure 20.8

Histograms

At first glance, histograms look very similar to bar graphs. Both graphs use vertical bars to represent their data. The height of a bar corresponds to the relative frequency of the amount of data in the class. The higher the bar, the higher the frequency of the data. A histogram, however, is a graphical representation of the **distribution** of numerical data. For example, Figure 20.9 shows the frequency of grasses of different heights in an area of tropical grassland. The height between 1.6 and 1.7 is the most frequently found.

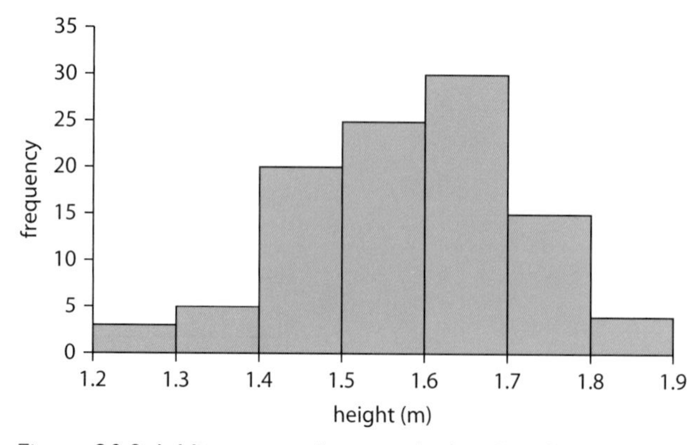

Figure 20.9 A Histogram showing the height of grasses in an area of tropical grassland

Triangular graphs

These are commonly used to compare the percentages of people who are working in **primary**, **secondary** and **tertiary** industry. They have **three** axes instead of the normal two. Figure 20.10 and the exam-style question that goes with it are typical of the way in which this type of graph may be used in an exam. A typical LIC will have a very high percentage of people working in **primary** industry – mainly poor subsistence farmers (such as Tanzania with 83% on Figure 20.10), while an HIC will have a very high percentage in **tertiary** (service) industries (such as the USA 72%, France 67% and Japan 58%). MICs, such as Nigeria and Indonesia, will be somewhere in between the two extremes as more of their population move into jobs in **secondary** (manufacturing and processing) and **tertiary** industries.

Exam-style questions

1 Explain why there are the differences shown on Figure 20.10 in the percentage of the labour force employed in the three different sectors (primary, secondary and tertiary) of the economy in the two groups of countries. [5]

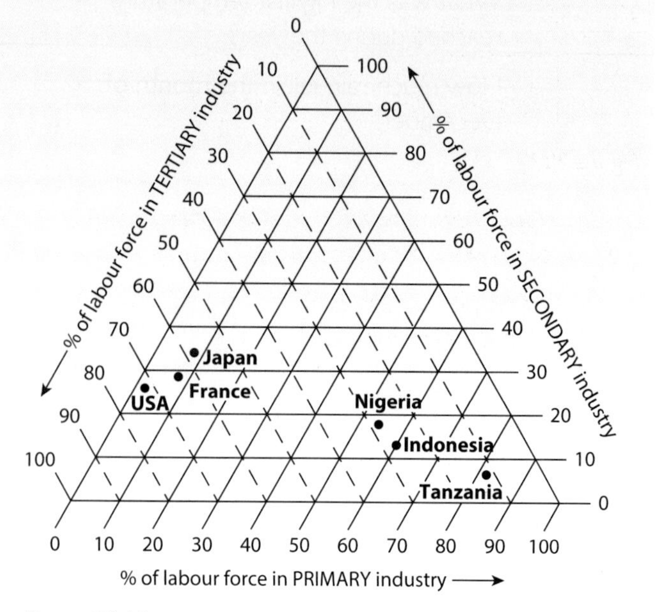

Figure 20.10

Divided bar charts

There are several options for displaying data. While Figure 20.10 makes use of a triangular graph, a vertical (Figure 20.11) or horizontal (Figure 20.12) divided bar graph could also be used for displaying similar data.

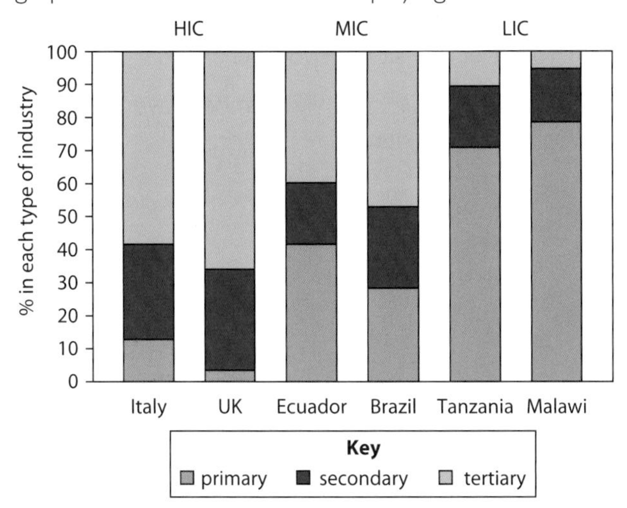

Figure 20.11 Triangular graph to show the relative percentages of the labour force employed in the primary, secondary and tertiary sectors of the economy of selected countries.

Exam-style questions

1 Figure 20.12, shows the employment structure of four countries in different stages of economic development.

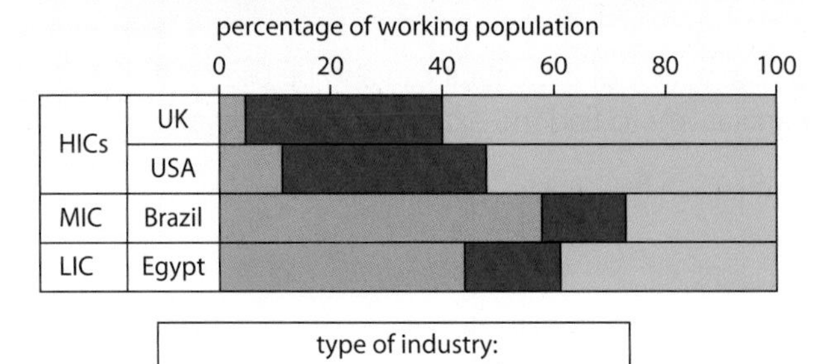

Figure 20.12

a Which country has the largest percentage of its workforce in tertiary industry? [1]

b What percentage of Brazil's workforce is in primary industry? [1]

c Using the information in Figure 20.12, describe what may happen to the employment structure of a country as it moves from a LIC towards a HIC. [3]

Pie charts

Pie charts are useful for showing the **size** proportions of different categories in a data set (Figure 20.13). They are visual diagrams that are easy to interpret. They often have a key to show what the different sections represent. The data to be displayed must first be converted into percentages and then into a proportion of 360° (i.e. the total number of degrees in a circle; this is done by multiplying the percentage values by 3.6). This gives us the number of degrees of the circle that a category takes up, which can then be plotted using a protractor. Pie charts generally do not tend to show exact figures, unless they are written on the diagram after the pie chart is constructed, meaning it can be hard to make detailed comparisons and quote actual figures.

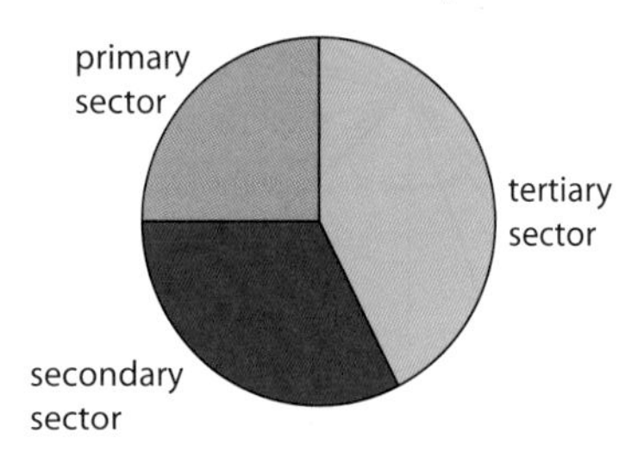

Figure 20.13 The proportion of primary, secondary and tertiary industry in a location

Proportionally divided circles

These are also pie charts, but they can be used to **compare** a particular set of **similar data in two areas**. For example, comparing the types of industry (P = primary, S = secondary and T = tertiary) in two countries which are very different in their total populations. Figure 20.14 shows the proportion of types of industry in the UK and Norway within the circles, while the size/area of the circles is proportional to the relative populations of each country.

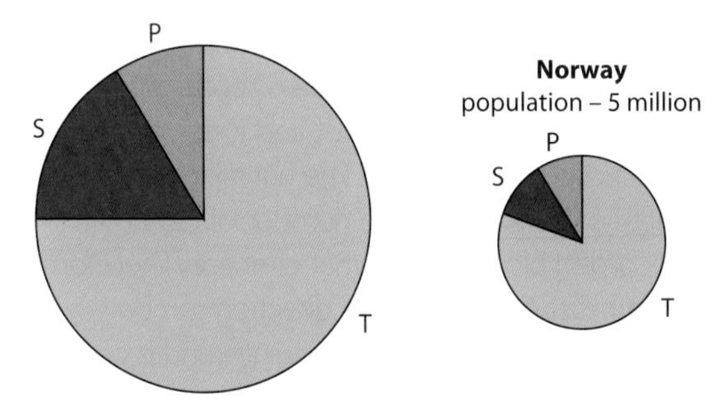

Figure 20.14 The proportion of types of industry in the UK and Norway

Exam-style questions

1 Study Figure 20.15. It shows the distance the people in a family have to travel from their house to reach the different services available in their settlement.

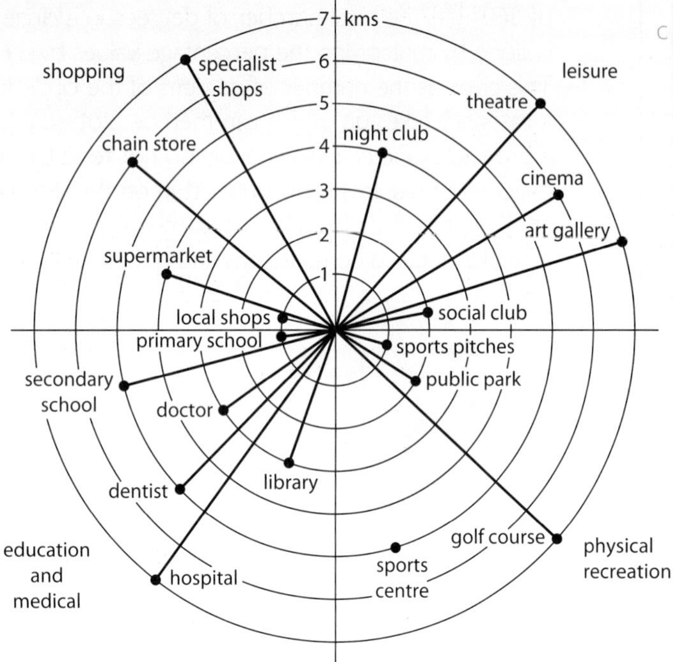

Figure 20.15

a In which of the sectors shown would you place a youth centre/club? [1]

b How far would people in the family have to travel to visit the sports centre? [1]

c How far would a family member have to travel to the dentist? [1]

Radial graphs

In radial graphs, the values extend out from a central point to show the relationship of each variable to the central point or location. They are useful because a number of different variables can be plotted on one graph, as more than one axis can be used (Figure 20.15). The most common form of radial graph is a **wind rose**, which shows the frequency of wind direction – the axes represent the four basic directions, North, South, East and West, or the eight basic directions (Figure 20.16). The number on each axis is the length of time, usually the number of days that the wind is recorded blowing from that direction. Therefore, the **prevailing**/most common, wind direction can be easily identified. In Figure 20.16, the prevailing wind is south west and the number of days it came from that direction is seven. During the same period, there were no days when the wind came from the north east. In a wind rose, the centre of the diagram is used to show the number of days (in this example, one) when there was no wind, i.e. it was calm. While the wind rose gives a good visual representation of the data, it can also be used statistically/quantitatively to find the exact number of days.

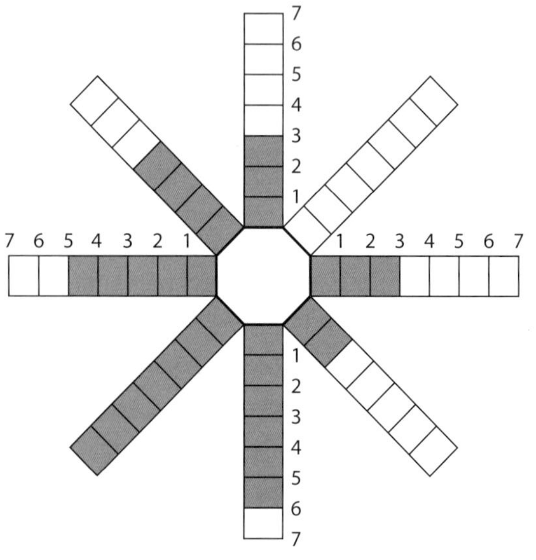

Figure 20.16 Wind rose

Scatter graphs

These are often used to show and compare a wide range of data. Scatter graphs show the **relationship between two variables** that are being investigated and any correlations between them. The independent variable of the investigation is plotted on the *x*-axis and the dependent variable is plotted on the *y*-axis (Figure 20.17), and the results are marked on the graph with a dot/cross. A line of best fit can then be drawn.

Scatter graphs can help highlight a correlation between two variables if it is present, and a large number of points can be plotted within a small space. However, the line of best fit can sometimes suggest a misleading relationship and it is not accurate due to the fact that it is drawn by eye. Scatter graphs are quick and easy to construct, but are much more useful when a statistical test is applied to the data.

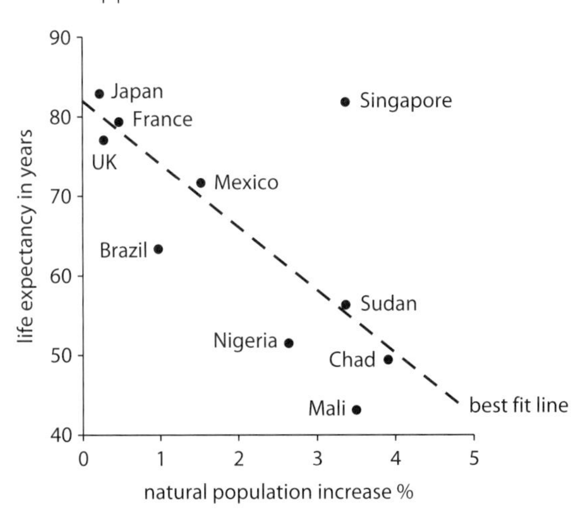

Figure 20.17 The best fit line that results when comparing the natural population increase (percentage) with life expectancy in years

Flow diagrams

These are often used to show **systems** with **inputs**, **processes** and **outputs.** The systems used are usually in the industry or farming questions, but they can be used in rivers, coasts and ecosystems.

Kite diagrams

Kite diagrams are useful for looking at patterns of zonation in vegetation such as Figure 20.18, which shows the number of species of three different plants in a transect through some sand dunes on a coast.

Kite diagrams have two axes, and in this way are similar to bar graphs. The *x*-axis has one or more categories and the *y*-axis shows the measurement sites. In Figure 20.18, the readings have been taken at 10-metre intervals. These graphs are most commonly used to show plant succession, with different plants forming the different categories on the *x*-axis. A central line is drawn up the graph from each category on the *x*-axis and at each measurement site, the percentage cover of each plant is plotted. However, the value is **divided into two**, with half of the result being plotted one side of the line and half on the other side, thus creating a '**kite**' shape.

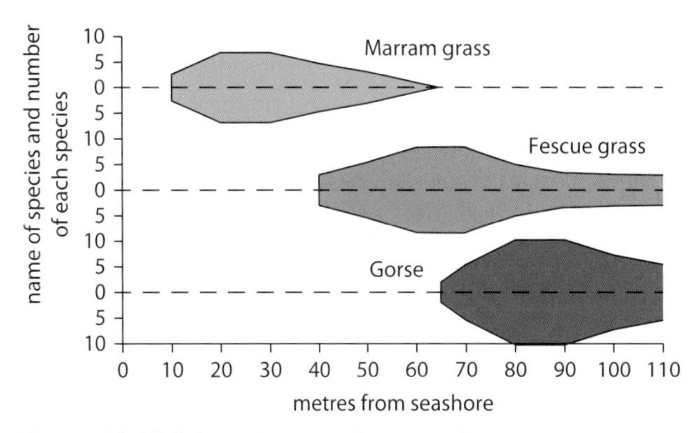

Figure 20.18 A kite diagram showing plant succession

Tables of data

Data tables may provide information on physical phenomena such as river discharges and flood events, size of earthquakes, on economic activities, population/settlement, agricultural and manufacturing output, etc. You may be asked to describe and analyse features and trends from the data provided in a table, or you may be asked to suggest an appropriate type of graph on which you could plot the data provided.

TERMS

correlation: the process of establishing a relationship or connection between two or more things

independent variable: the factor that is changed or controlled in a scientific study to test the effects on the dependent variable

dependent variable: the factor being tested and measured in a scientific study. The dependent variable is 'dependent' on the independent variable

line of best fit: a line through the centre of the results with an equal number of points above and below the line

kite diagram: a graph that shows the density or distribution of species that have been found along a transect

Written material

This may be extracts from books, periodicals and newspapers, and you will be expected to show an understanding of the material presented.

20.4 Photographic and pictorial material

Questions usually use oblique photographs. You should be able to describe the human and physical landscapes (landforms, natural vegetation, land-use and settlement) and other geographical features from photographs, aerial photographs, satellite images and GIS. You will only be asked to give simple descriptions. Questions may also use simple field sketches of physical and human landscapes to help in your understanding and geographical description and labelling (annotation). You will be expected to use a range of words in your descriptions. Therefore, it is just a matter of you remembering to look and name or describe certain obvious features.

TERMS

oblique photograph: a photograph taken with the camera axis inclined away from the vertical. It usually covers more ground area than a vertical photo taken from the same altitude

GIS – Geographic Information System: a system designed to capture, store, manipulate, analyse, manage, and present spatial or geographic data

field sketching: a drawing forming a fundamental part of any field work investigation. It is a simple qualitative technique which, done correctly, will support the data collected and enhance a study

In a **natural landscape**, you should be able to identify and use the following words and terms.

- high, low, steep, gently sloping, mountains, hills, cliffs, vertical slopes, bare rock, loose rock and scree
- **valleys:** small valleys, large valleys, V-shaped valleys – wide or narrow V-shaped valleys, tributary valleys, confluence (where two or more valleys or rivers meet), gorges, waterfalls, rapids, oxbow lakes, river cliffs, point bars (slip-off slopes)
- **rivers, streams:** wide or narrow river channels, fast or slow flowing, turbulent (white water), muddy/full of sediment, clear, lakes, ponds, reservoirs, well drained, poorly drained (marshy)
- **coastal:** beaches (large, long small, wide narrow), spits, bars, lagoons, tombolos, sand, pebbles, marshes (salt water), cliffs (high, low, steep, vertical, bare rock, vegetated, collapsed), wave-cut platforms, caves, arches, stacks, stumps, geos (caves where the roof has collapsed), cliffs – with cracks or faults, joints, bedding planes, layers of rock
- **vegetation:** forests, woods, trees (tall or small), shrubs, grass, bare ground, deforested, cleared
- **farmland:** fields (large or small), arable or pastoral farming or both, crops, hedges, shelter belts (trees grown on hedges to protect land from wind erosion), fences, ditches

In a human landscape:

- **cities and towns** may have shops, offices, banks, restaurants/cafés, hotels flats, apartments, residential/housing, multi-storey car park, bus station, tall/high/high rise/multi-storey buildings, low-rise buildings, modern/new/old buildings, flat roofs, steep roofs, high-density/low-density buildings, concrete/brick/glass, balconies
- **rural areas** may have villages, individual houses, farmhouses, single storey, two storey, etc., farms, farm land, fields, crops orchards, woods, forests.
- **streets and roads** may be narrow or wide, straight or winding, sealed or unsealed (sealed means covered in bitumen/tarmac, or unsealed not covered in bitumen/tarmac) pavements, motorways, roundabouts, cars, lorries/trucks, buses, coaches, cycles, motorbikes

Lastly, **cartoons** may be used to illustrate a geographical theme for you to interpret and analyse.

Alternative to coursework

21.1 Alternative to coursework

As an alternative to coursework, you may be set a series of tasks on issues relating to one or more of the syllabus themes previously covered. You are likely to be tested on the methodology of questionnaires, observation, counts and measurement techniques and will involve hypothesis testing appropriate to specific topics. Exam-style questions on common topics will be looked at in this chapter.

TERMS

methodology: a system of methods used in a particular area of study or activity

hypothesis: a proposed explanation for a phenomenon taking place

When you answer a hypotheses question that asks you whether you agree with the hypothesis or not, always give your opinion at the start of your answer before you give any supporting evidence. Your answer will usually be yes/no/to some extent. There is no need for you to copy out the hypothesis if you agree with it. It is important that you make a decision, say what it is and then provide the evidence for your choice. Make sure that you state your decision clearly. Do not be too vague, for example, try not to say 'this might be true' or 'this could be false'.

When you give figures in your answer make sure that you give the units, such as metres, millibars, etc. The style of writing numbers varies so it is very important that you write your numbers clearly, in particular, a figure 7 can look like a 1, a 2 can look like a 5 and a 4 can look like a 9.

Graphs and charts

If you are asked to shade a graph, use the same style of shading that is provided in the question and make sure that the pencil you use gives a good dark image – a 2B pencil shade will give a very good level of shading. Ensure that you understand the scales that are being used on graphs and diagrams and the significance of any data that has already been plotted for you. When completing pie charts or divided bar graphs, complete these in the order of the data given and in the order of the key. It is important to make sure that your shading matches the shading in the key, so if the question uses diagonal lines that slope down to the right, do not draw yours sloping to the left. When you have finished your answer, go back and check that you have fully completed all the graphs.

Command words

One simple instruction to follow is RTQ – Read The Questions carefully and identify the command word, such as the two command words **Describe** and **Explain** – say what it is like and give reasons for this. It is important to be aware that a question that asks **Why** requires you to give a reason and not a description. If you are asked to **Compare**, make sure that you make a comment on whether one set of values/data is higher, lower, and you do not just give a list of comparative statistics.

You will be tested on the processing, presentation and analysis of data. You must answer **all** the questions.

Questions can be taken from across the IGCSE Geography syllabus on a wide variety of topics and you will be assessed on your knowledge and application of the methodology used in a range of data-collection enquiry skills.

You are likely to be expected to show that you have some knowledge of fieldwork equipment, how it is used and fieldwork techniques. You should also be familiar with the different types of maps, tables and the graphs listed in the syllabus as this is an important element of this examination.

The topics that could be used include:

- changes in the weather in the school grounds
- downstream changes in the physical characteristics of a small river/stream
- the effects of different types of woodland on temperature and sunlight
- coastal processes

- investigating the action of longshore drift
- conditions in squatter settlements
- the characteristics of the CBD of a town
- traffic flows in and around a town centre
- land use in urban areas
- how a shopping area may change
- different types of housing area
- the environmental and economic effects of tourism
- the impact of tourism on a town.

21.2 Examples of exam-style questions

Weather study

TIP

A **weather study** is a relatively easy study to carry out in school grounds. It also allows you to get to know the various weather instruments and how to read, re-set, record and analyse their results. It is a study of a **microclimate**. It is surprising how climate can vary over a comparatively small area; even in an area as small as a school site. The microclimate can be important in determining how land use/type of crop is planned. Careful measurement of the weather variables are therefore of practical importance.

Weather studies/investigations require minimal equipment and much of the basic equipment, such as rain gauges, can be homemade. They can also generate large amounts of quantitative results so you have plenty to analyse.

Exam-style questions

1 Students were asked to study and measure elements of the weather every school day during a two-week period in their school grounds. A hypothesis was put forward for the students to investigate:

- **Hypothesis 1:** Precipitation increases as atmospheric pressure falls.
- **Hypothesis 2:** Temperature changes according to the direction which the wind is blowing from.

Over the two-week period, the students measured atmospheric pressure, temperature, precipitation and wind direction.

a Name the instruments that would be used to measure:

i Atmospheric pressure [1]

ii The highest and lowest temperatures [1]

The instrument used to measure the temperature was kept in a Stevenson screen (Figure 21.1).

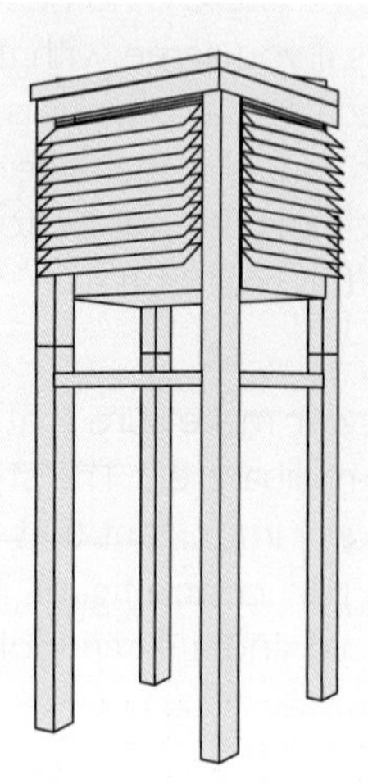

Figure 21.1

iii Name **four** characteristic features of the Stevenson screen shown in Figure 21.1 and for each of the characteristic features, give a possible reason for the feature. [4+4]

b Figure 21.2 shows a rain gauge.

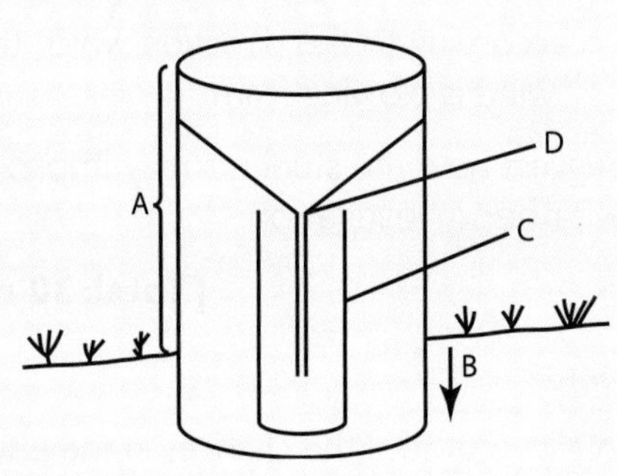

Figure 21.2

i Name the part of the instrument labelled **C.** [1]

ii Why should the top of the rain gauge (shown as **A** on the diagram) be well above ground level? [1]

iii Why should the rain gauge be sited on grass or loose gravel/shingle? [1]

iv Why is the rain gauge sunk into the ground (shown as **B** on the diagram)? [1]

v Why does the funnel (shown as **D** on the diagram) have a restricted opening? [1]

vi Suggest two factors which the students would need to consider in choosing a site for the rain gauge in the school grounds. [2]

c The students used a wind vane to observe wind direction. Suggest a good position to put a wind vane in the school grounds and explain your choice. [2]

d Suggest why the temperature was recorded daily at the same time of day, at 8.00 a.m. [1]

e Describe how the students could extend their study to measure the amount of cloud cover. [1]

The results recorded by the students are shown in Table 21.1.

f Using the data in Table 21.1, complete the graph and wind rose in Figure 21.3 by plotting the remaining data on precipitation (in mm), temperature (in °C), atmospheric pressure (in mb) and wind direction, starting with the data from 13 March. The first three days, 10–12 March have been done on each of the graphs and the wind rose. **[4]**

Date	Precipitation (in mms)	Temperature (in °C)	Atmospheric pressure (in mb)	Wind direction
10 March	0	23	1016	SE
11 March	0	24	1012	E
12 March	3	25	998	S
13 March	7	25	994	SW
14 March	0	28	992	SE
17 March	0	29	990	SE
18 March	6	24	992	SW
19 March	9	23	985	W
20 March	0	23	1008	SE
21 March	0	23	1016	E

Table 21.1

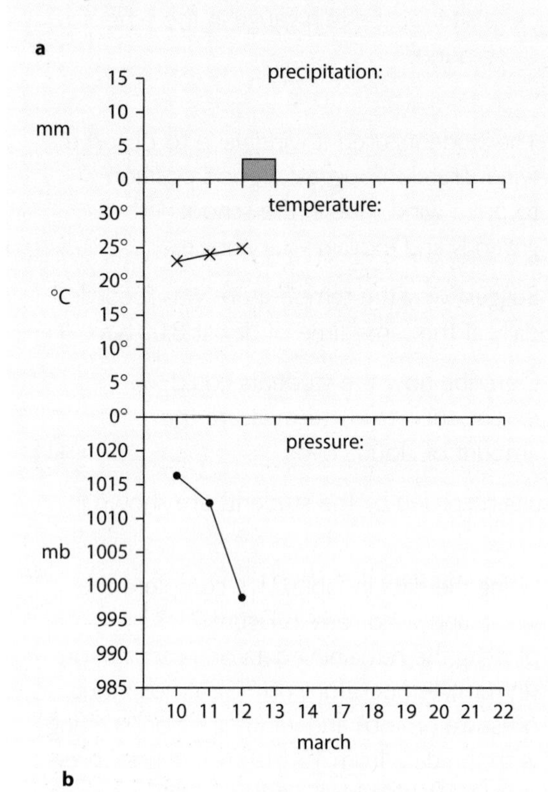

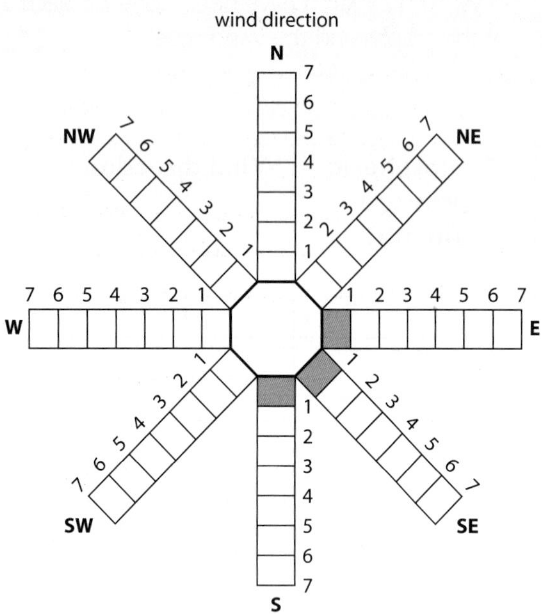

Figure 21.3 (a and b)

g Using evidence from Figure 21.3, what conclusion would the students make about:

i **Hypothesis 1:** Precipitation increases as atmospheric pressure falls. [2]

ii **Hypothesis 2:** Temperature changes according to the direction which the wind is blowing from. [2]

h Suggest how the students may improve the weather investigation. [1]

[Total: 30 marks]

River study

2 Students investigated the changes in the physical characteristics of a small river channel at three sites, **A**, **B** and **C**, as they walked from near the source down the river channel towards the mouth.

They measured three physical characteristics at each of the three sites: the width, depth and velocity of the water. This data was then used to investigate two hypotheses for their investigation:

- **Hypothesis 1**: The width, the depth and the velocity of the river will all increase as the distance from the source increases.
- **Hypothesis 2**: The velocity of the river will increase as the hydraulic radius of the river channel increases.

The **hydraulic radius** is a measurement which attempts to indicate how much friction there is between the river channel and the flow of the river. The hydraulic radius is calculated using the following formula.

$$\textbf{hydraulic radius} = \frac{\textbf{cross–sectional area (in}^2\textbf{)}}{\textbf{wetted perimeter (in)}}$$

To find out the data to put into this formula, the students had to calculate the cross-sectional area of the river channel at each site and the length of the **wetted perimeter**. The **wetted perimeter** is the total length, in, of the river channel bank and bed which the water in the river channel is actually in contact with.

To calculate the **cross-sectional area**, the students measured the width of the river channel and the depth of the river at several sites, at 0.5 m intervals across the channel. The results of sample measurements made at the three sites, **A**, **B** and **C**, are shown in Table 21.2.

a i Suggest **three** factors the students should have considered in their choice of the sites for their fieldwork. [3]

ii To prepare for this fieldwork investigation,

Distance across channel (metres)	0.5	1.0	1.5	2.0	2.5	3.0	3.5	**Average depth (in m)**	**Width of channel (in m)**	**Cross-sectional area (in m²)**	**Average time (in seconds) to travel 10 m**	**Velocity (in m/s)**	**Distance from site A (in m)**
Site A Depth of river (metres)	0.10	0.13	0.11					0.11	1.7	0.187	35.9	0.28	0
Site B Depth of river (metres)	0.12	0.13	0.15	0.15	0.17			?	2.8	?	31.6	0.32	1 000
Site C Depth of river (metres)	0.13	0.15	0.25	0.30	0.22	0.16	0.18	0.19	3.7	?	27.4	0.36	2 000

Table 21.2

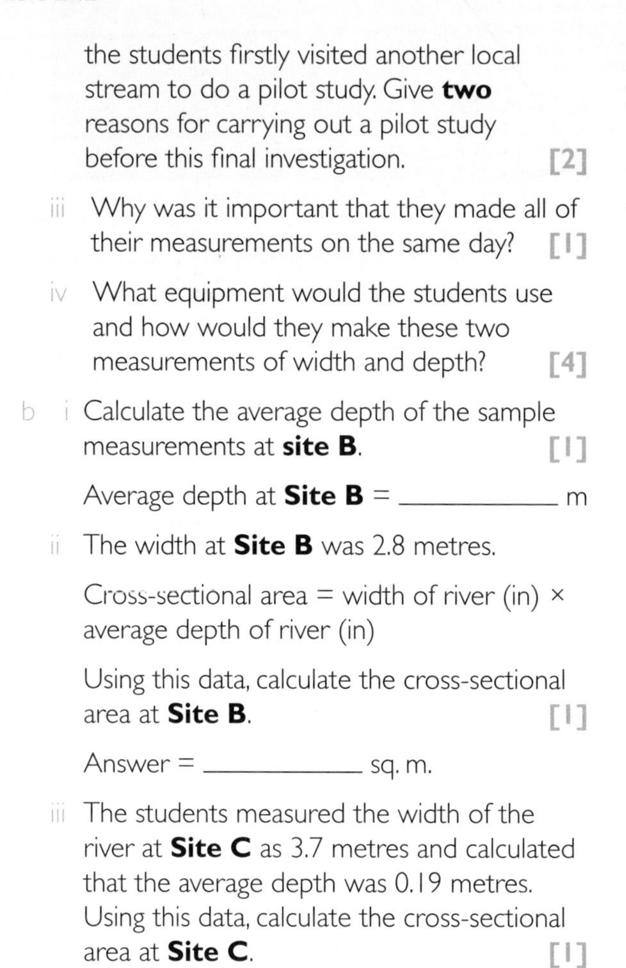

the students firstly visited another local stream to do a pilot study. Give **two** reasons for carrying out a pilot study before this final investigation. [2]

iii Why was it important that they made all of their measurements on the same day? [1]

iv What equipment would the students use and how would they make these two measurements of width and depth? [4]

b i Calculate the average depth of the sample measurements at **site B**. [1]

Average depth at **Site B** = ____________ m

ii The width at **Site B** was 2.8 metres.

Cross-sectional area = width of river (in) × average depth of river (in)

Using this data, calculate the cross-sectional area at **Site B**. [1]

Answer = ____________ sq. m.

iii The students measured the width of the river at **Site C** as 3.7 metres and calculated that the average depth was 0.19 metres. Using this data, calculate the cross-sectional area at **Site C**. [1]

Answer = ____________ sq. m.

The students then measured the **velocity** of the stream at each site. Using a stopwatch, they timed how long it took a float to travel a distance of 10 m down the river channel (Figure 21.4).

The float was released at three separate points in the cross section: quarter, half and three-quarters across the stream. Each measurement was repeated five times. The **average time** for each set of measurements was calculated and is shown in Table 21.2, along with the calculated **velocity**.

c i Why was the measurement repeated five times in each location? [1]

ii Explain how the velocity figures in Table 21.3 were calculated. [1]

d i Describe the link between the velocity and the depth results shown in Table 21.3. Give reasons for the relationship you have identified. [4]

ii Describe **two** possible problems in measuring the velocity that may cause the results to be unreliable. [2]

The students then measured the **wetted perimeter**. To measure the wetted perimeter, the tape was placed across the bed of the river at each site, starting and finishing at the water level on both banks. To make the method more

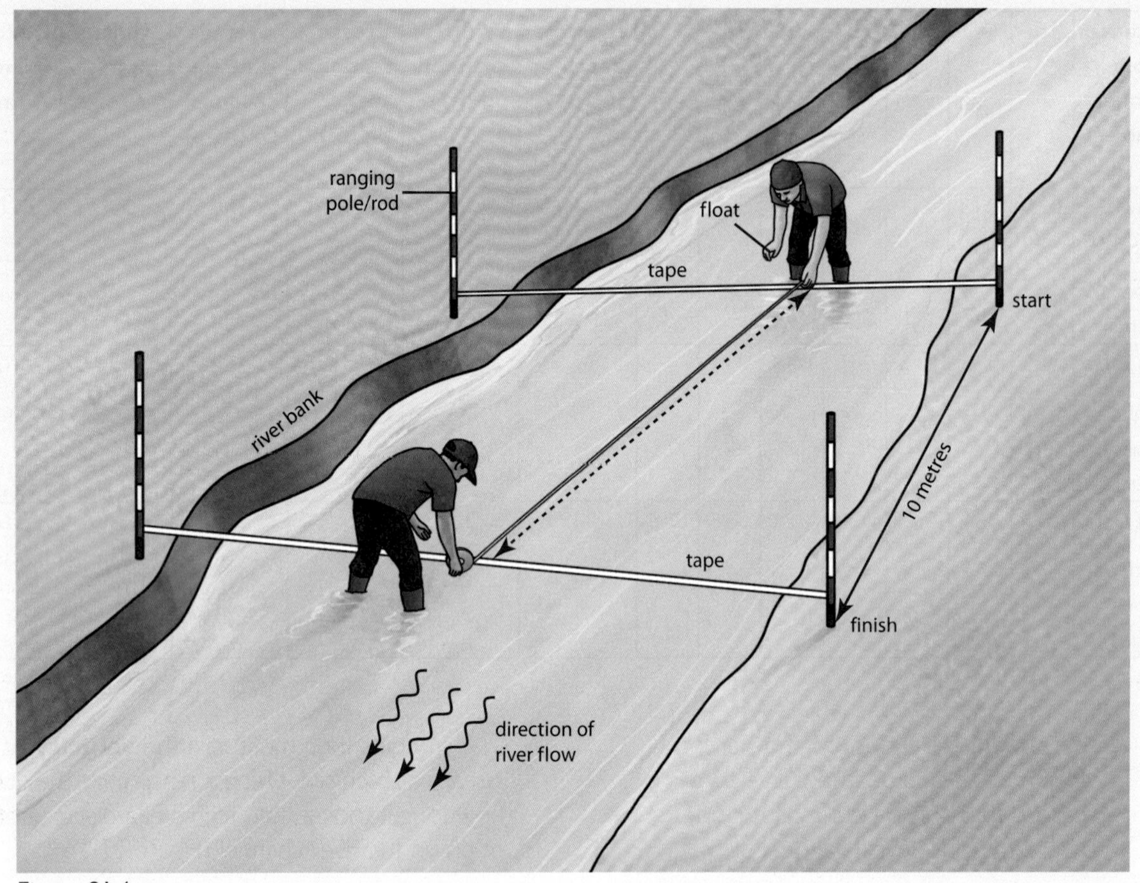

Figure 21.4

accurate, a student walked along the tape across the river.

iii Suggest **one** possible disadvantage of this method in a large river. **[1]**

The wetted perimeter results are shown in Table 21.3.

Site	Distance from site A (in m)	Cross-sectional area (in m^2)	Wetted perimeter – (in m)	Hydraulic radius $\frac{CSA}{WP}$
A	0	0.187	1.8	0.14
B	1 000		3.1	
C	2 000		4.0	

Table 21.3

You are now able to complete Table 21.3 with your cross-sectional area results.

e From this information, it is now possible to calculate the hydraulic radius for all three sites (to 2 decimal places). The hydraulic radius for site **A** has been done for you.

i Calculate the hydraulic radius for sites **B** and **C** and enter your results in Table 21.4. **[2]**

f When the students had collected and analysed this data they were then able to reach a conclusion for both the hypotheses. Using the data evidence that you now have, what conclusion do you think they reached for:

i **Hypothesis 1:** The width, the depth and the velocity of the river will all increase as the distance from the source increases. **[2]**

ii **Hypothesis 2:** The velocity of the river will increase as the Hydraulic Radius of the river channel increases. **[2]**

iii Suggest how the investigation of the hypotheses could be extended to improve the reliability of the results. **[2]**

[Total 30 marks]

Answers

Theme 1: Population and settlement

Chapter 1

Self-test questions 1.1

1 In the 20th century, population growth accelerated as birth rates fell (to about 20 live births per 1000) and death rates continued to fall (to about 15 deaths per 1000).

Birth rates fell due to: access to contraception – family size could be planned; increases in wages – families were better off and did not need as many children to work; increased urbanisation which changes the traditional values placed upon fertility and the value of children in rural society; urban living also raised the cost of dependent children to a family; a reduction in subsistence agriculture (where what you grow your family eats and there is little or no extra food to sell for cash) – not as many children were needed to work on the farms; an improvement in the status and education of women – increasing female education has led to more women getting jobs and following careers; working women have less time to raise children; this is particularly an issue where fathers traditionally make little or no contribution to child-raising (e.g. in the Ivory Coast); an increase in parental investment in the education of children.

Death rates fell due to: the availability of health/medical facilities; improvement in access to food; improved nutrition/less malnutrition; increased care and spending on older people; increased health awareness/improved health education.

2 Where few natural resources are available, such as fertile soils, an optimal climate for the growing of crops, sources of freshwater and minerals, population growth may be limited. Conversely, the potential for population growth will be increased where these resources are available.

Self-test questions 1.2

1 The reasons will include changes in the factors that impact on birth and death rates. Where both birth and death rates are high, there will be little change in population. Where birth rates remain high (due to a lack of contraception and improved health care, for example) and death rates fall (due to due to improved access to health care and food, increased care and spending on older people, for example), there can be an increase in population. Where both birth and death rates fall (due to increased access to contraception, decrease in the number of children needed to work on the farms, increased availability of health/medical facilities, and so on), population may fall.

2 Attempts to alter population growth can be pro- or anti-natal across the many different societies of the world and may involve one or more of the following practices: increasing access to contraception; increasing access to abortion; advertising campaigns putting forward the advantages of a smaller family; offering bonuses to those people who have smaller families; educating women about family planning; improving health care so that infant and child mortality rates drop, reducing the need to have more children.

A typical case study to use is China's One Child Policy, where the following measures were applied: couples were allowed only one child; men could not get married until they were 22 and women 20; couples had to apply to the authorities to get married and again when they wanted a baby; couples were rewarded for doing this by being given a salary/wage bonus (an extra 10%) and family benefits; priority in education/health facilities/employment/housing; those who did not conform lost these benefits and were given large fines; women who became pregnant a second time were forced to have an abortion and women who became repeatedly pregnant were sterilised; a 'workplace snooper' was employed by most factories and businesses who could grant permission for employees to have a child; the government advertised the benefits of small families such as having a greater amount of disposable income available.

Pro-natal policies were established in Singapore, where the incentives included: tax rebates for third children; up to four years' unpaid maternity leave for civil servants; pregnant women were to be offered increased counselling to discourage 'abortions of convenience' or sterilisation after the birth of one or two children; a public relations campaign to promote the joys of marriage and parenthood; a S$20000 tax rebate for fourth children born after 1 January 1988; for children attending government-approved childcare centres, parents were given a S$100 subsidy per month regardless of their income; third-child families were given priority over small families for school registration.

Exam-style questions

1 a There may be little or no birth control or family planning; having more children will mean that a family will have more help to farm the land, as they cannot afford machinery; in some societies, a large number of children is seen as a sign of virility; as so many children die when they are young, parents have several children to try to ensure that some will live and look after them in old age; religious beliefs – some religions encourage large families (e.g. Roman Catholics).

b As a response to poverty or environmental concerns; to improve the quality of life for a society; as a solution to over-population; to improve people's lives, giving them greater control of their family size.

c Improvements in the food supply brought about by higher yields due to improved agricultural practices and food storage and better transportation of food, which has helped to prevent death due to starvation. These agricultural improvements include crop rotation, selective breeding and seed drill technology. Significant improvements in public health care which has helped to reduce death rates (mortality), particularly for children. These included medical breakthroughs (such as the development of vaccinations) and, more importantly, improvements in water supply, sewerage, food handling and improved general personal hygiene which came from a growing scientific knowledge of the causes of disease and the improved education of mothers.

2 Over-population may be said to occur when the number of people living in an area exceeds the ability and the availability of the resources in the area to sustain them.

A named example, Bangladesh. Some specific reference to the high levels of traffic congestion, especially in cities like Dhaka, as there are too many cars/vehicles on the roads; farm/agricultural land on floodplains is being over-used/overgrazed, as on the Brahmaputra/Ganges/Meghna rivers; which may lead to a loss of soil fertility / lower crop yields; there is widespread deforestation as wood is cut for fuel/firewood on the steep, fragile mountain/hill slopes of Himalayas.

Chapter 2

Self-test questions 2.1

1 People may be **pushed away** from one area due to natural disasters, unemployment, lack of work, to escape from poverty, civil war, racial, political or religious intolerance, land shortages and famine or lack of food. They may be **pulled** to other areas which offer employment and higher wages, better housing and education opportunities, a higher standard of living and improved quality of life, greater racial, political and religious tolerance and the 'bright lights' of urban areas.

2 The positive impacts may include less pressure on land and resources in rural areas. Migrants may also send back remittances which improve the standard of living and quality of life of their families who remain. However, the negative impacts may include rural areas, from which people have migrated, can become depopulated; fewer farmers can lead to a drop in food production and fewer people to help with harvesting crops. Many of the people who migrate are male, often husbands and fathers, who leave their families behind for several months or years at a time. The burden of looking after the home and possibly a farm may fall on the remaining family members. The family members that remain may be elderly and find it difficult to cope. The people who leave may be economically active and skilled leaving skill shortages in health care, for example.

Self-test questions 2.2

1 a The European migrant crisis starting in 2015 was caused by several reasons. Many migrants were refugees, fleeing war and persecution or to avoid indefinite military conscription and forced labour. Some were economic migrants, moving away from poverty and a lack of jobs, many hoping for a better lifestyle and the prospect of employment.

b Destination countries receiving migrants may find it difficult with regard to housing, health care, education and in providing enough jobs for the migrants. Migrants may be very poor and can find themselves in squatter settlements, in very difficult situations with regard to a new culture and language and not having the education or skills relevant to the areas in which they find themselves. Many may find themselves exploited by employers and criminal elements.

Exam-style question

1 The positive effects include: provides more workers, either a skilled labour supply such as foreign doctors into the UK from Europe or cheap/unskilled labour on fruit/vegetable farms in the UK; it produces a multi-cultural society with increased cultural understanding; introduction of specialist amenities (e.g. restaurants and food outlets).

The negative effects include: possible racial conflict; increased pressure on employment as there are extra people looking for jobs; pressure on housing as more people need to be housed; pressure on the existing amenities and infrastructure.

Chapter 3

Self-test question 3.1

1 The wide base of the pyramid indicates that the birth rate is quite high; the rapidly sloping sides above the age of 30 show that the death rate was high in the fairly recent past. The base (0–4 years) of the pyramid indicates the birth rate is falling rapidly and that the death rate has stabilised at a low level. The percentage of young economically dependent people and young adults is much bigger than the percentage of elderly economically dependent age groups. The male : female ratio at the base of the pyramid is quite large, 1.11, and indicates the cultural preference for male children in India.

Self-test questions 3.2

1 a There are more people in the 70–75+ age group on the UK's pyramid compared to Somalia's pyramid. The UK pyramid has wider apex/top, while Somalia's is narrower.

b There is a greater percentage of population in younger age groups and a wider base to Somalia's pyramid.

2 You should make sure that you *compare* the countries and not write two discrete accounts in your answer. Also, do not simply repeat the figures without giving some reasons/interpretation. Your answer could include: there are a greater proportion of young dependents in Somalia; about 45% of the population in Somalia compared with less than 20% of that of UK; a greater proportion of old dependents in UK; about 15% of the population of the UK compared with less than 10% of that of Somalia; a greater proportion of total dependents in Somalia; over 50% of population Somalia compared with less than 40% of that of UK.

Exam-style questions

1 There will be a maximum of 3 marks available for LICs or HICs, so you must discuss both in your answer for a chance to gain full marks. For example:

In LICs: an extended family to look after children; children will be expected to look after elderly relatives; charities/aid organisations may be involved in helping.

In HICs: the government will support through the use of taxes/social security payments; people will be able to use state and/or private pensions; there will nursing and retirement homes.

2 a Your reasons can include: longer life expectancy; better treatment of diseases; universal immunisations and vaccinations available from birth; improved health care facilities; investment in care homes/retirement homes and services for the elderly; low birth rates and smaller families.

b Effects can include: an increasing percentage of elderly dependents; increased pressure/strain on the working population; higher taxation; more money to be spent on care/retirement homes/health care/facilities for the elderly; not enough workers for key positions; may be difficult to defend the country; an increased need to attract foreign workers; services for the young may become under used/uneconomical.

3 You can use any concerns resulting from rapid growth of population, in both LICs and HICs. You should refer to examples you have studied as these will be credited as development marks (maximum of 2 marks) if you link them with appropriate points. For example, rapid population growth could lead to the development of shanty towns/favelas, as in Rio de Janeiro/Nairobi.

Depending on the examples you chose you can use issues such as: over-population/not enough resources to go round; inadequate food supplies, leading to malnutrition and/or famine; poor/inadequate water supply/sanitation; a lack of work and an increase in levels of unemployment; an increase in the dependency ratio; an increase in poverty; a lowering of living standards/quality of life; need for more/lack of educational and health care facilities; overcrowded housing/lack of space for development/increased number of squatters; a decline of infrastructure; traffic congestion; air pollution; a lack of facilities for waste disposal/land pollution; the overuse of agricultural land/overgrazing; deforestation and/or the loss of natural vegetation and habitat.

Chapter 4

Self-test question 4.1

1 The factors can be divided into two groups – physical (natural) and human. The physical (natural) factors include climate, water supply, natural resources, relief/topography, natural vegetation and soils. The human factors may be social (including cultural), such as housing, health care, education and cultural opportunities; economic, such as transport and money (capital); or political, such as government investment in the infrastructure of an area, for example in roads, railways, airports and sea ports, and land reclamation; and can play a major role in deciding where industry, jobs, roads, railways, air and sea ports, housing, hospitals and schools are located.

Self-test questions 4.2

1 Use an example you have studied in class (the Revision Guide uses Mongolia, but your area doesn't have to be a country) and give a detailed, accurate answer, explaining why the area has a low population density. Make sure you include some facts that are specific to your chosen area. For example, it has few job opportunities and these may be limited to extensive farming of cattle, sheep or goats or jobs in forestry. It may have a very limited water supply and very high temperatures, summers in excess of 40 °C and low variable rainfall, less than 300 mm. It may contain large areas of highland, including many steep mountains which are unsuitable for building houses and factories, poor road and/or rail communications. It may lack other forms of infrastructure and services. It may lack government investment.

2 The physical/natural factors that encourage a high population density include: **Climate** – areas of high population density tend to be in, firstly, the temperate areas where there are not the extremes of temperature and there is adequate rainfall to provide a reliable source of water for both people and farming. For example, Western Europe, North Eastern USA, North East China and Japan. Secondly, the tropical areas, but not all – it has to be where year round high temperatures, reliable rainfall and fertile soils produce highly productive areas for farming. For example, the floodplains and deltas of major rivers, such as the Ganges in India and Bangladesh, the Nile in Egypt, the Mekong in Vietnam and the island of Java in Indonesia. **Water supply** – most people in the world obtain their water for drinking and farming from two sources –
rivers and lakes and from underground storages, aquifers, so areas of high population density are found where these reliable water sources are located. For example, the Nile in Egypt. Where there is a lack of reliable water, population densities are normally low. **Soils** – the most fertile soils are mineral rich and well drained, these are often found in river floodplains and deltas and in areas of fertile volcanic soils. Where water is available, either naturally or by irrigation, areas of fertile soil can support high population densities – such as the island of Java in Indonesia. **Relief (topography)** – population densities tend to be lowest where land is high and steep, and highest where land is low and gently sloping or flat. Most of the world's population tends to be found in the lower areas of the world – around the coastlines and on river floodplains and deltas. **Natural resources** – areas with abundant natural resources, such as sources of energy and fuel, minerals, especially metallic ores like iron ore, provide employment and income and high population densities.

Exam-style questions

1 Use an example you have studied in class (the Revision Guide uses Mongolia, but your area doesn't have to be a country) and give a detailed, accurate answer, explaining why the area has a low population density. Make sure you include some facts that are specific to your chosen area. For example, it has few job opportunities and these may be limited to extensive farming of cattle, sheep or goats or jobs in forestry. It may have a very limited water supply and very high temperatures, summers in excess of 40°C and low variable rainfall, less than 300 mms. It may contain large areas of highland, including many steep mountains which are unsuitable for building houses and factories, poor road and/or rail communications. It may lack other forms of infrastructure and services. It may lack government investment.

2 Due to the physical geography, such as: where a supply of water is available in an arid area, such as an oasis in a desert; or where rivers flow through an arid area; a dry area in otherwise marshy area/wetland; a valley in otherwise highland/area of steep slopes.

Due to the human geography, such as: a mining settlement/an area producing of oil/natural gas; the growth of a tourist resort; a route centre/junction of major highways; towns which are of strategic importance or market towns; new towns set up as a result of government policies.

Chapter 5

Self-test questions 5.1

1 The 'site' describes the area of land actually covered by the buildings in a settlement. The factors include its height, or range in height, in; whether it is flat, gently or steeply sloping; its aspect (the compass direction it faces).

2 The main factors relating to the situation which can determine a settlement's future growth include: the presence of flat or gently sloping land that is easy to build on; a good defence site; near a reliable source of water; a site to avoid flooding; having building materials nearby; a supply of fuel; fertile land for growing food; a sheltered site; good transport links.

Self-test questions 5.2

1 The size of a settlement's sphere of influence will depend on a number of factors, including: number and type of services it provides; transport facilities available to the settlement; level of competition from surrounding settlements.

2 The analysis of the nature and characteristics of the settlements can aid the planning and organisational processes for the future provision of services in the area, such as health care, schools, public transport, recreation and retail services.

Exam-style questions

1 It is where settlements are arranged in rank order according to their size/importance/services. In any area of settlements, there will be more low-order settlements (hamlets/villages) than high-order settlements (towns/cities). More services/variety of services are found in larger settlements/high order than small settlements/low order. High-order services are found in cities/high-order settlements. High-order settlements/towns/cities will have a larger sphere of influence, and vice versa. High-order settlements/towns/cities have a larger threshold population, and vice versa.

a A **nucleated** settlement is circular in shape with the buildings mostly concentrated around a route centre; while a **dispersed** settlement is where individual buildings are spread out across a landscape.

b Low-order services include: small shops which provide convenience goods; Post Office, small church/chapel, mosque, temple; public house; community hall; bus stop.

High-order services include: specialist shops/department stores; cinemas/theatres, large secondary schools/colleges/university; hospital; financial and insurance services, such as banks; estate agents; large supermarkets.

Chapter 6

Self-test questions 6.1

1 Rural villages may expand to form suburbanised villages, adding new housing estates and provide more services for the village and surrounding rural community – new schools, cafés, hairdressers, and health care facilities. In other areas, many of the new inhabitants may be commuters and shop in large supermarkets on the edge of towns and not use a village's shops and services which may be forced to close. Social impacts include close/tight-knit village communities losing their community spirit as people move in.

2 The rural–urban fringe has proved to be an attractive location for new housing, science and business parks, retail parks and hypermarkets, hotels, conference centres, road developments, sports centres and stadiums, landfill (rubbish) sites and sewage works. All these put pressure on the land in the fringe area.

Self-test questions 6.2

1 Many of the urban areas in both HICs and LICs have large areas of old, traditional industry just outside and adjacent to the CBD. As industry has changed and developed, these inner-city sites became unsuitable for modern industry which need large, flat, easily accessible sites. Many areas become derelict and abandoned. Urban regeneration schemes have attempted the rehabilitation of substandard urban areas by renovating buildings or demolishing and replacing them with new ones, changing and updating their functions. Urban river and coastal industrial sites have been cleaned up to provide new residential and business sites.

2 Integrated and coordinated mass transit systems have been introduced into many large urban areas to address transport problems, such as the combined over and underground railway system forming Dubai's metro system. This has involved the setting up of transport hubs (transport interchanges) served by adequate car parks to allow park-and-ride schemes to operate where people can leave their cars in car parks and be taken in to their place of work or to shop by frequent buses and the metro system. Public transport services operate for at least 18 hours per day. Congestion charging may be introduced into certain central areas in Dubai. Discounted travel tickets, Nol Silver cards, for commuters who can use them for all forms of public transport in the city, including bus, metro or water in Dubai. Some major companies/employers provide workers with a free (or lower priced) public transport subsidy or seasonal-ticket as a standard employee benefit. Restricting parking in the city centre and fining or towing away offenders.

Exam-style questions

1 The question in this case does not ask for examples, but they can be used to illustrate the reasons why urban authorities may want to carefully control any development in the rural fringe area. The reasons include: to prevent the development of urban sprawl on the edge of the town or city; to protect any agricultural land on the urban edge; to protect and provide

areas of open space around the urban area for the recreational use of the people in the urban area, possibly through the creation of green belts/wedges; to prevent the possible merging of neighbouring towns/cities, which could lead to the formation of conurbations; to prevent the development of unplanned squatter settlements on the edge of the urban area.

2 Reasons may include: it provides a large/extensive area for development not often available in the urban area; it allows for a spacious store layout and a large area for car parking; it is less likely to suffer from congestion than inner urban areas; it gives the possibility of having room for future expansion; land in the rural–urban fringe may possibly be cheaper than land in the urban area; access may be easier by either being located beside an outer ring road or by creating new access roads; the new development will be close to a large residential area and therefore a large number of potential customers.

Chapter 7

Self-test question 7.1

1 The **push** factors causing people to move away from rural areas include: the mechanisation of farming activities and a shortage of alternative jobs resulting in a lack of employment opportunities in rural areas and high levels of rural unemployment; the trend for large landowners to take back the land of their tenant farmers to grow cash crops for export; high infant mortality in many rural areas due to a lack of clean water, electricity, sewerage and health care; natural disasters, such as periodic droughts in north-east Brazil; a lack of infrastructure, such as adequate roads, education facilities, retail opportunities/shops.

Self-test questions 7.2

1 Reasons include: the poverty of the people and the fact that many are unemployed/working for low wages often means that they cannot afford to buy houses, and/or the houses are too expensive, or the rents are too high; the city/local authorities are overwhelmed by the rapid natural increase/inward migration of people; a lack of city/local authority investment in housing; inadequate housing or not enough houses; many migrants are unqualified/lack relevant qualifications/are uneducated; migrants may build houses themselves but lack money/materials to build adequate structures; there is overcrowding/lack of space near their workplaces.

2 The problems include: overcrowding – these settlements have high population densities; devastating fires – fires can spread quickly as there is little or no room between the tightly packed buildings and there are very few points of access for fire engines and their equipment; overpopulated – the settlements do not have enough resources to support their large and growing populations; intense competition for any form of employment; poor sanitation and limited health care services and facilities can lead to the spread of disease; poor access to education services and schools; a lack of space in the settlements – the newest and often poorest arrivals may be forced to live on the worst quality land, often the steeper, less stable slopes, wetland/marshland, or land most prone to flooding; a lack of infrastructure – access to services can be poor; public transport is limited and connections to the water, sanitation and electricity supply can be limited and sometimes dangerous.

Exam-style questions

The Curitiba and Rio de Janeiro case studies can be used to answer the following questions.

1 Curitiba is the capital and largest city of the Brazilian state of Paraná. The Curitiba Master Plan was set up, with the following results.

- It opened the world's first Bus Rapid Transit (BRT) system in 1974 and its huge popularity saw the population of the city shift from car travel to bus travel. It was estimated that it caused a reduction of about 27 million car trips in the city per year. Previously, 28% of BRT passengers had travelled by car.
- A planned transportation system consisting entirely of buses and including lanes on major streets devoted to a BRT system. A new road design to minimise traffic. Concentric circles of local bus lines connect to five lines that radiate from the centre of the city in a spider web pattern. On the radial lines, triple-compartment buses in their own traffic lanes carry 300 passengers each. They go as fast as subway cars, but at one-eightieth of the construction cost, on exclusive bus lanes in the main roads. Five of these roads form a star that converges on the city centre. Land farther from these roads is zoned for lower-density development, to pull traffic away from the main roads.
- Compared to eight other Brazilian cities of its size, Curitiba uses about 30% less fuel per capita, resulting in one of the country's lowest rates of air pollution. The buses stop at Plexiglas tube-shaped stations where passengers pay their fare, enter through one end of the tube, and exit from the other end. This system eliminates paying on board, and allows faster loading and unloading, less idling and air pollution, and a sheltered place for waiting.
- BRT's 1100 buses make 12 500 trips every day, serving more than 2 million passengers, 50 times the number from 20 years ago. People in the city now spend only about 10% of their income on travel, considerably below the national average. Nobody lives more than 400 metres from a bus stop and passengers are charged one price regardless of distance. Some employers subsidise their employees who use it; 80% of travellers use it.

2 Possible strategies include: creating urban growth boundaries, parks and open space protection; improved planning and increased expenditure to promote public transportation; reversing and changing government policies that help create sprawl; revitalising already developed areas through measures such as attracting new businesses, retailing, reducing crime and improving schools. Managing rapid urban growth in a sustainable way in Curitiba can also be used in this answer, explained in the answer to question 1.

3 Reasons include: the poverty of the people and the fact that many are unemployed/working for low wages often means that they cannot afford to buy houses, and/or the houses are too

expensive, or the rents are too high; the city/local authorities are overwhelmed by the rapid natural increase/inward migration of people; a lack of city/local authority investment in housing; inadequate housing or not enough houses; many migrants are unqualified/lack relevant qualifications/ are uneducated; migrants may build houses themselves but lack money/materials to build adequate structures; there is overcrowding/lack of space near their workplaces.

Theme 2: The natural environment

Chapter 8

Self-test questions 8.1

1 On the global distribution map, volcanoes are found in three possible types of location. Firstly, where two tectonic plates move away from each other – such as the Mid-Atlantic Ridge; secondly, where two plates move towards each other – along the Andes mountain range, running north-south along the coast of South America; thirdly, over thinner, weaker areas towards the centre of a tectonic plate, called hot spots, such as Hawaii in the Pacific Ocean. Most of the world's earthquakes are found where tectonic plates move towards each other at destructive/convergent plate boundaries, such as in Japan and Chile, and collision boundaries (e.g. the Himalayas) or conservative boundaries (e.g. California) where they may rupture, causing an earthquake. The largest number, 90% of them and 81% of the largest in size, take place along the Pacific Ring of Fire.

2 There are four types of plate boundary movement and each produces different results. **Constructive/divergent plate boundaries**, where two plates move away from each other, has both volcanic activity and earthquakes on these margins but they tend to be relatively gentle in comparison to the other margins. **Destructive/convergent plate boundaries**, where plates made of heavier oceanic crust move towards plates made of lighter continental crust and the heavier oceanic crust is forced down under the lighter continental crust, forming a subduction zone, where the oceanic crust sinks, melts and forms magma. This may rise to the surface and emerge as lava to form very explosive volcanoes. These margins often generate very powerful earthquakes. **Conservative plate boundaries**, where plates slide past each other, may become locked together and pressure builds up until they tear apart, producing very powerful earthquakes, but they do not produce volcanic eruptions and land is not created or destroyed. **Collision plate boundaries,** where two continental plates converge/move towards each other. These lighter plates are not dense enough to sink into the mantle, but, when the plates collide, they can produce powerful earthquakes, but they do not produce volcanic eruptions.

Self-test questions 8.2

1 The actual number of deaths caused by earthquakes will depend on a combination of factors: the strength of the initial earthquake and the aftershocks that may follow it; the depth of the earthquake – many earthquakes take place deep in the crust, below 150 km, so much of their energy is absorbed by the crust above them; distance from the epicentre – as the shock waves spread away from the epicentre, they become weaker; the geology of the rocks in the area – loose sedimentary rocks may liquefy and cause buildings and structures to sink in to the ground, while solid/harder rocks will normally provide the safest foundations for buildings; building construction materials and designs – steel-framed buildings are better able to absorb movement than concrete-framed buildings; the space between buildings – as buildings sway, they may hit each other if they are built too close together and become damaged; number of storeys – in a tall, high-rise building, shock waves become amplified as they move up the building, which can cause them to sway and collapse; the density of population living in an area – a densely populated urban area is likely to suffer many more casualties and damage than a low-density rural area; the time of the day when the earthquake occurs – at night in residential areas most people will be inside their homes and asleep.

2 People live in these volcanic areas for a number of reasons: volcanic soils are often very fertile and yields of crops are high – in the case of Mount Etna in Italy, the fertile volcanic soils support extensive agriculture with vineyards and orchards spread across the lower slopes of the mountain and the broad Plain of Catania to the south; people can obtain hot water for heating and also generate electricity from the volcano using the hot steam to produce geothermal power; volcanoes provide raw materials such as sulphur, zinc, gold and diamonds, which can be mined and sold; volcanoes can attract tourists, local people can get jobs as tour guides and earn money from tourism – for example, the people who live in towns near Mount Etna in Italy such as Messina and Catania, can earn money from renting accommodation to tourists; governments, as in Italy, can set up volcanic and earthquake prediction equipment and then local people may feel more secure in living in high risk areas; many people have lived near volcanoes and earthquake zones all their lives, where they are close to their family and friends and their place of work and, also, many just cannot afford to move away to another area.

However, volcanoes can produce a wide variety of hazards that can injure and kill people, and destroy properties: large explosive eruptions can endanger people and property hundreds of kilometres away and even affect global climate; some volcanic hazards, such as landslides and fumaroles, can occur even when a volcano is not erupting; heavy ash fall can collapse buildings, while minor ash fall can damage crops, electronics and machinery; volcanoes can emit large volumes of gases during eruptions; deaths caused directly by lava flows are uncommon because most flows move slowly enough so that people can move out the way, but lava flows can bury homes and agricultural land under tens of of hardened rock; powerful lahars can rip up and carry trees, houses, and huge boulders miles downstream.

Exam-style questions

1 Volcanoes are landforms, often the size of a mountain or hill, typically conical in shape, having a crater or vent through which lava, rock fragments, hot gases and steam are, or have been, erupted through the Earth's crust.

2 There are three main types of volcano based on their shape and what they are made of.

- **Composite or strato volcanoes:** these are made from a combination/composition of both lava and ash, often in alternating layers as both lava and ash come out of the vent during an eruption. They form on destructive plate boundaries where oceanic crust has melted as it is subducted. The lava forces its way up through the crust and emerges as a violent explosion, such as Mount Etna in Sicily, Italy.
- **Shield volcanoes:** these are made from lava only and form on constructive plate boundaries or at hot spots, like Mauna Loa in Hawaii, where lava appears at the surface as two plates pull apart. They form large volcanoes, with much more gently sloping sides, sometimes hundreds of kilometres across because the lava that forms these is alkaline and very runny and travels a long way on the surface before cooling and solidifying.
- **Dome volcanoes:** these are also made from lava only but their lava is acid and thicker and cools quickly. It does not flow very far before solidifying and so these volcanoes are steep sided and high, such as Mount St. Helens in the USA.

3 Mudflows or lahars are a mixture of melted snow and ice from the top of the volcano, often combined with rainwater, which mixes with ash and volcanic debris and runs off the volcano at speeds of 30–50 kilometres per hour and can travel more than 80 kilometres. Some lahars contain so much rock debris (60–90% by weight) that they look like fast-moving rivers of wet concrete. These flows are powerful enough to rip up and carry trees, houses and huge boulders miles downstream. Farther downstream, they flood valleys and entomb everything in their path in mud.

4 The ways include: ground movement which can cut lines that cross faults: for example, tunnels, highways, railroads, power lines and water and gas pipes; shaking can damage older buildings; liquefaction when shaking turns what looked like solid ground into mud; aftershocks can cause the final collapse of the structures damaged by the main earthquake; subsidence, where the land drops during an earthquake, can damage tunnels, roads, railways, power and gas lines and harbours; in flat coastal areas; a drop in the land can allow the sea to permanently invade and destroy forests and farm land.

Chapter 9

Self-test questions 9.1

1 **Waterfalls** may form by **differential erosion** – where a band or layer of more resistant rock runs across the river channel. The softer, less resistant rock is eroded at a faster rate, by hydraulic action, abrasion and solution, causing a drop in the river bed – a waterfall; a **drop in sea level**, leaving the mouth of the river suspended above the lower sea level; an **earth movement** causing a drop in the river bed along a fault line.

Oxbow lakes may form as a meander develops; its 'neck' becomes very narrow; with time, the river may break through the neck, often during high flow conditions; resulting in a section of the river channel being isolated from the main river channel as an oxbow lake.

2 Where the river loses energy, it will start to drop/deposit its bedload and suspended load. This may take place in the river channel, on its floodplain or on its delta. The heavier material will be deposited first and as the river continues to slow down and become shallower, the smaller material will be deposited. The river's velocity/speed will slow and deposition may take place when: the gradient of the river channel decreases / becomes less; the river channel bed becomes rougher and shallower; the river meets a large, static body of water such as the sea or a lake and is slowed/halted.

Self-test questions 9.2

1 Rivers flood when the water in their channel reaches the top of their banks (bankfull discharge) and then flows across its floodplain. This normally occurs in two ways: when there has been a short period of heavy, torrential rain which the land cannot absorb quickly enough and overland flow may take place (when the water cannot infiltrate fast enough) , causing flash floods, when too much water arrives too quickly for the river to transport away; when there has been a long period of steady rain and the soil and rocks below the ground are full (saturated) with water so that any more rain that falls cannot infiltrate the soil and so it is forced to stay on the surface of the ground and then flows quickly towards the river channel and fills the river channel.

2 Rivers present a number of potential hazards through flooding. These include: destruction of crops and buildings; floods can also kill people and animals; floods can be sources of water-related diseases because they provide a breeding ground for certain animals that spread disease, such as the mosquito which can spread malaria and dengue fever, and the bilharzia snail; polluted water from flooding can also spread diseases such as cholera and diarrhoea; destruction of bridges and road/rail communications and infrastructures.

3 The methods include:

- **Dredging:** where the bedload is dug out by diggers and there is greater channel capacity as a result. This sometimes means that the soft channel sides may be eroded and collapse into the river channel, so the channel sides are strengthened either by concrete or stone.
- **Wing dykes:** these are walls built out from the side of the river made out of concrete and stone on one side only of a river channel. The aim is to force the river into a smaller area which means it flows faster and carries away the bedload and alluvium and therefore does the job of dredging without the use of machines. These are used on the Mississippi river.
- **Two-stage channels:** this is where the top of one side of the channel is cut away to increase the amount of room in the channel for flood water.
- **Building embankments or artificial levées:** these are high banks of soil, clay, sand and gravel built beside the river channel to increase the volume of the channel. Sometimes these are built further away from the river channel to make the capacity even bigger.
- **Straightening the river channel:** this involves cutting out meanders so that floodwater can flow away much more quickly.

- **Holding dams:** these dams are built in the upper sections of rivers and hold back floodwater from melting snow or heavy rainfall from Monsoons, which they can then release after the flood threat is over. The Mississippi river in the USA has over 200 holding dams.
- **Check dams:** these trap the river's bedload as it is being transported down the river in a flood so that the bedload does not fill up the river channel further down the river and reduce its capacity to carry water. They can be emptied by diggers when they fill up.
- **Overflow channels or spillways:** these allow water to flow away from the main river channel.

Exam-style questions

1 a Watershed

b Where two river channels join or where a tributary river joins the main river channel.

c Hydraulic action is where the weight and force of the water flowing in the river removes particles of rocks from the river channel's bed and sides.

Abrasion is where the river's bedload (boulders, pebbles, gravel, sand and silt) rolls, bounces and collides with the channel bed and sides, removing particles of rock from the channel as it does so.

d Two processes from the following four can be used:

- **Traction:** the larger, heavier material that makes up the river's bedload (boulders, pebbles and gravel) is rolled along the river bed.
- **Saltation:** the lighter material that makes up the river's bedload (gravel, sand and silt) is bounced along the river bed.
- **Suspension:** the smaller, lighter material that makes up the river's suspended load (clay-sized particles, less than 0.02 mm in diameter) is carried/suspended by the river.
- **Solution:** dissolved material that makes up the river's solute load is moved by the river in solution.

2 The Nile, Mississippi, Ganges, Brahmaputra and Indus rivers are relevant examples. The Indus, for example, is the key water resource for Pakistan's national economy. It accounts for most of the nation's agricultural production, supports many heavy industries and provides the main supply of drinking water in Pakistan.

The advantages/opportunities of living on a floodplain include: the alluvium that has been deposited on them during floods provides extremely fertile soils which mean that they are often very important for agriculture – such as the river Nile in Egypt; the water from the river can be used for irrigation, allowing land that lacks water to be used for agriculture; larger rivers are very important route ways for transport and communications (by the rivers themselves and by roads and railways built on the flat land on the floodplain); they provide large areas of flat land that can be used for building houses and industry.

The difficulties of living on its floodplain include: their liability to flood, which destroys crops and buildings; floods can also kill people and animals; the damage done to agriculture through the Indus flooding has included damage to irrigation infrastructure; the prolonged flooding/inundation of large areas of cultivated land as a result of this flooding has resulted in massive losses in the agriculture sector; the alluvium that makes up the floodplain and delta is not very stable for building so foundations need to be carefully constructed; the floodplain is densely populated and so there is great competition for space.

Chapter 10

Self-test questions 10.1

1 You can use three of the following four processes of coastal erosion:

- **Hydraulic action:** the weight and force of a wave crashing against a cliff removes particles of rocks from the cliff. It also includes the process where air is trapped by a wave in a crack in the cliff and the enormous hydraulic pressure this creates opens up the crack further, which weakens the cliff.
- **Abrasion (sometimes called corrasion):** boulders, pebbles, shingle and sand are picked up by a wave and thrown against the cliff. This constant collision removes particles of rock from the cliff and wave-cut platform.
- **Solution (sometimes called corrosion):** some minerals (mainly the carbonate minerals found in rocks, such as limestone and chalk) are put into solution by the weak acids found in sea water.
- **Attrition:** this process does not actually erode the cliff but it is the process that breaks up the boulders, pebbles, shingle and sand on the beach. It takes place when the rocks on the beach are rolled up and down the beach by swash and backwash. As this happens, they collide with each other and become smaller and rounder as a result.

2 These features form in narrow, rocky headlands. The sequence starts when a vertical crack, or fault, in the headland is eroded by hydraulic action, abrasion/corrasion and solution/corrosion to form a **cave.** As the cave enlarges, it may erode all the way through a headland to form a natural **arch**. Over time, the roof of the arch may become weakened and will collapse. This leaves an isolated pinnacle of rock, called a **stack**.

Self-test questions 10.2

1 The conditions that are required for the development of a coral reef are controlled by seven limiting factors.

- **Temperature:** the mean annual temperature has to be over 18 °C. The optimal temperatures for them is between 23–25°C.
- **Depth of water:** coral reefs can only grow in depths of water less than 25 m.
- **Light:** the shallow water allows light for tiny photosynthesising algae, called zooxanthellae. In return for the corals providing the algae with a place to live, these tiny algae provide the corals with up to 98% of their food.

This is an example of a symbiotic relationship – an ecological relationship which benefits both sides.

- **Salinity:** corals can only live in seawater, and they can tolerate seawater of high salinity.
- **Sediment:** sediment clogs up the feeding structures and cleaning systems of corals. Cloudy water also reduces light penetration in the water, reducing the light needed for photosynthesis.
- **Wave action:** coral reefs prefer areas of high energy wave action. This ensures freshly oxygenated water. It helps to clean out any trapped sediment. It brings in microscopic plankton – a food source for the corals. In areas that are too exposed, however, corals may be easily destroyed.
- **Exposure to air:** corals die if they are exposed to air for too long. They can only survive and grow, therefore, at the level of the lowest tides.

2 People protect a coastline when something of economic value is threatened by coastal erosion, such as a settlement, an industrial area, a port or an important transport link. Decisions on the methods to be used are planned using a cost-benefit analysis which considers the social and economic aspects of the proposed method of protection. The benefits of different methods (new businesses or jobs and savings in lives and property) are divided by the costs of building and maintaining it. Environmental Impact Assessments (EIA) may also be used to assess the effects any protection method will have upon an area. This is especially important along coastlines as attractive scenery and ecosystems are valuable tourist and ecological assets. Several protection methods, such as the hard engineering methods, are often very expensive, may not be sustainable and can sometimes cause damage to other parts of the coastline through beach erosion/scour and preventing the movement of sediment by longshore drift. Where a soft engineering method such as beach nourishment is planned, care needs to be taken to use the same type and size of sand, otherwise it may be removed by wave action if it is too small and will need to be replaced, at further expense. Therefore careful planning has to take place before any method is put in place.

Exam-style questions

1 a It is a harder, more resistant area of land projecting out to sea.

b Two taken from: cliffs, wave-cut platforms, caves, arches, stacks, bays.

c If there are weaknesses in the rocks forming a coastline, such as sections of softer, less resistant rock or fault lines, differential erosion will take place. As the softer, less resistant rock is eroded at a faster rate through differential erosion, it will form a bay, leaving the harder, more resistant rocks projecting out to sea as headlands.

2 A named area can include: St Petersburg Beach in Florida, USA; Geraldton on the coast of Western Australia; Studland in Dorset, southern England; all of which experience wind action, where smaller grains of sand can be moved by the wind and can form sand dunes at the back of a beach. For this to happen there needs to be: a large sand beach to supply the sand; a strong, onshore wind to firstly dry out the sand and then transport it inland; an obstruction to trap the sand, such as seaweed at the top of the beach – the strand line. The sand will accumulate (build up) in to a small dune, about 1 m high, called an embryo dune. Pioneer species of plants, such as marram grass, will colonise the small dune. The roots and stems of these plants will trap more sand and speed up the process of deposition so that the sand builds up in to bigger mobile or yellow dunes. As this process continues, the sand dune will increase in size and height to become fixed or grey dunes. Figure 10.2 summarises these processes and features.

Chapter 11

Self-test questions 11.1

1 Weather is defined as being 'the state of the atmosphere at any particular moment in time'.

Climate can be defined as 'a description of the averages and extremes of weather variables of an area over an extended period of time' of any location.

2 **Rain gauge:** located away from buildings and trees which might affect their collection of rainfall; top of the rain gauge needs to be over 30 cm high to avoid surface water running in to the rain gauge and the spray from rain splash entering them as rain drops hit the ground; the base of the rain gauge is sunk in to the ground so that it is not easily blown or knocked over; normally located on either grass or gravel which absorbs the impact of falling rain drops and stops rain splash.

Cup anemometer: placed in the open; away from buildings or trees that can alter the wind speed reading; normally placed on a tall mast on top of a building; 10 above ground or building level; so that the reading is not altered by buildings/trees/shelter.

Self-test questions 11.2

1 Wind from the south-east; wind speed 8–12 knots

2 a 4 oktas cloud cover

b wind direction south

c wind speed 3–7 knots

d snow showers

e current air temperature 16 °C

f cirrus cloud

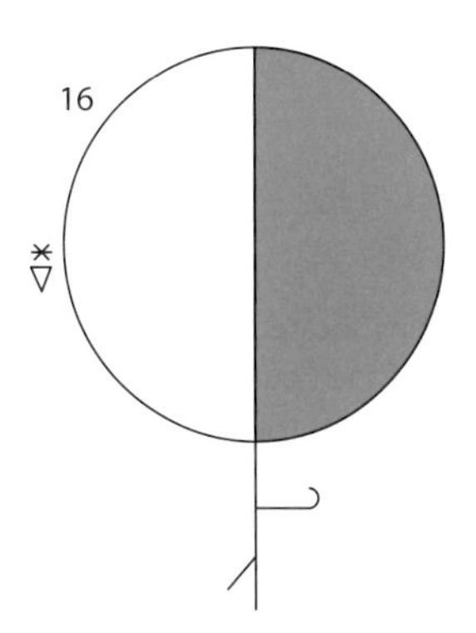

Exam-style questions

1 a Barometer (or aneroid barometer) and maximum and minimum thermometer.

b Present reading 1016/1017 millibars or 30 inches; past reading 1006/1007 millibars or 29.6–29.65 inches.

c Reading by:

- **barometer** – using the black marker; reading is read off the outside scale; in millibars or inches
- **max and min thermometer** – reading taken from the bottom of the two markers.

Instruments reset by:

- **barometer** – move the red indicator so that it rests above the black indicator
- **max and min thermometer** – using a small magnet; drag the metallic indicators down the glass tubes until they come into contact with the top of the mercury in both glass tubes.

2 a So that shade temperatures are taken; so that air temperatures are taken; so that wind does not affect the reading.

b Factors in locating: it needs to be in an open space to minimise the effects of buildings and trees; in the Northern Hemisphere, the door of the screen should always face north so as to prevent direct sunlight on the thermometers when it is opened; opposite applies in the Southern Hemisphere.

Reasons for characteristic features: double louvered/slatted sides allows air to circulate freely around the instruments, but not blow directly on them; standard height for the instruments in the screen; about 1.25 m above the ground; double roof to provide a layer of air between the two roofs; helps insulate the screen from the heat of the Sun; painted white to reflect the Sun's radiation.

Chapter 12

Self-test questions 12.1

1 a Hot season/summer is in April, May, June, July, August, September and October.

b Most rainfall is seen in April, 29.4 mm.

c Least rainfall is in June and September.

d The warmest month is August, 35.1 °C.

e The coldest month is January, 14 °C.

f Several factors can be used, including: located in latitudes from 15° to 30° north and south of the Equator where **atmospheric pressure is high** as the air is usually descending and therefore warming. Rainfall cannot occur where air sinks; many deserts are a **long distance from oceans, seas or large lakes** which means that they receive little rainfall; some deserts have **prevailing winds that blow over large areas of land (offshore)** and so there is no source of moisture; some deserts are in areas of **rain shadow** – where moist air is blocked by tall mountains; **relief/topography of the land** can impact on the hot desert climate. Low areas on the inland/lee side of mountains can be extremely dry.

Self-test questions 12.2

1 Climate is the most important factor in determining where tropical rainforests can be found, including: a low range of monthly average temperatures, normally only a 2–3 °C difference between the hottest and coldest months; constant high temperatures, an average (mean) monthly temperature of about 26 °C with no month below 18 °C; daytime highs of about 30 °C dropping to 23 °C at night, giving a daily/diurnal range of about 7 °C; high rainfall, 1750–2500 mm annual rainfall, evenly distributed throughout the year, with the driest month above 60 mm.

2 Case study of an area of tropical rainforest: for example, the Sinharaja rainforest in south west Sri Lanka. The threats to the reserve are typical of those faced by tropical rainforests globally. There are four main threats:

- Habitat change and loss as a result of the expansion of human settlements, agricultural land, including illegal tea cultivation in the peripheral areas of the rainforest, gem mining and related infrastructure such as road networks. Many of these activities involve deforestation and the cutting down of valuable tropical hardwoods are major threats.
- The over-exploitation of the flora and fauna species by the killing of animals for bush meat, the killing of elephants to supply the ivory trade, the export of ornamental fauna and flora, timber felling/illegal logging and the illegal trading of economically valued species.
- Invasive alien species – species whose introduction and/or spread outside their natural habitats threatens biological diversity.
- Pollution and climate change resulting in the extinction of endemic/indigenous animals and plants.

Self-test questions 12.3

1 Climate is the most important factor in determining where hot deserts can be found, including: average monthly temperatures of over 29°C in their hot summer season, but 10°C in their cool winter season; daytime temperatures of over 38°C, can reach 50°C in summer, but capable of falling as low as 5°C at night and so can have a large diurnal/daily range of temperature; very little precipitation – deserts are defined as areas with an average annual precipitation of less than 250 mm; isolated, irregular rainfall events, often less than 100 mm a year; low humidity of 25–30%.

2 Over 90% of the total land area of Kuwait now suffers from some form of desertification, and 44% of its hot desert ecosystem is severely or very severely degraded. Human interference has been the main threat to the natural hot desert ecosystem in Kuwait through a combination of activities, including: over-grazing, mainly by camels, sheep and goats;

quarrying of sand and gravel; off-road vehicles; oil production operations and their infrastructure; overhunting; poor law enforcement

Exam-style questions

1 For tropical rainforests: of all the threats to the protection of Sinharaja in Sri Lanka and other rainforests like it, it is the socio-economic threats relating to the people and organisations found in the immediate vicinity of the reserve that are perhaps among the most important and pose the main threat of clearance. Land being cleared by local people for farming and cultivation is the biggest threat. This is closely followed by licenced timber contractors who open up routes and roads to facilitate their logging operations on the edge of the reserve which, unfortunately, makes the reserve more accessible to illegal timber operations. Illegal gem mining is also a serious problem, organised by wealthy gem dealers from outside the Sinharaja region. This needs to be stopped. Finally, the lack of a uniform land-use policy and the large number of governmental and semi-governmental agencies involved in land-use planning in Sri Lanka are the major administrative constraints in evolving a suitable protection plan for the rainforest in Sinharaja.

   For hot deserts: in many of the world's hot deserts, nomadic pastoralists graze large herds of camels, goats, sheep and small groups of donkeys. In many areas, such as Kuwait, the livestock have overgrazed the land and the natural vegetation has been removed, exposing the sand to severe wind erosion, often producing major sand storms that can cover hundreds of square kilometres. The domesticated herds also compete for food with wild animals, such as gemsboks in Namibia.

2 The creation of the Sinharaja Forest Reserve and its designation as a National Park, a Biosphere Reserve and a World Heritage Site have all contributed to the future conservation and careful management of this tropical rainforest. A conservation plan has been set up between the Sri Lankan government and the international non-government organisation IUCN (International Union for Conservation of Nature). To protect the reserve, a scheme of zonation and management has been set up for areas both inside and outside the reserve. In the areas outside and surrounding the reserve, essential forest products for sustained use are grown to meet the needs of the local people and eliminate their former dependence on resources within the reserve.

   Other strategies include: establishing a 3.2 km wide buffer zone round the reserve, protecting the core area and using the buffer zone for various uses; relocating illegal settlements and villages to areas outside the reserve.

3 Deforestation, where the rainforest is logged for its valuable timber such as mahogany and teak, means that the habitat for thousands of species of plants and animals disappears and delicate food webs and food chains are destroyed, resulting in some animals becoming extinct. Deforestation also takes away the habitat for the indigenous peoples.

   The consequences include: extinctions and loss of biodiversity of plants, insects, animals, indigenous peoples; habitat fragmentation which disturbs the animals' habitat and may force them to enter habitats which are already occupied; soil erosion occurs when trees and plants are removed and rainwater washes the nutrients in the top soil away; climate change (more carbon dioxide is released into the atmosphere, thus increasing the effects of global warming); pollution – ground, water and air pollution from mineral extraction.

Theme 3: Economic development

Chapter 13

Self-test questions 13.1

1 Traditional measures of development use indices such as GNP (Gross National Product), GDP (Gross Domestic Product) or GNI per capita (Gross National Income per person). Other ways include the Human Development Index (HDI) and the Multidimensional Poverty Index (MPI).

2 A number of factors normally combine to produce these differences between countries. Two factors can be taken from: **geographical location and physical environment** – countries found in the interior of continents, without access to the sea, have generally developed more slowly than countries with access to the sea and the world's oceans, as it affects their ability to trade; **size of country in area** – many small countries have developed more slowly than large countries; **climate** – polar and tropical countries have developed more slowly than those countries with a more equable, temperate climate; polar/sub-polar and tropical climates often produce more infertile soils which can be difficult to farm sustainably and productively; **economic policies** – many of the HICs have applied economic policies which have encouraged strong economic growth. The setting up of Trade Blocs, such as the European Union (EU) or the North American Free Trade Association (NAFTA), have seen their collective economies grow compared to many countries that have developed alone. Many LICs have often found it hard to market their products to these rich, large markets because of the regulations, laws, import taxes and tariffs placed on their products; **stable governments** – many European countries, the USA, Canada, have democratic political systems which they believe encourages economic growth. Political unrest can impact on vital infrastructure development and can adversely affect economic and social development; **population policies** – governments can encourage or discourage higher birth rates to affect population growth. A growth in population can provide economic benefits, such as a large and productive workforce and an increasingly larger and richer market within the country. However, some LIC economies have been unable to expand fast enough to keep up with their population growth, and have not put in place successful population policies, leading to high levels of unemployment, especially among younger people, and high levels of poverty.

Self-test questions 13.2

1 LICs have a high proportion/percentage of their workers in primary industry. In LICs, a large proportion of people work as subsistence farmers in the primary agricultural sector as this is the only way they can make a living. In subsistence farming, people are often poor in economic terms because they eat the crops and animals they grow and rear and there is often

little surplus crops/animals to sell for cash in local markets. The same situation is common for coastal and lake-side fishing settlements.

There is also a small percentage of workers in secondary industry in LICs as it is often craft-based, such as carpentry or the making and repair of simple tools. Many LICs have a small educated workforce as the lack of adequate infrastructure, such as an efficient transport network of roads and railways, is not attractive to foreign companies/multi/transnationals. Jobs in the tertiary/service sector often require a higher level of education, which is not available to most people, many of whom may only have been able to complete primary education before being required to work on the family farm or business for little or no personal income.

In HICs, the percentage of primary jobs drops dramatically as agriculture is more mechanised. HICs are also able to attract inward investment by foreign companies/multi/transnationals (TNCs). TNCs set up new businesses providing employment for the farm workers who have either left the rural areas due to the mechanisation of agriculture, or a lack of paid employment. The foreign companies are often attracted by a mixture low wage rates and relaxed labour and environmental laws. There will also be a growing local market for manufactured goods and the national infrastructure of roads, communications, etc. will be developing. The growth in their economy also leads to local companies expanding, as in Malaysia and South Korea.

The number of jobs in the tertiary sector will also increase as the growing and more complex economy requires more services. The products of the secondary manufacturing sector need to be transported around, so there will be more jobs in transport services. New businesses and factories will need a whole range of tertiary service jobs in finance, marketing and advertising and cleaning. Developing the provision of secondary and higher education and an improved health service is often given priority in an expanding economy, so the tertiary sector employment in these areas also increases.

In HICs, the proportion/percentage of jobs in the secondary sector falls because many manufacturing jobs have moved to MICs where the labour costs are lower. Higher labour costs, computer automation and the use of robot technology mean that only a few workers are needed for the control room and maintenance of the machines.

2 Japan's Toyota Motor Company is one of the world's largest TNCs. It sold 10.18 million vehicles globally in 2016 in over 170 countries and had 53 manufacturing companies in 28 countries on six continents. In 2016, Toyota had over 344 100 employees worldwide and it was the 14th largest company in the world by revenue. Toyota has a major impact on the economies of countries where it invests. Its US operations alone directly employ over 28 500 people, worth nearly $2.3 billion to the US economy. However, Toyota indirectly contributed to the support of more than 365 000 other jobs and provided over $12 billion to the total US economy in 2016.

Several of Toyota's manufacturing locations have taken advantage of government policies aimed at attracting the investment and jobs that a Toyota facility will bring to a country or region. Toyota does not simply have high-end functions in HICs and low-end production functions in LICs. It has Research and Development centres in the USA, Germany, France, UK, Spain, Belgium, Thailand, China and Australia, in addition to its Japanese facilities.

Toyota has developed two new production systems which have helped to increase its profits and efficiency. Firstly, a lean manufacturing or Just-In-Time, JIT, system. Just-in-time (JIT), or lean, manufacturing is a method of industrial production aimed primarily at reducing flow times within production system as well as response times from suppliers and to customers. It has now been adopted by many other TNCs. It has allowed the company to reduce the number of parts it has to store on a factory site and produce only the exact number of parts it needs, based on actual customer vehicle orders and it minimises waste production. Secondly, the Jidoka (roughly translated as 'automation with a human touch') system. When a problem occurs on the production line, the equipment stops immediately, preventing defective products from being produced. It also allows workers to identify the cause of the problem and prevent its recurrence.

Exam-style questions

1 a One from mining, quarrying, farming, fishing and forestry.

b One from those industries that process and manufacture the products of the primary industry, the raw materials, such as iron and steel making, processing food, assembling the component parts made by other secondary industries such as car assembly.

c One from those industries that provide a service or skill such as jobs in education, health care, retailing, office work, transport and entertainment.

2 The tertiary sector has grown considerably in the HICs. In the UK, the financial service industries, such as banking and insurance now provide over 30% of the country's GDP and employ large numbers of highly educated people. The research and development jobs of most multi/transnational companies, along with their marketing and advertising jobs, are still based in HICs while manufacturing is based in MICs. The aerospace industries, such as Boeing in North America and Airbus in the EU, are highly concentrated in HICs where they can have access to a highly-educated workforce and the research facilities of leading universities.

Employment in secondary industries in the HICs has been contracted because of two main reasons. Firstly, many manufacturing jobs have moved to MICs where the labour costs are lower. Secondly, high labour costs and an ageing population (shortage of workers) in HICs mean that automation has increased, with computer-controlled robots doing much of the work on the assembly lines of car factories.

3 For HICs, the advantages may include: an increase in high-level forms of employment in the field of research and development; decreased labour costs; lower taxes; less strict employment and environmental laws and regulations. The disadvantages for HICs include: the closure of manufacturing factories; loss of jobs; increasing unemployment.

For LICs, the advantages may include: more people will be in employment; they will receive higher wages than in their existing work; an improvement in skills; improvements in transport infrastructure – to roads and railways; improvements in service infrastructure, e.g. electricity, water supply and sanitation; rise in the standard of living; improvements in public services such as health care and education; the multiplier effect / cumulative causation – there will be positive effects on other areas of the economy – people will be able to buy more goods in the shops; they may want to extend and improve their houses, they can afford to send their children to school. The disadvantages for LICs may include: exploitation of the labour force; low pay for workers; long hours at work for workers; poor working conditions in many factories; loss of rural land/ farmland for the building of new factories; increased pollution to the local area from some factories, e.g. air, water, noise and visual pollution.

Chapter 14

Self-test questions 14.1

1 There are several types of large-scale commercial farming that can be used in this answer. Plantation farming is one of the most common and involves the growing of one crop (called monoculture), often over very large areas. Examples include sugar cane, bananas, rubber, tea, coffee and pineapples. Other types of large-scale commercial farming include the growing of cereals, such as wheat, maize and barley, and the rearing of livestock animals such as cattle and sheep.

2 Commercial sheep farming in Australia is found on very large farms in marginal areas – areas where other animals and crops would not be as successful or as profitable due to physical and human factors. They are therefore often found in areas of low rainfall, high temperatures and poor quality grazing. Per hectare, sheep farming has very low **inputs** of capital – much of the land that is used is of relatively small value as it cannot often be used for arable farming, so it is cheap to buy. Farms may need up to 25 hectares of grazing land per animal as grazing land is so poor.

Processes – it takes very few people to both look after large numbers of sheep as they can be left out in the fields all year round, or gathered together for shearing (the actual shearing is often done by groups of skilled shearers who move from farm to farm); applying any pesticide to their fleece and antibiotics to overcome any pests. The **output** of this industry is that it produces about 620 000 tonnes of meat and 575 000 tonnes of wool in a typical year.

Self-test questions 14.2

1 The Green Revolution refers to a set of research and development of technology transfer initiatives occurring between the 1930s and the late 1960s that increased agricultural production worldwide, particularly in MICs and LICs. The initiatives resulted in the adoption of new technologies, including the development of High Yielding Varieties (HYVs) of five of the world's major cereal crops – rice, wheat, maize, sorghum and millet.

2 The reasons why people are suffering from hunger is usually a combination of natural and human factors.

Many of the natural factors are extreme climate events, including:

- **Drought and unreliable rainfall:** in much of East Africa, including the north of Kenya, Sudan, Somalia and Ethiopia.
- **Tropical hurricanes/cyclones/typhoons:** when these storms hit, with their high winds, torrential rainfall and storm surges (where sea levels are raised by the high winds and then surge inland flooding low-lying areas like southern Bangladesh), they can devastate farm land and crops.
- **Floods:** though often associated with tropical cyclones/ typhoons/hurricanes, they usually result from heavy rainfall, often associated with monsoons or El Niño events. Crops can be destroyed by flooding, as well as the effect on whole communities who may lose their houses, belongings, animals, roads.
- **Pests and diseases:** there are many pests and diseases which can prey on crops, for example, locusts, and diseases, such as mildew. Without the expensive pesticides and sprays to deal with them poor rural communities may suffer severe crop losses and food shortages.

The human factors include:

- **War:** Darfur in the south of Sudan, for example, has suffered from many years of civil war.
- **Increasing population:** as a result of rising population more land is having to be cleared of forest, less fertile, marginal land brought in to use, and farms divided into much smaller units among farmer's children as they grow up and have their own families to feed. As less land becomes available, food shortages result.
- **Inability to invest capital and improve infrastructure:** the lack of capital available in LICs, particularly in rural areas, means that schemes to improve agricultural production, improve food storage, improve roads and transportation for distributing farm products are very hard to initiate and develop.
- **Volatile global food prices:** between 2001 and 2016, the price of many globally important crops rose dramatically.

Exam-style questions

1 Nomadic farming is where people keep animals like sheep, goats, cattle and camels and move over large areas, grazing their herds. Sedentary farming is where the farm is located in a permanent location.

2 a Irrigation is the artificial application of water to the soil, usually for assisting in growing crops. In crop production, irrigation is mainly used in dry areas and in periods of rainfall shortfalls, but also to protect plants against frost. Irrigation also helps to suppress weed growing in rice fields.

Drainage may improve crop production as land needs to be well drained to allow most plants to grow and not find their roots waterlogged. Flat land is easy to plough but may become waterlogged and flooded. Putting in drainage ditches/canals can therefore improve production.

b The problems include:

- Salination where the overuse and poor irrigation practices have led to increased salt content in the soil, reducing the productivity of the land. Irrigation salinity is caused by water soaking through the soil level, adding to the groundwater below. This causes the water table to rise, bringing dissolved salts to the surface. As the irrigated area dries, the salt remains.
- There is increased competition for water, from individual farmers, communities and even countries.
- Over-extraction of water can lead to the dropping and depletion of underground aquifers.
- Ground subsidence (e.g. as in New Orleans, Louisiana, USA) may occur as water is removed for irrigation from the underground aquifer.
- Over-irrigation because of poor distribution may lead to water pollution.

3 By introducing appropriate technology (technology which is suited to the level of wealth, knowledge and skills of local people and is developed to meet their specific needs) to rural areas. For example:

- The building of small earth dams and wells to provide water for basic irrigation projects.
- Simple methods of soil conservation, such as planting trees to make shelter belts to protect soil from wind erosion in dry periods, contour ploughing or building low earth or stone walls along the contours of a slope, to stop the runoff of rainwater and allow it time to enter the soil, helping to prevent soil erosion and increasing the amount of water in the soil and making it available for crops.
- Tied ridging where low walls of soil are built in a grid of small squares which stops rainfall runoff and again allows water to drain into the soil. Crops such as potatoes and cassava are grown on the soil walls.
- Strip or inter-cropping, which has alternate strips of crops being grown, at different stages of growth, across a slope to limit rainfall runoff as there is always a strip of crops to trap water and soil moving down the slope.
- Tier or layer-cropping, where several types and sizes of crops are grown in one field to provide protection from rainfall and increase food and crop yields. For example, the top tier, or layer, may be coconut trees, below this may be a tier of coffee or fruit trees, and, at ground level, vegetables or pineapples.
- Improved food storage which allows food to be kept fresh and edible for longer periods of time and protected from being eaten by rats and insects and affected by diseases.

Chapter 15

Self-test questions 15.1

1 An industrial system has things that go into it (inputs), they are then worked upon and changed (processes) and manufactured goods come out (outputs). Inputs can include raw materials, energy, transport, labour, money (capital investment) and government policies. Processes can include the processing of raw materials, assembling component parts, packaging and administration. Outputs can include the finished products, the profits and waste.

2 A manufacturing industry or processing industry often involves the manufacture of raw materials, into another product by manual labour or machines. The iron and steel industry is usually located on large, flat sites to accommodate the very large buildings, storage areas for raw materials and to give plenty of room for future expansion. They are often beside deep water ports to bring in their bulky raw materials – iron ore, coal and limestone and have good road and rail transport links to distribute their finished goods. The raw materials are put into blast furnaces from where the molten iron is processed, extracted and refined into steel. The steel may then be rolled into sheets, slabs or coiled wire and then distributed from the iron and steel works to be used in other industries, such as ship building, railways and trains, the automobile industry' pipelines and the steel framework for may large buildings.

Self-test questions 15.2

1 In deciding where to locate a factory, or industrial zone, a number of factors need to be taken into account. These factors can be put into two groups –
physical factors and human and economic factors. The **physical factors** include: **easy access to raw materials** – the iron and steel industry uses large amounts of bulky raw materials which would be much easier to locate near the source of the raw materials or at a location where they can be cheaply transported to; easy **access to cheap sources of power** – a location on or near a major coalfield was a perfect location, for example, South Wales in the UK and the Ruhr in Germany; a **site that is cheap to buy or rent** and is easy to build and expand on, a large, flat site is easier to build on and for a factory to expand on in the future; a **geographical situation** that allows easy transport routes to be set up: a good, natural route way such as where valleys meet (called a confluence), rivers meet or a port, provide the ways in which raw materials, component parts or finished goods can be transported easily to and from factories.

The **human and economic factors** include: the availability of labour – some industries need large numbers of relatively unskilled workers, such as some types of farming, other industries need relatively few workers but they must be highly skilled, such as the IT industry; **availability of capital** (finance/money) to invest in the factory; a **market** where the products of the factory can be sold: the size and the location of a market have now become more important than raw materials, large HIC cities like New York, Paris and London provide very large markets; availability of **cheap, fast and efficient transport**:

transport costs can make up a larger proportion of production costs and therefore finding the cheapest forms of transport for moving raw materials and finished goods is very important; **government policies** affecting the location of industry; **economies of scale** –
a business with many small factories may not be very profitable compared to those which have just one large factory location and so many businesses have closed their smaller plants and built larger ones to put all the industrial processes in one site; **changes in technology** – the use of robotic machines run with computer software to do repetitive jobs has transformed many factories, for example car assembly plants; **living and working environment** – many businesses now look at attracting workers by offering them better living environments.

2 The iron and steel industry in South Wales in the UK has undergone enormous change through time, reflecting the changing global factors influencing the location of this particular industry. In the early 19th century, the industry was located where it could find its three essential raw materials – iron ore, coal and limestone. Another factor in favour of their location was the fact that the valleys in which they were located led down to the coast where iron products could be transported and then exported easily.

By the 20th century, many of the smaller iron and steel works had closed down. The reasons for this were that the raw materials became exhausted –
the iron ore deposits and the easily accessible coal deposits were used up and so these now had to be imported. The works inland suddenly found themselves at a real disadvantage as the iron ore and coal had to be unloaded at the coast and put on trains to be transported to the works. This added enormously to the production costs and made the inland works non-competitive as their products became much more expensive than those at coastal works.

By the 21st century, two huge fully integrated sites were built on the South Wales coast. These were on large, flat sites to accommodate the very large buildings, over 1 km long and to give plenty of room for future expansion. The main factor governing the future of both plants was the world price for steel and the demand for steel.

Exam-style questions

1 A footloose industry is one which is not dependent on a particular location, unlike many other types of industry which need to be beside their raw materials, ports, fast and easy transport. The reasons for them being footloose are that: they use small, light component parts; their products tend to be small and light; they use electricity as their power source; they need a small, skilled labour force; they are non-polluting and so can locate in or near residential areas

2 The growth of high technology industries in MICs provides a number of benefits for the companies: there is a large and expanding market for goods in many MICs; there is a large and educated workforce available and so the Hi Tech companies provide new employment opportunities; the labour supply is relatively cheap compared with Europe and North America; the low cost of land for building on; government help and assistance in the setting up of new industries. This can all bring benefits to both the people and the economy of the MICs: more people will be in employment; they will often receive higher wages than in their existing work; there will be an improvement in skills; the economy will benefit from improvements in transport infrastructure – to roads and railways and improvements in service infrastructure, e.g. electricity, water supply, and sanitation; the people will see a rise in their standard of living; improvements in public services such as health care and education; the multiplier effect / cumulative causation will improve the overall economy.

3 Many national governments and the EU have a wide range of policies which they can use to encourage industries to move to particular locations. A common policy is to decentralise their own government departments. In the UK, the government has moved the Royal Mint and the Passport Office away from London to South Wales and parts of the Foreign Office to Milton Keynes and the Pensions Department to Newcastle. They can also provide incentives such as lower company taxes, subsidised wages, lower rents and improved infrastructures such as improving roads and railways. All these can attract new industry to an area.

Chapter 16

Self-test questions 16.1

1 Reasons include: a rise in incomes which gave people, after they had paid for their basic needs, 'disposable' money which they could spend on leisure activities and tourism; increased leisure time caused by a shorter working week, flexitime, paid annual holidays, earlier retirement with a pension; increased mobility as a result of private car ownership, improved roads, a decrease in the cost of air travel combined with greater numbers of airports, the expansion of budget airlines, like Air Asia, Ryanair, Jazeera, Jet, Air Dubai and Easy jet and the increased numbers of flights to a wider range of destinations (in 1970, there were 307 million airline passengers; by 2011 this had reached 2.75 billion); increased media coverage by television, magazines and the internet of different holiday destinations and types of holiday; governments have used major sporting events such as Winter and Summer Olympic Games, World Championship Athletics, Football, Rugby and Cricket World Cups to advertise tourist opportunities; increased international migration encouraging more people to visit relatives and friends abroad.

2 Tourism provides many economic, social and cultural advantages for the people who live in LICs.

- **Economic:** increased employment opportunities in the many jobs created directly and indirectly by tourism; this can help reduce migration, especially from rural areas, as employment can be found in small cafés, hotels, souvenir shops, tour guides, local taxis, etc. Many of these jobs will be in the informal sector, which helps the people of LICs in particular.
- **Social and cultural:** an increased understanding of different peoples, cultures and customs; increased cultural links with other countries; increased foreign language skills for both visitors and hosts; increased social and recreational facilities for local people; the preservation of traditional heritage sites and customs.

Self-test question 16.2

1 The impact of tourism on the Kenyan environment. The Kenyan National Park environments are fragile and sensitive – both the natural landscapes and the animals that live in them. As a result, they can be changed and damaged by the thousands of tourists who visit them every year.

When visiting the Park as part of a Safari (meaning journey), most tourists will take one of the many tourist buses to get into the Park and close to the animals. Minibuses are meant to keep to well-defined trails but sometimes drivers may go off the tracks to get closer to animals so that tourists can get better views and photos. This can increase the driver's tips and increase their incomes. During the wet season the tracks can also get very muddy so drivers drive outside them and widen them, some end up 50–60 metres wide as a result. The National Parks and Game Reserves are often part of the traditional grazing areas of the nomadic tribes. These tribes move their animals over very large areas so that they do not overgraze any area. However, the Park boundaries stop the tribes using quite large areas of land, which means that they have smaller areas to put their animals on and the land can be overgrazed as a result. This means that the tribal people lose income and see their traditional environment become degraded. Many tribal people therefore have to live in more permanent settlements earning money from selling products they make or from putting on dance performances for tourists. Recently, the Kenyan government has worked more closely with the tribal peoples to give them a share of the tourist income.

Exam-style questions

1 Ecotourism is a sustainable form of tourism which allows people to visit natural environments and traditional cultures while enabling local people to share in the economic and social benefits of tourism. At the same time, however, measures are taken to protect the natural environment, the local way of life and the traditional culture.

Ecotourism is encouraged in the National Parks in Kenya and the protection of the environment is carried out in several ways: restricting tourist numbers to both the Parks and to certain areas of the Parks; a limited number of tourist firms are licensed to use the Parks and their activities are regulated in several ways (for example, minibuses are not allowed within 25 metres of animals); limiting or preventing the destruction of natural vegetation and habitat that is cleared for tourist development; ensuring that any building developments are low level and made out of local materials and in local styles; using local labour in as many activities as possible and provide training for local people; educating tourists with regard to the environmental and conservation issues in the Parks.

2 Tourism also brings many economic, social and cultural disadvantages, which affect the people who live in LICs.

Economic: seasonal unemployment – if people come for summer sun or winter skiing, the rest of the year may mean few or no tourists and little or no employment; leakage of tourist income – airlines, hotels and tourist activities in LICs are often foreign owned which can mean that 60–75% of tourist income may either never come to, or may leave, an LIC; many tourists may spend most of their money in, have most of their meals in, and do trips organised by their hotels so, although the companies and individuals that own the hotels benefit, they have little impact on the wider local economy; many of the jobs provided by tourism in an LIC are low paid and low skilled; many of the higher skilled and better paid jobs will be taken by foreigners; some locations may become over dependent on tourism, so, should a natural or human disaster occur, they may have little alternative income; water shortages caused by tourist complexes, hotels and golf courses using large amounts may lead to local farms and villages not having enough; traffic congestion and pollution from litter, increased sewage, etc. especially at 'honeypot' sites where there are large numbers of tourists in one location.

Social and cultural: the demonstration effect – local people may copy the actions of some tourists in different dress, diets, habits, and, possibly, alcohol and drug abuse and so their traditional values may be abandoned; an increase in prostitution and the development of 'sex tourism'; young people may truant from school to work in the informal tourist industry and earn money as unofficial guides or selling souvenirs; people leave family farms to work in the tourist industry and this makes it more difficult to run the farms without their help; people may be moved from their houses and land to make way for tourist developments. Local landowners may sell large areas of land and coastline to non-local or foreign buyers who may then deny access to local people; house and land prices may rise as non-locals buy them and put them out of the reach of local people.

3 Named area: the Kenyan Coast.

Tourists are attracted both by the beautiful beaches along the coast of Kenya, fringed by coconut palms, blue skies, warm climate and clear warm blue seas and the traditional Swahili culture of the Kenyan coast. In addition to the beaches, Kenya has tropical coral reefs which are a major attraction. The thousands of tourists that visit the coast bring several advantages. Most of the people who live on the coast are in small fishing villages – the people depend on the coral reefs for fishing as a source of food and to sell the fish as an income. The fishermen can greatly increase the income by using their boats to take people on trips to see and dive on the reefs. They also have a whole new market for what they catch in the new hotels and restaurants; hotels, restaurants and other facilities also provide hundreds of new jobs, from building to maintenance, cleaning and cooking, making furniture, etc.; jobs in cleaning and looking after tourists can add enormously to the local people's incomes; the wide range of water activities provides many new jobs; local crafts people also have a much bigger market for their goods as well as local farmers who can provide the food for the hotels and restaurants; shops now provide a wide range of goods which greatly increase local incomes; transport also provides many jobs with coach and bus companies needing drivers, engineers and cleaners.

Chapter 17

Self-test questions 17.1

1 A non-renewable source of energy is one that is either finite or non-sustainable. This is because their use will eventually lead to them running out. Non-renewable sources include fossil fuels such as coal, oil, natural gas and peat. A renewable energy source is one that can be used continually without running out – it is a sustainable resource. Wind, water, geothermal, wave, tidal, biogas, biofuels (like ethanol) and solar energy are examples.

2 It can be very dangerous if there is a nuclear accident and radioactive materials are released in to the environment, as happened in 1986 when a reactor at Chernobyl, in the Ukraine, exploded and in 2011, when the Fukishima reactor exploded in Japan when over 200 000 people had to be evacuated from the immediate area and the long-term effects are at present unknown; nuclear waste can remain dangerous for several thousand years and so there are problems in storing it; the cost of shutting down (decommissioning) nuclear reactors is very high and there is constant debate as to who will pay for this – national governments or electricity companies.

Self-test questions 17.2

1 There are disadvantages to the **Three Gorges Dam**; many of these also apply to other HEP sources. They include: the dam was very expensive to build; the reservoir lake covers large areas of natural habitats and farmland; it has destroyed these wildlife habitats; people have been forced to move (1.3 million people in the Three Gorges Dam area in China) and whole towns and communities have disappeared along with historical and archaeological remains; it may trap sediment carried by the river and gradually fill up; there is a possibility that it may collapse or be damaged, by earthquakes, for example, although this is rare, a dam collapsed in Indonesia in March 2009 killing 55 people and 200 000 people living below the Oroville dam in California, the tallest dam in the USA, were evacuated in February 2017 when a spillway appeared in danger of collapse; the reservoir behind the dam has created a large area of still water, ideal for mosquitoes to breed – malaria may appear; the visual impact of the dam and its reservoir has changed the look of the natural landscape.

2 Demand is now outstripping supply in many areas of the LICs where it is the main fuel source; people, especially women, are having to travel further away from their homes to cut and collect the fuelwood; much of it is usually burnt inside or close to houses and as these are not properly vented, it is a major air pollutant for many families and currently accounts for 1.5 million deaths from respiratory illnesses in LICs every year; deforestation also leads to increased soil erosion and a decrease in water quality as soil gets into rivers, streams and lakes; in turn, this can increase flood events in both size and number as there is a lack of interception of water on the valley sides so that surface runoff reaches the rivers more quickly and in larger amounts; plus, river channels become filled with sediment so that they cannot contain as much floodwater and so flood more easily.

Exam-style questions

1 a A fossil fuel is either finite or non-sustainable; their use will eventually lead to them running out and their exhaustion. Non-renewable fossil fuels include coal, oil, natural gas and peat.

b The advantages of using HEP include: renewable, clean and non-polluting; cheap (after the initial cost of the dam); dams also help with flood control; provide water for the local population and for farming (irrigation) and industry; they can also be stocked with fish and support a local fishery; they can be used for recreation and attract tourists; the new source of electricity may attract manufacturing industry and new jobs will be created.

2 Advantages of using nuclear power include: it is not a bulky fuel – 50 tonnes per year for a power station compared to 540 tonnes of coal per hour for a large coal fired thermal power station; nuclear waste is very small in quantity and can be stored underground; it does not produce greenhouse gases, does not produce carbon emissions or contribute to acid rain; it stops countries relying on imported oil, coal and gas; there are relatively large reserves of uranium available in the Earth; nuclear power stations have relatively low running costs.

Disadvantages of using nuclear power include: it can be very dangerous if there is a nuclear accident and radioactive materials are released in to the environment; nuclear waste can remain dangerous for several thousand years and so there are problems in storing it; the cost of shutting down (decommissioning) nuclear reactors is very high and there is constant debate as to who will pay for this – national governments or electricity companies.

3 In the area of Coral Bay in NW Australia, wind power and solar energy are being introduced. Both forms of energy can be used in off-grid locations in remote areas, such as Coral Bay. It is a small, remote community on the northwest coast of Western Australia not connected to the main electricity network. It now has a new wind–diesel system comprising of three wind turbines and a diesel power station. The wind turbines can be lowered in case of cyclones, to which the area is very prone. The system is clean, renewable, sustainable, pollution free and cheap.

Solar power can provide electricity by either generating it by providing heated water for a thermal generator or by using photovoltaic cells which convert sunlight to electricity. It is commonly used to provide hot water, and thermal energy for cooking, to many of the buildings and homes in the area. Again, it is clean, renewable, sustainable, pollution free and cheap.

Chapter 18

Self-test questions 18.1

1 The methods include: **surface freshwater** – structures such as dams may be used to store water and if properly designed and constructed, dams can help provide a sustainable water supply; **groundwater** accounts for greater than 50% of global freshwater and is a major source of water, it can be accessed by digging wells, or drilling boreholes, down to the water table and

either pulling up the water using a container from the bottom of a well or pumping it up; **artesian basins** are found where the groundwater underground is confined or trapped in an area, a well or borehole is dug/drilled and water is pumped/drawn up to the surface; **rainwater harvesting**, collecting water from precipitation, is one of the most sustainable sources of water supply since it is easier to control and manage compared to other surface and groundwater sources, and it directly provides water of drinking quality; **reclaimed water**, or water recycled from human use, can also be a sustainable source of water supply. It is an important solution to reduce stress on primary water resources such as surface and groundwater; **desalination** has the potential to provide an adequate water quantity to those areas of the world that are freshwater poor, including small island states; **bottled water** is where mostly private companies provide drinking water in a bottle for a cost, in some areas, however, bottled water is the only reliable source of safe drinking water for a population.

2 Sustainable sources of water supply are those sources which are renewable in the long term. Renewable water supplies include surface water and groundwater. **Surface freshwater** is unfortunately limited and unequally distributed in the world. In addition, pollution from various activities into rivers and lakes leads to surface water that is not suitable for drinking, therefore, treatment systems must be put in place. If properly designed and constructed, dams can help provide a sustainable source of water supply. **Groundwater** accounts for greater than 50% of global freshwater and is a major source of water. The water emerging from some deep groundwater sources may have fallen as rain on the land surface many decades, hundreds, thousands or in some cases millions of years ago. If so much water is pumped out of the groundwater source that the water table falls below the bottom of a borehole or well, it will dry up. This is a frequent occurrence in many parts of the world where groundwater is over-used, or where the groundwater has not been recharged with precipitation as a result of long periods without rainfall or in areas experiencing droughts.

Self-test questions 18.2

1 There are several reasons, both natural and human, why some areas have a shortage of water whereas others can have a surplus, they include: amount of precipitation received; amount of evaporation/evapotranspiration taking place; temperature – the higher the temperature, the greater the amount of evaporation/evapotranspiration taking place; type of land use, for example natural forests compared to farming – farming may consume large amounts of water for irrigation; level of economic development – richer HICs may consume larger amounts of water; population density – more densely populated areas may use more water; presence or absence of water-bearing rocks or aquifers may mean that more or less water is available to people; how close people live to rivers will affect how easy it is to obtain water; political decisions – where a river flows through several countries, they may not agree on how much water each country may take – the Nile in north east Africa, for example.

2 The California State Water Project provides additional water to approximately 25 million people and about 303 500 hectares of irrigated farmland. The project is a water storage and delivery system of reservoirs, aqueducts, power stations and pumping plants. It has involved the construction of 21 dams and almost 1300 km of canals, pipelines and tunnels. Its main purpose is to supplement the already available water supply in California, which was not capable of meeting the needs of the state. It does this by storing water and distributing it to 29 urban areas and farm land in California. About 70% of the water provided by the project is used for urban areas and industry in Southern California and the San Francisco Bay Area, and 30% is used for irrigation in the Central Valley.

In this particular example of water resources, there are several strategies that can be put in place for California and elsewhere:

- Reduce the leakage of water from pipes and aqueducts as well as the loss of water by evaporation from aqueducts. This would stop up to 25% of losses.
- Recycle water that is used in industry and from sewage (grey water). This does not have to be treated to the same standard as drinking water and can be used to irrigate gardens and golf courses and flush toilets.
- Reduce water subsidies. At the present time, farmers in the south west of the USA only pay 10% of the actual cost of the water they use for irrigation. The federal government subsidises the rest of the cost. Farmers may then look to use irrigation water more efficiently – such as using drip irrigation systems which are 100 times more efficient than flood irrigation. LIC farmers also point out that US farmers have an unfair advantage in markets as their product is heavily subsidised and can make LIC products uncompetitive.
- Grow less water-dependent crops rather than rice and the fodder crop, alfalfa.
- Several cities in California are looking at using desalination plants to produce water

Exam-style questions

1 In Malawi, an LIC in southern Africa, water is often obtained either from wells dug by hand or collected, from distant rivers and lakes. However, the supply can be unreliable. Sometimes the well itself can be a source of disease and they need to be covered or protected to stop animals falling into them. To improve supply, several schemes are now being introduced, including:

- Gravity-fed schemes where there is a spring on a hillside, or from a river or lake further up a valley. The water can be piped, or small channels can be constructed from the spring/river/lake down to the villages/fields.
- Boreholes can require more equipment to dig, but they can be created quickly and usually safely. They may require a hand pump to bring the water to the surface.

In addition to locating new supplies of water, strategies are being introduced to reduce the need for water. These include: collecting rainwater landing on the roofs of buildings; recycling waste water to use on crops; improving irrigation techniques; growing crops that are drought resistant or less dependent on a constant water supply.

2 The Nile Basin Initiative is a regional intergovernmental partnership that seeks to develop the River Nile in a cooperative manner, share substantial socio-economic benefits and promote regional peace and security. It includes all the countries through which the Nile flows – Burundi, Democratic Republic of the Congo, Egypt, Ethiopia, Kenya, Rwanda, South Sudan, Sudan, Tanzania and Uganda. Eritrea participates as an observer.

The positive effects are that the project provides a means for multi-country dialogue and information sharing, as well as joint planning and management of water and related resources in the Nile Basin. These countries have a total population of 437 million, of which 238 million (54% of population of basin countries) live within the Nile Basin. The project also plans to enable the Nile Basin to successfully address ever-growing challenges and pressures. These include: climate change, which is expected to increase the likelihood of extreme events (prolonged droughts and floods) and will adversely affect the food, water and energy security of the basin countries; high population growth rates; the demands of faster economic growth across the basin.

There are several negative effects. Water experts believe there is not enough water in the river to meet the various irrigation goals of the Nile basin nations. Adding to potential water stress, many large hydropower dams are being built or considered and some countries in the region are acting alone to secure their water supplies and needs at the expense of other countries in the basin.

3 In general, LICs and MICs will have most of their water used in agriculture and relatively little in industry or domestic use. Bangladesh, Kenya and Egypt have agriculture as a large part of their economy so a large percentage of their water is used for that purpose.

HICs tend to have a higher percentage for industrial use. There are exceptions. The USA, Australia and Japan are HICs, but still have a high amount of water used for agriculture because it is still an important activity.

The variations are greatest when comparing the rich HICs and the poor LICs use of water. For example: a larger amount of water could be used for agriculture and irrigation in LICs because people are more dependent on the land; a larger amount could be used for industry in HICs, as there is more industry and more factories in developed countries; a larger amount could be used for electricity generation (HEP) in HICs as richer countries consume more electricity, or more may be used for cooling in thermal power stations.

Chapter 19

Self-test questions 19.1

1 Several processes contribute to soil erosion: water erosion when water flows over the surface of the ground and removes small particles of soil; wind erosion, when dried soil particles are exposed to high winds and are picked up/entrained; the over-abstraction of groundwater, which may lead to soils drying out and being easily removed by heavy rain and wind, or resulting in there being a lack of water in the soil for crops to germinate and grow; salination is an increasingly common problem in semi-arid areas where high rates of evaporation draws groundwater up to the surface of the soil; climate change will probably increase the problem of soil degradation as an increase in the number and severity of floods will cause more water erosion. Also, an increase in the number and severity of droughts will cause more wind erosion.

2 Desertification is caused primarily by human activities and climatic variations.

Burkina Faso, in the Western Sahel of Africa, in an attempt to remedy some of the problems caused by desertification, with the help of aid agencies such as the Eden Foundation (Oxfam) and USAID, intervened both to supply the equipment needed for sustainable farming and to educate local people and communities on sustainable farming techniques. Farmers were taught how to use **drip irrigation** techniques, 'using far less water than flood irrigation,' reducing the amount of water lost to evaporation by the traditional flood irrigation method, where much water was lost as it was spread in large quantities across the whole field. To raise crop production, **microdosing** was introduced, involving the application of small, affordable quantities of fertiliser with the use of a bottle cap, either during planting or 3–4 weeks after germination. This technique maximises the use of fertiliser and greatly improves crop productivity.

Self-test questions 19.2

1 The impacts of the enhanced greenhouse effect (EGE) may include:

- Average global temperature has risen by around 0.7 °C since the early 20th century. This may not sound like much, but some regions will experience a much more extreme response than others.
- The amount of Arctic sea ice, snow cover and the volume of glacier ice globally have all decreased and surface ocean temperatures have increased.
- The warmer ocean temperatures can kill large areas of coral reefs. The phenomenon of global coral bleaching events that have occurred in recent years where the corals are killed leaving their bleached skeletons, is caused by ocean warming (93% of the EGE heat is absorbed by the oceans).
- Rising sea levels – threatening low lying countries such as the Maldives, Bangladesh and the Netherlands
- Changing climatic patterns – a pole-ward shift of climatic belts and the impact this will have on farming crops and animals.
- More extreme weather events such as storms, hurricanes, floods and droughts.

2 Industrial waste is released into rivers, lakes and seas because it is an easy/quick/convenient method of disposing of many forms of waste; it is a cheap method of waste disposal; environmental protection laws and regulations, in LICs in particular, are weak; laws/regulations are not enforced; economic growth is given preference over the quality of the natural environment.

Exam-style questions

1 a Deforestation (when people clear forests, woodlands and shrub land) to obtain timber, fuelwood and other products is currently far in excess of the rate of natural re-growth. This has become an increasing problem in semi-arid environments, where fuelwood shortages are often severe.

In rainforest areas, deforestation can be carried out for several reasons:

- Logging for valuable timber, like mahogany and teak. This also badly impacts the habitat for thousands of species of plants and animals and destroys the delicate food webs and food chains that exist in the natural environment. It also takes away the natural home for many indigenous peoples
- Plantation agriculture, where the forest is cleared to create huge farms for growing plantation crops such as sugar cane, oil palms, bananas and pineapples
- Cattle ranching to meet the growing demand for beef and burgers from HICs and MICs in particular
- For the construction of new settlements, often to provide land for small-scale farmers
- Over-cultivation, where low income farmers attempt to increase the yield from their land
- Rising populations are forcing farmers into farming more marginal areas on desert fringes, where they would not normally choose to farm
- Overgrazing (when too many animals are kept on the same land for too long and the animals east too much of the grass/fodder plants so that they cannot re-grow and die) is resulting in a decrease in the vegetation cover, and this is a major cause of wind and water erosion.

b The rise in carbon dioxide started with the clearing of forests – this is often done by burning the forests, which increases the amount of carbon dioxide in the atmosphere and it also has an additional impact as the trees convert carbon dioxide to oxygen. Industrialisation since the mid-19th century has put large amounts of carbon dioxide into the atmosphere from the burning of fossil fuels. Emissions from internal combustion engines and jet engines.

2 Industrial activity poses several threats to the natural environment. These include: the burning of fossil fuels can release carbon dioxide into the atmosphere, a greenhouse gas, causing global warming. Burning fossil fuels such as coal and oil can release sulphur dioxide into the atmosphere causing acid rain as the sulphur dioxide joins with rainwater to form dilute sulphuric acid which threatens vegetation and makes soils, rivers and lakes more acid, threatening their flora and fauna, such as fish. Nitrous oxide from the exhausts of vehicles is harmful to humans.

3 **Overgrazing** (when too many animals are kept on the same land for too long and the animals eat too much of the grass/fodder plants so that they cannot re-grow and die) is resulting in a decrease in the vegetation cover, and this is a major cause of wind and water erosion.

Over-cultivation, where low-income farmers attempt to increase the yield from their land. Instead of allowing the land to rest for a period of time by not planting crops and allowing the soil to recover and regain nutrients, farmers are continually planting, often the same crop in the same land every year and the soil is losing fertility.

4 Pollution is the introduction of contaminants into the natural environment that can cause an adverse change to the environment. Pollutants, the components of pollution, can be either foreign substances/energies or naturally occurring contaminants.

5 Aircraft, vehicular traffic (highway and off-road), and the rapid increase in the use of machines and other technologies are all associated with noise pollution. Like many other forms of pollution, noise appears to disproportionately affect poor and disadvantaged communities. Heathrow airport in London has 1340 average daily movements of aircraft landing or taking off. It is surrounded by suburban housing, business premises and mixed-use open land to the north and south, suburban housing and business premises to the east, and several large reservoirs, mixed-use open land, housing and business premises to the west. The area affected by noise pollution around the airport is estimated to be an area of 105 km^2 and a population of 270 100 is affected by the noise of aircraft landing and taking off.

Chapter 20

Exam-style questions: Population pyramids

1 The population pyramid is triangular in shape and has the concave shape typical of a LIC. The wide base indicates a high birth rate and the sloping, concave sides show that the death rate is high in all age groups. It also shows that life expectancy is low and infant mortality rates are very high, the latter shown by the marked reduction in population between the 0–9 age groups. The number of young, economically dependent people is very high. With so many people under 15, the population may continue to increase for many years.

2 a i There is a larger percentage of the HIC population above 65 years of age

ii There is a larger percentage of the LIC population under 5 years of age

b In the HIC, 34% of the population are economically dependent compared to 51% in the LIC; 18% are below 15 and 16% above 65 in the HIC compared to 44% under 15 and 7% over 65 in the LIC.

c In an LIC, the economically dependent population is most likely to be supported by their family both below 15 and over 65, including contributing toward education below age 15, whereas in a HIC, the government may also be providing support for the under 15 age group in health care and education and then in health care and social housing care after age 65 especially.

Exam-style questions: Line and bar graphs

1 a 26 °C

b 260 mm

2 Small annual range of temperature; 2 °C, between 24 and 26 °C; precipitation in every month; varying between 160 and 260 mms; wettest season between November and January.

3 High temperatures throughout the year; between 24 and 26 °C.

Exam-style questions: Triangular graphs

1 HICs have low percentages in primary, 2–8% and higher percentages in secondary, 26–35% and tertiary, 57–73%, compared to the LICs – primary, 55–83%, secondary, 6–18%, and tertiary, 12–28%.

Exam-style questions: Divided bar charts

1 a UK

b 58/59%

c Percentage in primary will drop, from over 40 to below 8%, while secondary will increase in percentage from 15 to 30% and tertiary from 40 to over 60%.

Exam-style questions: Radial graphs

1 a Leisure

b 5 km

c 5 km

Chapter 21

Exam-style questions

1 a i Barometer

ii Maximum and minimum thermometer / Six's thermometer

iii Four explained features from: double-louvered sides – allows air to circulate freely around the instruments but not blow directly on them; the screen has to be at, or very near, 1.25 m above the ground/surface – a standard height for the instruments to measure the air temperature, not the ground/surface temperature; allows temperatures to be compared accurately with those measured in earlier years and at different places around the Earth; the screen allows shade temperatures to be measured –
if sunlight falls on the glass thermometers the temperatures will increase; a double roof –
to provide a layer of air between the two roofs to insulate the screen from the heat of the Sun; painted white – to reflect the Sun's radiation; siting in an open space – to minimise the effects of buildings and trees; the door of the screen should always face north in the Northern Hemisphere / south in the Southern Hemisphere – to prevent direct sunlight falling on the thermometers when it is opened. (4 features + 4 reasons = 8 marks)

b i Collecting jar / container, measuring cylinder

ii The top of the rain gauge needs to be over 30 cm high to avoid surface water running into them and the spray from rain splash entering them as rain drops hit the ground.

iii So that the impact of rain drops is absorbed, and does not bounce/splash up into the rain gauge.

iv So that it is not / less likely to be blown or knocked over.

v To reduce any possible loss of the water that is collected in the collecting cylinder by the process of evaporation.

vi They have to be located away from buildings and trees; which might affect their collection of rainfall; through sheltering the rain gauge or cause water to drip into it from trees or buildings.

c They must be placed in the open, away from buildings or trees that can alter the wind speed. They will normally be placed on a tall mast on top of a building, or 10 metres above ground or building level.

d To give a constant set of readings, with one day / 24 hour period between readings at the same time of day.

e By measuring the amount of cloud covering the sky above the weather station – measured in oktas, where one okta is one eighth cloud cover – one eighth of the sky is covered/ obscured by cloud.

f Completed graphs and wind rose are shown in Figure 22.2.

g i **Hypothesis 1:** The hypothesis is correct for the dates 12, 13 and 18 March but on the 19, it is raining as atmospheric pressure is rising.

ii **Hypothesis 2:** Temperature changes according to the direction which the wind is blowing from. The hypothesis is incorrect. For example, for the three days 19–21 March, the wind is blowing from three different directions and the temperature remains the same, 23°C. Numerous examples of different temperatures from similar wind directions to use in answer.

h Could be taken over a longer period of time / different times of the year to extend data set. There is a gap of two days in the middle – a weekend, 15 and 16 March, from where data is missing – which could mean important data / trends are missing.

[Total: 30 marks]

2 a i The factors the students may consider in their choice of the sites for their fieldwork include: the depth of water; the speed/velocity/strength of the water in the river channel; avoiding stretches of the river near waterfalls/ rapids; the fact that it might flood; the possibility of encountering wild/dangerous animals; finding locations that are easily accessibility/easy to get to; ensuring sites are open to the public and, if it is private land, that

a

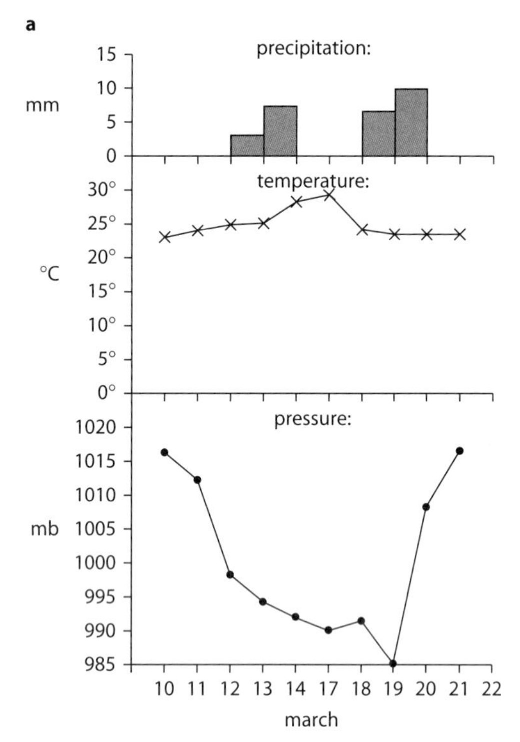

b

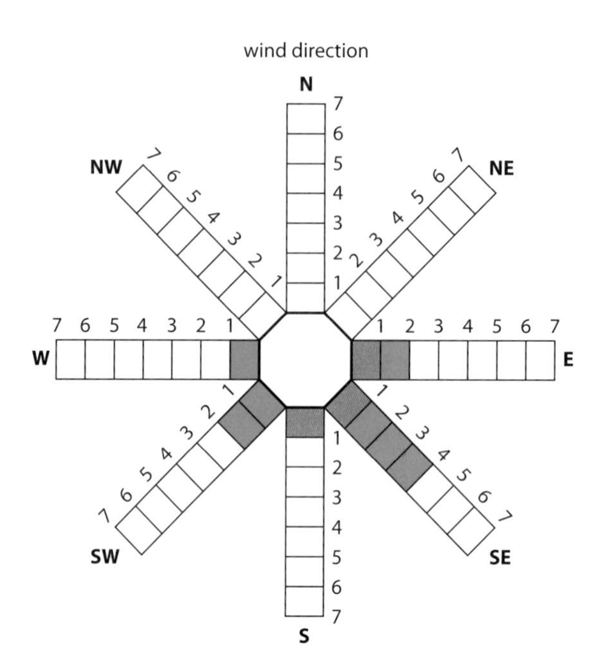

Figure 22.2 (a and b)

permission has been granted for access; the distance from source and the distance between sites/locations; avoiding sites/locations that may be affected by human impact/dams/weirs/structures/buildings/houses.

ii Reasons for carrying out a pilot study before the final investigation may include: practise fieldwork techniques / gain experience and confidence in methods and techniques / use of equipment; test fieldwork equipment; experience of working on a project as a team; working out the number of people needed on the techniques; ascertain the time needed to carry out the techniques and the possibility of the team doing several techniques at the same time; whether teams are too small/large for the techniques to be practised by everyone or carried out accurately and efficiently.

iii Important that all of their measurements are taken on the same day because: ensures consistency in results; the speed/velocity and the depth/width/river conditions may change if taken over several days; the weather conditions/rainfall might change, whereas, river discharge/height/width readings, if taken on the same day, should stay the same.

iv Equipment the students would use and how to make the two measurements of width and depth include: ranging rods/poles/sticks/bamboos to be put on each bank of the river; tape measure/string/rope stretched between the ranging rods/poles/sticks/bamboos across the river to measure width; if string/rope used, stretch it across river then measure it; metre rule(s) to measure river depths at measured/fixed points across the channel.

b The answers have been also been entered in Table 22.1, and are:

i Average depth at **Site B** = 0.12 m

ii Cross-sectional area at **Site B** = 0.336 sq. m.

iii Cross-sectional area at **Site C** = 0.703 sq. m.

c i Measurement repeated five times in each location to give an average reading and avoid a single anomalous reading / gives a more accurate / more reliable / more representative / average result.

ii Velocity figures were calculated by: dividing the distance in, in this case 10, by the average time taken, in seconds, to give the average velocity in per second.

d i The link between the velocity and the depth results indicates that, as channel depth increases, velocity increases. The reasons for the relationship is the reduction in the influence of friction; from the channel / river bed; on the velocity of water; as the river increases in depth. Deeper water; can overcome the frictional drag; of the river/channel bed; so there is a faster flow of water.

ii If a flow metre is used, it may not be correctly positioned in the water. For example, it may not be fully below the surface of river / may not be fully submerged; The propeller must be facing upstream; it must be held in the water for a specified time; the reading on the flow metre may be affected by an object(s) which is restricting normal flow / such as the presence in the water of water plants / rocks etc.; there may be student error in timing; if a float is used, it may be affected by plants/rocks and the effects of wind on the float.

Distance across channel (metres)	0.5	1.0	1.5	2.0	2.5	3.0	3.5	Average depth (in m)	Width of channel (in m)	Cross-sectional area (in m^2)	Average time (in seconds) to travel 10 m	Velocity (in m/s)	Distance from site A (in m)
Site A Depth of river (metres)	0.10	0.13	0.11					0.11	1.7	0.187	35.9	0.28	0
Site B Depth of river (metres)	0.12	0.13	0.15	0.15	0.17			0.12	2.8	0.336	31.6	0.32	1 000
Site C Depth of river (metres)	0.13	0.15	0.25	0.30	0.22	0.16	0.18	0.19	3.7	0.703	27.4	0.36	2 000

Table 21.3

iii The water in the river channel is deep; there is a fast-flowing / strong current of water; the water current may pull the measuring tape out of position / pull it downstream; the tape measure may not be long enough; it may be more dangerous to work in a large river.

Table 22.2 has to be completed using answers already given in (b) (ii) and (b) (iii) – added below:

e i The hydraulic radius for site **B**, 0.11, and site **C**, 0.18, have been entered in the table.

f i **Hypothesis 1:** The width, the depth and the velocity of the river all increase as the distance from the source increases so the hypothesis is **true**. Width increases from 1.7 to 2.8 to 3.7; depth increases from 0.11 to 0.12 to 0.19 metres; velocity increases from 0.28 to 0.32 to 0.36 metres per second. Site **B** readings are not necessary to gain the extra mark, but may be included to give a full answer.

ii **Hypothesis 2:** The velocity of the river will increase as the hydraulic radius of the river channel increases is **partly true**, as it is true when comparing sites A and C, but there is an anomaly at Site B because although the velocity has increased, the hydraulic radius has decreased. A reason is not needed here but the gradient of the river channel may have increased / and/or plant/rock obstructions may have increased slowing the velocity of the river, but these factors/ variables are not being measured/observed here. This can be mentioned in the next question on possible improvements.

iii The investigation could be extended to improve the reliability of the results, using more sites; comparing different rivers; carrying out investigations/ fieldwork at other times of year; taking more velocity measurements; use of a flow metre instead of floats for measuring velocity; measuring gradient, noting the presence/absence of vegetation.

[Total 30 marks]

Site	Distance from site A (in m)	Cross-sectional area (in m^2)	Wetted perimeter – (in m)	Hydraulic radius CSA WP
A	0	0.187	1.8	0.14
B	1 000	0.336	3.1	0.11
C	2 000	0.703	4.0	0.18

Table 21.4

Index

A

B

C

D

E

F

G